Design Rules

Carliss Y. Baldwin and Kim B. Clark

Design Rules

Volume 1. The Power of Modularity

The MIT Press
Cambridge, Massachusetts
London, England

This book was set in Times Roman by Graphic Composition, Inc. on the Miles 33 typesetting system and was printed and bound in the United States of America.

Library of Congress Cataloging-in-Publication Data

Baldwin, Carliss Y. (Carliss Young)
Design rules / Carliss Y. Baldwin and Kim B. Clark.
p. cm.
Includes bibliographical references and index.
Contents: v . 1. The power of modularity
ISBN 0-262-02466-7 (alk. paper)
1. Electronic digital computers—Design and construction—History.
2. Modularity (Engineering) 3. Computer industry—History.
4. Industrial organization. I. Clark, Kim B. II. Title.
TK7885.A5B35 1999
621.39—dc21 99-16204
CIP

To our loving and understanding families

Contents

Part III Design Evolution via Modular Operators

Part IV Modular Clusters

Acknowledgments

There are many people who, knowingly or unknowingly, contributed to the writing of this book. We can only acknowledge a few, but we are grateful to all who helped us along the way.

First, we owe a great debt to our teachers, John Lintner, William Abernathy, Robert Merton, Alfred Chandler, Robert Hayes, and Michael Jensen. They showed us how structures match functions, and how solutions reflect problems across a vast range of human activity. Their influence pervades our work.

Our project to study the computer industry began in 1987. Initially we planned to take only a few months to gather data and arrive at insights, from which we could draw relevant conclusions. Instead we found that each insight led to another question, and then another and another. There were many times when our spirits flagged, and we asked ourselves whether we could possibly be on the right path. At those times, a number of people encouraged us to continue and offered corroborative insights or evidence. Those who gave us encouragement at critical times included Andre Perold, Michael Harrison, Sam Wood, John Roberts, Rick Ruback, Krishna Palepu, Jennifer Francis, David Upton, Rene Stulz, Diane Lander, Martha Amram, Mal Salter, Richard Tedlow, Joyce Tedlow, Nathalie Blais, James Blais, Rebecca Henderson, Chris Argyris, Brian Hall, Anita McGahan, Ashish Nanda, Marco Iansiti, Gerald Zaltman, Eric von Hippel, Michael Kelly, Dan Raff, Gabriel Szulanski, John Sutton, Nancy Koehn, and Kent Bowen.

We also thank Steve Eppinger and Donald Steward, who generously explained their work to us, and shared their individual views on the design structure matrix.

As the work progressed to the point of working papers and draft chapters, a number of our colleagues were kind enough to give us comments and feedback. Their critiques, which we took to heart, have made the work substantially better than it would otherwise have been. We are especially grateful to George Baker, Mal Salter,

Karen Wruck, Richard Tedlow, Chris Argyris, Anita McGahan, Tarun Khanna, Josh Lerner, Michael Tushman, Karl Ulrich, Rebecca Henderson, Nitin Joglekar, and Brian Kernighan.

We would also like to acknowledge our debt to our students, Steve Spear, George Westerman, Richard Bergin, Glen Taksler, Sonali Shah, and Seki Obata. Their questions, comments, and criticism have never failed to show us when our arguments were weak, and our expositions murky.

We have had the best of help and support. As research assistants, we were fortunate to work with Jack Soll at the start of the project, and Joanne Lim toward the end. Todd Pulvino and Partha Mohandas helped to define the set of firms that eventually were incorporated into our database. Jim Schorr has worked to update the database each year as new financial records become available.

Andrea Croteau and Jean Cunningham made sure that with all else that was going on, we still had time to meet and to work on this project. Without their intervention, we would never have approached completion. Andrea and Jean also took charge of the last stages of manuscript preparation, thereby bringing order to chaos.

We are grateful to the people at MIT Press, especially Victoria Warneck, Katherine Almeida, Yasuyo Iguchi, Chryseis Fox, Jean Wilcox, Bill McCormick, and Jud Wolfskill.

Although many people have helped us along the way, one person stands out as our prime support and champion. Barbara Feinberg, our editor, began to work with us when Kim became dean of Harvard Business School. A few weeks later, we threw away our first draft and started over. She, more than anyone, understood the nature of the project. She knew what was needed, and told us what we had to do. Without her, the book would have been quietly abandoned long ago, brought down by its complexity and inherent contradictions. We are most grateful to Barbara for her patience, understanding, and encouragement throughout this time.

Finally, we thank our families—Randy and Sue, our parents, children, and grandchildren. It is an understatement to say that they have taught us to see the world in new ways. Their multifarious, ever-changing encounters with daily life—their delights and annoyances—are the core experiences from which this book has sprung. As small recompense for the joy they bring us, we dedicate this book to them.

Der Tag ist schön auf jenen Höhn.
[The day is beautiful upon those heights.]
—Friedrich Rückert

1 Introduction: Artifacts, Designs, and the Structure of Industries

As the twentieth century draws to a close, we live in a dynamic economic and commercial world surrounded by objects of remarkable complexity, sophistication, and power. It is a world where the speed of business is rapid and accelerating, where new technologies create new products at once stunning and useful, and where the pace of change has created an environment that is fraught with risk and reward. This much has been widely recognized and has inspired any number of articles in the popular press about the "late twentieth-century economy."

Much of this discussion has focused on emerging products and technologies in information, materials, electronics, biology, and medicine. But change is not limited to products and the technologies needed to make them. Think of developments in the last few years in financial services, energy, autos, communications, food, retailing, computers, and entertainment. In these and many other industries, changes in products and technologies have brought with them new kinds of firms, new forms of organization, new ways of structuring work, new contracts and relationships, new ways of bringing buyers and sellers together, and new ways of creating and using market information.

These changes in products, technologies, firms, and markets are not a passing phenomenon, like froth on the waves, caused by shifting economic winds. These are fundamental changes driven by powerful forces deep in the economic system, forces which moreover have been at work for many years. The developments we read about in the headlines and watch on the nightly business report are the most visible effects of a process with deep roots. They are manifestations of forces that are profoundly affecting the nature of enterprise, the character of work, the products and services we use, even the fabric of daily life.

An understanding of the forces driving change is crucial to comprehending the opportunities and the risks that change creates. But in our search for understanding we must be prepared to dig deep, for the forces that matter are rooted in the very nature of things, and in the processes used to create them. What we see around us is not the result of some deus ex machina working outside of our influence or control. Human beings, working as individuals and in groups, create the new technologies, the new forms of organization, the new products and markets. To be sure, the consequences of their actions are not always intended or even anticipated. But the "things" themselves—the tangible objects, the devices, the software programs, the production processes, the contracts, the firms and markets—are the fruit of purposeful action. They are "designed."

Hence the phrase "deeply rooted in the very nature of things" is not merely a slogan, but is the touchstone of our work. It is in the deep structure of technologies, products, firms, and markets that we seek understanding of the forces driving change in our economic world. Simply put, this book is about the process of design and how it affects the structure of industry. Design is the process of inventing objects—our "things"—that perform specific functions.[1] These objects, the products of human intelligence and effort, are called "artifacts."[2]

Each of us is surrounded by thousands and thousands of artifacts. The physical artifacts in our world include kettles (invented circa 8000 B.C.), automobiles (circa A.D. 1900), and computers (circa A.D. 1950). The kettle sits on a stove, the automobile sits in a garage, and the computer sits on a desk—all artifacts. There is a clock in the stove, a radio in the car, and many, many files in the computer. Taken as a whole, these objects create a rich and colorful backdrop for our lives.

Some artifacts are intangible. Rules, customs, systems of law, and property rights are artifacts, as are treaties and contracts. Organizations, including the firms that make kettles, stoves, automobiles, and computers, are artifacts. Games and strategies are artifacts, as are computer programs and engineering designs. Science, the knowledge we use to understand the world, and technology, the knowledge we use to create things, are artifacts too.

Artifacts evolve. This is one of the most important characteristics of "the very nature of the things" and is a theme that will carry throughout our work. Neither technologies nor the products that embody them emerge full-blown, all at once. They develop over time. The same is true for the firms that create products, and the markets that support them. Indeed, products and technologies and firms and markets evolve

---

1 Alexander (1964).

2 Simon (1969).

interactively over time to create a complex adaptive system, which we call "an industry." Some transitions in this process are sharp and striking, but much of what we see is the cumulative result of small changes. Indeed, barely perceived changes in the designs of common artifacts occur from day to day. Major and minor modifications in these designs accumulate over decades and half centuries, until eventually this process of evolution transforms our institutions, our economies, our culture, and the texture of our daily lives.

Computers

Any attempt to understand powerful and deep-seated forces must find a point of entry, a window into the complex phenomenon. Although design and industry evolution are at work across the economic landscape, we have chosen to study this phenomenon and its implications for firms and markets in a single context, albeit one where the underlying forces are especially salient. Our context for this work is an artifact of the second half of the twentieth century: the electronic computer.

A computer is a physical artifact made of plastic, glass, and metal.[3] It is also an intangible artifact, whose essence lies in meanings assigned to flickering patterns of electrical current deep within the structure of the machine. Emerging on the scene in 1944, computers, their programs, and their patterns of use—all artifacts—have developed in astonishing ways. Where once a roomful of vacuum tubes, programmed via a switchboard, were used to calculate angles for aiming heavy artillery,[4] today a handful of chips, using a myriad of stored programs, can calculate numbers, format text, generate pictures, compare outcomes, and make decisions. The results can be delivered anywhere on earth, displayed on paper or on a screen, subjected to further processing, and stored for future use.

The complexity of even the smallest computer today is impressive. The complexity, power, and diversity of all computers seems close to miraculous. From a relatively inauspicious start, how could these manmade things have become so powerful, so versatile, and so ubiquitous in such a short space of time?

Questions like this have been asked many times in the past. In answer, many have pointed to Moore's law—the dramatic rate of increase in the number of circuit

[3] Before 1950, a "computer" was a person, and "computing" was a job description.

[4] Aspray and Burks (1987); Dyson (1997).

elements that can be packed on a "chip."[5] Indeed, ever-increasing chip density is an important fact, for it explains why computer processing speeds are high and increasing, while costs are low and decreasing. But the amazing trends in speed and cost are only part of the story. By themselves, they cannot tell us why there are today literally thousands of different pieces of computer hardware, millions of programs, and tens of millions of specific uses for computers. Low cost explains why ordinary people can afford to buy powerful computers, but not why they want them. It explains why small children are allowed to play on computers, but not why they find them fascinating.

Computers are fascinating, interesting, and delightful to human beings because they are complex. Most of us are not especially intrigued by their raw speed or low cost. It is the many things computers do, and the many different ways they can be configured, that makes them interesting and useful. And it is the ability of computers to fulfill idiosyncratic, even whimsical desires (like the desire to see colored fish "swimming" on a screen) that causes these artifacts to surprise and delight us.

For all of these reasons the computer is a powerful lens through which to observe and study the evolution of designs, and the development of an industry. But telling the story of the development of this artifact and its industry is only part of what we are trying to do. We are also seeking to construct a theory that explains the story, that is, a theory of design and industry evolution. The story and the theory are closely intertwined: the story will motivate and ground the theory in technical and commercial reality, and the theory will inform and sharpen the story, and extend its lessons beyond the context of a single artifact and industry.

Throughout these two volumes we shall move back and forth between the story of computers, and the conceptual framework of our theory. All the while we will be building our understanding of design and industry evolution. Before we set out on this intellectual journey, however, it will be useful to have a map of the terrain ahead. Therefore we begin this work with an overview of the story and the theory, and then proceed to a detailed exposition.

---

[5] In 1965, Gordon Moore predicted that the number of circuit elements on a chip would double every eighteen months. His prediction has held to a remarkable degree, though lately the doubling rate may have slowed. Different components of the computer appear to have different characteristic rates of improvement. "Moore's law," Intel Corporation; no date; viewed 4 December 1998; available at http://www.intel.co.uk/intel/museum/25anniv/hof/moore.htm; Hennessy and Patterson (1990); Pfister (1998).

Complexity and Modularity

Complexity is at the heart of the story of electronic computers. The broad outlines of this story are familiar: the design of room-sized calculating machines made of vacuum tubes in the 1940s; the emergence of IBM and its mainframe computers in the 1950s and 1960s; the arrival of the minicomputer in the 1970s; the personal computer in the 1980s; the Internet in the 1990s. However, within this broad sweep lies an important story about how human beings created an incredibly complex array of artifacts and made them economical and beneficial for use throughout society.

Human beings are limited in their capacities to learn, think, and act. Luckily, complex artifacts—like computers—do not have to be the product of one person's hands or mind. If an artifact can be divided into separate parts, and the parts worked on by different people, the "one person" limitation on complexity disappears. But this implies that techniques for dividing effort and knowledge are fundamental to the creation of highly complex manmade things.

The word "complex" in this context has the following commonsense meanings:

complex *adj.* [L. *complexus,* past participle of *complecti,* to entwine around, fr. *com-* + *plectere,* to braid] **1a:** composed of two or more separate or analyzable items, parts, constituents, or symbols . . . **2a:** having many varied parts, patterns or elements, and consequently hard to understand fully . . . **b:** marked by an involvement of many parts, aspects, details, notions, and necessitating earnest study or examination to understand or cope with.[6]

These definitions clearly contemplate a spectrum ranging from the starkly simple to unimaginably intricate. If we think of arraying artifacts along this spectrum, two interesting points arise as we move from simple to complex: (1) the point at which an artifact can no longer be made by a single person; and (2) the point at which an artifact can no longer be comprehended by a single person. Crossing into the first region requires a division of labor; crossing into the second requires a division of the knowledge and effort that go into creating a design.[7]

Divisions of effort and knowledge cannot take place in a vacuum, however. The essence of a complex thing is that its parts are interrelated. The different pieces must

---

[6] *Webster's Third New International Dictionary of the English Language.* Springfield, Mass.: G. & C. Merriam, 1964.

[7] There is, as yet, no generally accepted, formal definition of complexity in the field of economics. We think it is useful to think of the capacity of a single human being as a "natural unit" of complexity for both knowledge and effort.

work together, and the whole must accomplish more than any subset of its parts. Coordinating mechanisms are needed to channel effort and knowledge toward useful, attainable, and consistent goals.

Therefore a complex artifact—like a computer—is necessarily embedded in complex systems, whose purpose is, first, to divide effort and knowledge, and second, to coordinate tasks and decisions. As the basic artifact becomes more complex, moreover, the number of tasks and decisions that surround it will increase. To respond to these demands, the coordinating systems must themselves grow. The coordinating systems' efficiency in dealing with higher levels of complexity in turn will have a huge effect on the economics of the artifact—the cost of designing, making, and using it, and the extent to which it diffuses through human society.

In these two volumes, we will argue that computers were able to evolve into very complex and diverse forms because the coordinating systems surrounding them were very efficient. The systems we have in mind arose at two levels—first, the level of engineering design, and second, the level of economic organization.

At the level of engineering design, computers proved amenable to an approach we call "modularity in design." Under this approach, different parts of the computer could be designed by separate, specialized groups working independently of one another. The "modules" could then be connected and (in theory at least) would function seamlessly, as long as they conformed to a predetermined set of *design rules.*

Modular design rules establish strict partitions of knowledge and effort at the outset of a design process. They are not just guidelines or recommendations: they must be rigorously obeyed in all phases of design and production. Operationally, this means that designers may not solve the problems of module A by tweaking parameters affecting module B.

Because of the severe constraints it imposes, full-fledged modularity is never easy to achieve in practice. However, when implemented faithfully, modularity greatly reduces the costs of experimenting with new designs. With modularity enforced, it is possible to change pieces of a system without redoing the whole. Designs become flexible and capable of evolving at the module level. This in turn creates new options for designers, and corresponding opportunities for innovation and competition in the realm of module designs.

Modular Designs of Computers

Before the mid-1960s, computer designs were not modular. The first "truly modular" computer design was IBM's System/360, a broad, compatible family of computers introduced in 1964. Measured in terms of its total market value, System/360 was without question the most successful line of computers ever introduced by a single company. But even as this system was sweeping the marketplace, its huge base of users and IBM's high prices provided a target of opportunity for designers (many of them former IBM employees) who knew how to exploit the options embedded in its modular design.

As modular designs spread through the industry (a process that began around 1970), the structure of the computer industry began to change.[8] New firms entered and new institutions emerged to take advantage of the opportunities inherent in the new designs. As a result, between 1970 and 1996, the industry changed from being a virtual monopoly into a large "modular cluster," which today comprises more than a thousand publicly traded firms and many more hopeful startups.

Plate 1.1 is a picture of how the industry developed. It graphs the capital market values of different sectors of the computer industry from 1950 to 1996 in 1996 inflation-adjusted dollars.[9] To construct this chart, we collected data on the market value of all firms in the Compustat database in the sixteen four-digit SIC codes that make up the computer industry. The market values of about 1500 firms (in round numbers) are represented in the data. Of these, 991 were "alive" at the end of 1996. We aggregated their end-of-year market values into sixteen subindustries by SIC code. In addition, we broke out the market values of three individual companies (IBM, Intel, and Microsoft), and a group of five Japanese firms (NEC, Fujitsu, Hitachi, Toshiba, and Kyocera).

IBM dominated this industry for forty years: it is the blue "mountain range" that forms the backdrop of the chart. As the chart shows, during most of this period, it was by far the largest firm in the industry.[10] Since the late 1980s, Microsoft and Intel have been the largest firms (by market value) in their subindustries. They have also jointly inherited much of IBM's former market power in the personal computer sector. Finally, Japanese computer firms have very different structures from those in the United States—they are large, integrated, and much more diversified than their U.S. counterparts. Rather than lump them in with the U.S. firms, we chose to treat them as a separate group.

The relations among firms in different subindustries shown in plate 1.1 are quite complicated. The vast majority do not make stand-alone products, but instead make

---

[8] The change in structure occurred in the United States, but not in Japan or Europe (Smith 1997; Anchordoguy 1989).

[9] We used the consumer price index to deflate our numbers. Using other indices does not affect the results in any significant way.

[10] In 1995, IBM was for the first time surpassed in market value by Microsoft Corporation.

parts ("modules") that perform specialized functions within larger computer systems. The many disparate modules are tied together by complex webs of design rules. New modules and new design rules are being created almost every day.

The Changing Structure of the Computer Industry

As the chart shows, there was a time when the computer industry was highly concentrated. For five years after the introduction of System/360, IBM's total market capitalization was in excess of $150 billion (in 1996 dollars). But IBM's halcyon days were short-lived. The company did not fulfill the expectations implicit in this valuation: in 1973, its value dropped precipitously, and it never reached those highs (adjusted for inflation) again.

As IBM's value was dropping, whole new subindustries were emerging. As indicated, the firms in these subindustries did not make whole computer systems: they made parts of larger computer systems. In many cases, their products obeyed System/360's design rules, hence they were "plug-compatible" with IBM equipment, and could be used in IBM installations.

Table 1.1 shows the order in which subindustries emerged on the scene. In constructing this table, we defined the start date of a subindustry as the first year in which Compustat reported six or more firms in a given four-digit SIC category. (We imposed the "six-firm" constraint because we wanted to know when each sector became "economically important" enough to attract more than just a handful of firms.)

Ten of the sixteen subindustries crossed the six-firm threshold between the years 1970 and 1974. This was the heyday of IBM plug-compatible peripheral companies (the companies that sold compatible modules of IBM systems at greatly reduced prices). As figure 1.1 shows, these were also the years of IBM's greatest value decline.

In fact the emergence of these subindustries was the first hint that a significant change in industry structure was underway. But change came very slowly: the computer industry remained highly concentrated, and IBM remained dominant for another twenty years. Although its share of total value continued to erode through the 1970s and 1980s, as late as 1990 IBM accounted for close to 35% of the capital market value of the entire computer industry.

Indeed in the mid-1980s, following the spectacular success of the IBM PC, it seemed that IBM would reverse its value-share decline. Events took a surprising turn, however. First, through a series of strategic mistakes, IBM lost control of the design of personal computers. A few years later, in 1991–1993, the mainframe market, the mainstay of IBM's product line, collapsed. Both these events, moreover, were the

Table 1.1 Dates of emergence of new subindustries in the computer industry

Code	Category definition	Start date[a]
3570	Computer and Office Equipment	1960
3670	Electronic Components and Accessories	1960
3674	Semiconductors and Related Devices	1960
3577	Computer Peripheral Devices, n.e.c.	1962
3678	Electronic Connectors	1965
7374	Computer Processing, Data Preparation and Processing	1968
3571	Electronic Computers	1970
3575	Computer Terminals	1970
7373	Computer Integrated Systems Design	1970
3572	Computer Storage Devices	1971
7372	Prepackaged Software[b]	1973
3576	Computer Communication Equipment	1974
3672	Printed Circuit Boards	1974
7370	Computer Programming, Data Processing, and Other Services	1974
7371	Computer Programming Services	1974
7377	Computer Leasing[c]	1974

[a]Start date is the first year in which six or more are present in the category.
[b]This category had six firms in 1971, dipped to five in 1972, and went back to six in 1973.
[c]The database does not pick up many of the leasing firms of the 1960s.

result of the evolutionary dynamics of computer designs, whose roots can be traced back to the modularity in the design of System/360.

What happened in the 1980s and early 1990s was that the forces of modular design evolution, which had been at work on the periphery of the computer industry for almost twenty years, moved to center stage. At this point, economic power ceased to depend on "extensive control" of all elements of a computer's design, and instead came to depend on "intensive control" of critical modules within a set of complex and evolving design architectures.

In these two volumes, we want to explain how and why the computer industry changed from a quasi-monopoly into a large "modular cluster." We believe that a true understanding of this phenomenon depends on understanding the evolution of computer designs not just over the last ten years, but over the last fifty years. In particular, one of most the important forces shaping the evolution of these designs was the drive toward modularity.

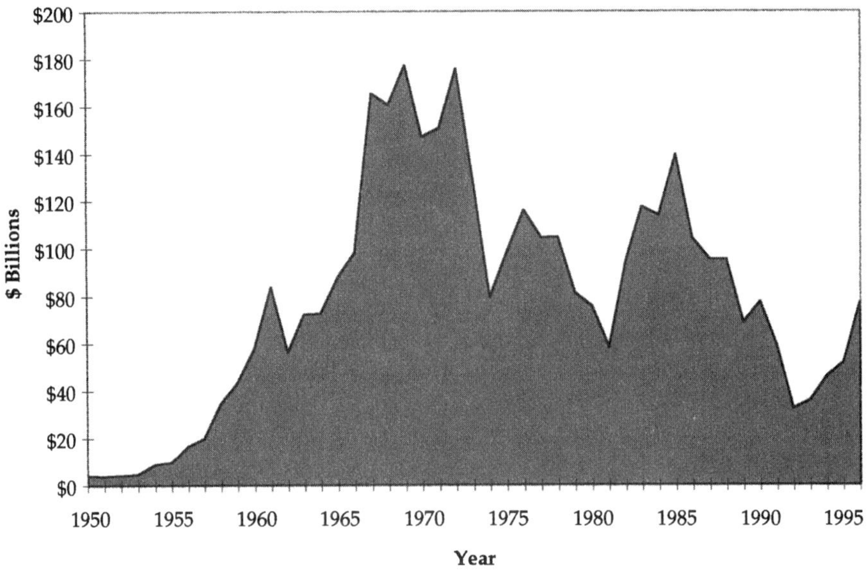

Figure 1.1 Market capitalization of IBM corporation in constant 1996 dollars.

Thus our plan in these two volumes is to look at fifty years of computer designs, and develop a theory of design and industry evolution. The framework we shall use to construct this theory is John Holland's framework of complex adaptive systems.[11] In the next section, we describe the foundations of the theory and the plan of the first volume.

An Overview of the Theory

A theory about a phenomenon generally begins with a list of structures that constitute the subject matter of the theory. The theory then describes the context or environment in which the structures exist. Finally, it depicts a set of changes to the structures. It is the task of the theory to explain how and why those changes occur. Theories of "complex adaptive systems" (of which this is one) follow the basic structures-

[11] Holland (1992).

context-changes template, but in addition, they are combinatorial: complex structures are built up of simple structures and complex changes occur as a result of combinations and sequences of simple changes.[12]

We follow the template of complex adaptive systems in developing our own theory of design and industry evolution. Following chapter 1 the remainder of this volume is divided into four parts. In part I (chapters 2–5), we develop the language of our theoretical framework and the basic facts about computer designs that motivate and ground the theoretical work. In part II (chapters 6–8), we look at how modularity emerged as a set of design principles in the computer industry between 1944 and 1960, and how those principles were applied to create System/360, the first modular computer family. In part III (chapters 9–13), we explain how modularity in design multiplies and decentralizes valuable design options, thereby making possible the process of *design evolution.* Then, in part IV (chapters 14 and 15), we explain how in the 1970s, the multiplication and decentralization of design options led to the emergence of a new industry structure for the computer industry, the form we call a *modular cluster.*

Part I: Structures, Context, and Operators

Part I bounds the scope of our theory and defines its fundamental elements. Specifically, in chapter 2, we define the structures of interest: these are *artifacts, designs,* and *design processes.* Designs and design processes, we will argue, have microstructures; that is, they can be broken down into smaller units of analysis. For designs, the "micro" units are design parameters; for design processes, the "micro" units are design tasks. We argue that at the level of parameters and tasks, designs and design processes are fundamentally isomorphic. As a result, the physics and logic of an artifact, expressed in its design parameters, affect the process of designing it in deep and unavoidable ways.

Moreover, among all possible variants, we will concentrate on modular designs and design processes as our particular concern in this investigation. They are the focus of chapter 3. Here we define modularity as a particular pattern of relationships between elements in a set of parameters, tasks, or people. Specifically, modularity is a nested hierarchical structure of interrelationships among the primary elements of the set.

---

12 Ibid.

The characteristic pattern of nested hierarchical blocks, which is the essence of modularity, may arise in many different contexts. For the purposes of this work, we distinguish modularity in the design of an artifact from other types of modularity, specifically modularity in production, and modularity in use. Finally, we discuss how modularity in design may be achieved via the creation and enforcement of a comprehensive set of *design rules.* We will give an example of an actual and important *modularization* in the computer industry—Mead and Conway's reconceptualization of the architecture of chip designs, described in their book *Introduction to VLSI Systems.*[13]

In chapter 4, we look at the broader economic context in which designs and design processes are lodged. These contexts range from primitive societies to advanced market-based economies. Economies with capital markets, we argue, offer large, direct rewards to value-creating enterprises, and commensurately large incentives for human beings to cooperate for the purpose of creating economic value. These perceived rewards in turn operate on designs "like a force." Metaphorically, they "pull" designs in the "direction" of higher market value.

Advanced market-based economies also offer their members a powerful and efficient set of contracting tools that can be used to organize collective enterprises. These *contract structures* can be overlaid on design and task structures, thereby defining the firms and markets of an industry. Finally, advanced market economies provide technologies that can be used to direct and coordinate effort within collective enterprises. These *guidance mechanisms* include technical knowledge about how an artifact works, and managerial knowledge about how to motivate individuals and manage groups.

This brings us to our third fundamental step in theory-building. Ours is a theory of how designs change. The crux of our theory is that modularity in design—an observable property of designs and design processes—dramatically alters the mechanisms by which designs can change. A modular design in effect creates a new set of *modular operators,* which open up new pathways of development for the design as a whole.

In chapter 5 we define six modular operators, which form a basic repertory of actions that can be applied to any modular design. They are

1. *splitting* a system into two or more modules;
2. *substituting* one module design for another;

---

[13] Mead and Conway (1980).

3. *augmenting*—adding a new module to a system;
4. *excluding* a module from the system;
5. *inverting* to create new design rules;
6. *porting* a module to another system.

Many complex changes in modular designs (we argue) can be represented as some combination of these operators.

Part II: The Creation of Modularity in Computer Designs

Chapters 2 through 5 give us a framework for thinking about designs, modularity, and changes in modular designs over time. With the framework in hand we are ready for a closer look at the story of computers, which we take up in part II (chapters 6–8). We begin in chapter 6 with a look at the "premodular" period of computer design, which lasted from 1944 to approximately 1960. Here we find the predecessors of modularity in the very early designs of stored-program computers, in the development of microprogramming to control the basic operations of the machine, and in early concepts for the design and packaging of electronic circuits.

Chapter 7 then goes on to describe the creation of the first "truly modular" computer family—IBM's System/360. The breadth of System/360 was made possible by a new theory of how to design a computer. The architects of the system first partitioned the design via an explicit set of *design rules,* and then created a management system that strictly enforced those rules. The result was a powerful and flexible computer family, which was a spectacular financial success. Part of that success lay in the technical accomplishments of the design, but part lay in IBM's capacity to capture the economic value inherent in the design.

From the very outset of the System/360 project, the architects of the system believed that a modular design would create significant potential value for IBM. However, realizing that value required the development of specific contract structures and specific strategies in finance, marketing, operations, and competition. The contract structure that IBM developed—which we label "extensive control"—is the subject of chapter 8. The power of that structure was evident in the financial and competitive impact of System/360 on IBM and its rivals, which we also discuss in chapter 8.

Part III: Design Evolution via Modular Operators

The introduction of System/360 was a watershed in the history of computer designs. It not only created a broad, compatible family of products for IBM but it opened up a whole new development path in which computer modules could begin to evolve along independent trajectories. Through *splitting,* modules could be subdivided; through *substitution,* they could be improved. Through *augmentation,* new modules could be created; through *exclusion,* new systems could be tried. Through *inversion,* redundant activities could be consolidated into single modules; and through *porting,* systems could be linked via common modules.

What moved designers to explore these possibilities, and induced investors to fund their efforts, was (and still is) the desire to create or capture economic value. With value as the force propelling change, and the modular operators as the means of effecting it, the design of a complex artifact could evolve in unplanned but nonetheless coordinated ways. The result was a "complex adaptive system," whose elements were constantly being modified, giving rise in turn to modifications of other elements. No architect had to give permission for these changes to take place; the possibilities were inherent in the modularity of the design itself. Like the walls of a house, the design rules provided a framework that allowed the working out of many different possibilities, leading to ever-changing versions of the design.

This process of design evolution is the focus of part III (chapters 9–13) of this volume. Chapters 9, 10, and 11 are the central chapters of the book. In them we combine the basic elements—structure, context, mechanisms—into a theory of design evolution. Our theory is that modular designs evolve via a

- decentralized search
- by many designers
- for valuable options that are embedded in
- the six modular operators.

When a design becomes "truly modular," the options embedded in the design are simultaneously multiplied and decentralized. The multiplication occurs because changes in one module become independent of changes in other modules. Decentralization follows because, as long as designers adhere to the design rules, they are free to innovate (apply the modular operators) without reference to the original architects or any central planners of the design.

In chapters 10 to 13 we connect the modular operators (the mechanisms of change) to valuations in the capital markets (the reason and reward for change). Using splitting and substitution as our first examples, we show that the modular operators create options for designers, options that have great potential economic value. In the case of System/360 we estimate that the option value of splitting alone may have led to a fivefold increase in value over previous designs. But there is more. By combining a modular design structure (splitting) with independent, parallel experimentation on modules (substitution), we find that designers can create as much as a twenty-five-fold increase in system value.

The analysis of splitting and substitution in chapter 10 focuses on symmetric designs wherein all modules have the same degree of complexity (require the same number of design tasks). But the essence of the remaining four modular operators—augmentation, exclusion, porting, inversion—lies in creating asymmetries in modules. Chapter 11 thus expands the theoretical framework to include differences in complexity, visibility, and technical potential across modules. The expanded model gives us the ability to identify differences in the value of different modules, and highlights the potential evolutionary impact of modular operators that create and exploit those differences.

Chapters 12 and 13 extend the analysis to the asymmetric operators. Our goal in these chapters is to characterize the options associated with each of the operators, defining both their benefits and their costs. In the process of modeling the value of the operators, we describe some of the common trajectories of modular designs, as well as some of the fallacies that trap designers as they make a transition from non-modular to modular designs. We look at the operators in natural pairings—augmenting and excluding (chapter 12) and inverting and porting (chapter 13)—and we illustrate each pair of operators with examples from the history of computer designs.

Specifically, in chapter 12 we add an important strand to our story by focusing on Digital Equipment Corporation (DEC). The rise of DEC and the development of the minicomputer marked an important episode in the evolution of computer designs. We describe how DEC used the exclusion and augmentation operators to create a modular family of computers, without the massive investment cost IBM incurred with System/360. Moreover, the modular structures and clean interfaces that DEC developed and the "minimalist" design of its early computers encouraged investment in DEC's architecture by third-party developers of hardware and software. Thus we will see in DEC's story a new effect of modularity: the structure of the design influencing the structure of firms and markets in the surrounding industry.

Following our discussion of DEC, we turn in chapter 13 to the last two modular operators: inversion and porting. The fact that we come to them last is symbolic of the fact that these operators only come into play after a modular design has been elaborated through repeated application of the other operators. Once that occurs there will be opportunities to take low-level, recurring components and make them visible and standard (inversion). There will also be opportunities to take a specific module and make it work in other systems (porting). We use the story of Unix, an operating system created at Bell Laboratories in the early 1970s, to illustrate the inversion and porting operators.

We shall see that, as in the case of DEC, Unix spurred many coinvestors to write software or develop hardware around its design rules. Unix stimulated third-party coinvestment because its design was both modular and portable. Developers could not only write applications that ran under Unix, but could add to the modules of Unix itself. Moreover, the Unix operating system was portable, and thus could run on many different hardware platforms.

As it happened, Unix evolved without central control, and with almost no value accruing to AT&T, the firm that technically owned it. The contrast with System/360 and IBM is therefore quite sharp. Both System/360 and Unix were examples of modular designs of great utility and inherent economic value. However, their divergent economic histories demonstrate that the value of an evolving modular design may be distributed among potential claimants in many different ways.

Part IV: Modular Clusters

By definition, the value created by a new design will be split in some fashion between designers, producers, investors, and users. The way value is split in turn will depend on

- the nature of the contracts that bind designers and investors together;
- conditions of entry to the design process;
- the anticipated structure of one or more specific product markets;
- the patterns of competition that ultimately prevail in those markets.

Complicating matters further, each of these elements may change over time as the design and its value evolve. Still there is likely to be a close relationship between the structure of the design and the economic structures—firms and markets—that emerge to realize it. These economic structures in turn are the focus of the fourth and final part of this volume.

In the computer industry, modularity in design gave rise to a new industry structure, which we call a "modular cluster." A modular cluster is a group of firms and markets that "play host" to the evolution of a set of modular designs.

Chapter 14 begins the process of mapping the forces that govern the design-industry structure connection. We first look at how a modularization affects the number of workgroups required to realize the economic potential of a set of designs. We argue first that the *multiplication of options* accompanying a modularization increases the number of workgroups and the resources that will be invested in the design and production of that particular class of artifacts. Second, the *decentralization of options* (the other consequence of modularity) changes the incidence of transactions and agency costs, making it possible for work to be distributed across many separate firms. The firms in turn will be connected by markets for goods, information, labor, and capital. The totality of firms and markets is what we are calling a "modular cluster." This form becomes a potentially viable industry structure following a large-scale modularization of a set of artifact designs.

In fact, a modular cluster did emerge in the computer industry of the United States following the creation of System/360, and the dissemination of the underlying principles of modularity to computer designers. Table 1.1 already showed that ten new subindustries "became important" early in the 1970s. In chapter 14, we provide additional data on entry and the distribution of market values by sector. These data show that the majority of the new entrants to the industry made what we call *hidden modules.* These could be tangible pieces of hardware, like disk drives or DRAM chips; or software; or a combination of hardware and software. Hidden modules have essentially no value as stand-alone products, but are useful and valuable parts of larger computer systems. However, in order for their products to work, the designers of hidden modules must know and obey the *design rules* laid down by the architects of those larger systems.

When hidden modules become objects of exchange in product markets, competition among the companies that design and make those modules becomes an engine of change in the greater economic system. In chapter 15, we model the dynamics of hidden-module competition, using as background the development of plug-compatible peripheral devices for System/360. We analyze a two-stage investment game among a variable number of players when coordination is imperfect, and valuation is costly. We show that there are many equilibriums associated with this game; moreover, an evolving, modular design will generate many opportunities to play, each with a different characteristic value function, and (possibly) a different set of players.

Thus, we argue, if the processes of design and production of hidden modules are dispersed over a large number of firms, design evolution in artifacts may give rise to a dynamic process of entry and exit into submarkets by individual firms. The firms' entry and exit decisions will be motivated by local, imprecise calculations of value. Hence their moves will not be perfectly coordinated; moreover, the returns to a firm in any submarket may depend on the outcomes of technologically risky bets. Under these circumstances, we think it is unlikely that the industry as a whole will converge to a single stationary equilibrium. Instead ex ante and ex post equilibria are likely to vary across submarkets and over time.

This process of entry and exit into submarkets is capable of transforming markets and industries through the cumulative combination of many local moves. It is a type of *industry evolution* that mimics the ongoing evolution of the underlying artifact designs.

The analysis of hidden module competition take us to the year 1980, and the end of volume 1. In volume 2, we investigate how the interrelated processes of design and industry evolution operate once a modular cluster has formed. Explaining how and why the computer industry developed as it did from 1980 to the present requires further development of the theory. In particular, in volume 2 we will focus on the infrastructure needed to support a modular cluster, the generic strategies used to capture value in modular design hierarchies, and the dual roles played by valuation technology and the financial system in shaping design and industry evolution within a modular cluster.

We now turn to the task at hand. We aim to provide an explanation of how designs change within a complex, adaptive economic system. We begin in chapter 2 by looking at the microstructure of designs.

Part I Structures, Context, and Operators

2 The Microstructure of Designs

Artifacts: Their Structure and Functions

Understanding the structure and functions of an artifact and how they differ is an essential first step in building our theory. This chapter is our starting point. In it, we look closely at designs and design processes. We delve into the "microstructure" of artifacts and designs to lay bare their essential organization. It is in these microscopic elements of structure that we find the critical mappings that link changes in designs to changes in organizations, markets, and industries.

In chapter 1, we introduced and defined the concept of an artifact. To this we need to add the related concepts of *design, design parameters,* and *design tasks.* A "design" is a complete description of an artifact. Designs in turn can be broken down into smaller units, called *design parameters.* For example a red mug and a blue mug differ in terms of the design parameter "color." They may differ in other dimensions as well—for example, height, width, and material. The *tasks* of design are to choose the design parameters, which the artifact will embody.

Next we need the concepts of *design structure* and *task structure.* The design structure of an artifact comprises a list of design parameters and the physical and logical interdependencies among them. (The concept of design parameter interdependency is explained below.) The task structure of a design process comprises a list of tasks that need to be done to complete the design, and the connections between the tasks. We will show that *the design structure of an artifact and the task structure of the design process are isomorphic.* Thus the structure of a design, which is dictated by physical and logical constraints on the artifact, affects the tasks of design in deep and unavoidable ways.

To increase the precision of our thinking, we will describe design and task structures using a tool called the design or task structure matrix (DSM or TSM).[1] With the help of this tool, we will define a set of generic design and task structures and describe how, with sufficient effort and knowledge, one kind of structure may be converted to another.

To make these concepts concrete, we will introduce a real thing, but one less complicated than even the smallest computer.[2] Therefore, let us begin by examining a simple artifact from everyday life—a coffee mug. In fact, imagine there are two mugs on the table next to you as we describe them. Turning to the mugs before us, let us contemplate their similarities . . . and differences. Both are hollow cylinders with bottoms and handles. One holds twenty ounces of liquid; the other holds eight. One has a cap; the other doesn't. The mug with the cap is made of hard plastic; the other is made of a glazed ceramic material. The ceramic mug is midnight blue and displays the Jayhawk emblem of the University of Kansas; the plastic mug, with the cap, has a pattern in black, brown, and white and bears the logo of the Au Bon Pain chain of restaurants.

These are the most obvious characteristics of each mug. But there are not-so-obvious characteristics, too. Although both mugs are hollow cylinders with bottoms (so the liquid doesn't fall out), there are subtle differences in the shape of the vessels and their handles. The plastic mug, moreover, is constructed with a layer of air between two walls for better thermal insulation. The cap on this plastic mug is designed to fit snugly into the top of the vessel, and has a small handle that can be grasped. The cap also has two holes, one large and one small.

Below this level, invisible to the user, are things that the producer of the mug needs to know. What is the exact plastic or ceramic material? What are the precise spatial dimensions of each mug? Is the color intrinsic to the material, or applied to the exterior? What are the emblems, and how are they applied?

An artifact, such as a mug or a computer, has both a structure (what it is, how it is constituted) and functions (what it does, what it is for). The structure of a mug, for

---

[1] This tool was first developed by Donald Steward (1981a, b) and extended and refined by Steven Eppinger (1991), together with colleagues and students at MIT. Additional citations are found in the relevant sections below.

[2] This section draws heavily on a number of different strands of literature. For related work, see the decomposition of a kettle by Alexander (1964), and of a room fan by Henderson and Clark (1990). Readers may also consult Ulrich's description of contrasting architectures for a truck (1995a), Kelly's chapter, "In the Library of Form" (1994), and Dawkins's (1989) account of the construction of biomorphs.

example, includes all the specifics of its composition—its height, diameter, and handle dimensions, as well as its material, color, decorations, and so forth. The primary function of a mug is to hold liquid, but it may have other functions as well (it can hold pencils, for instance). A computer (including all its files and programs) has a much more complex physical and logical structure than a mug. Its primary function is to process and transform information (data) in accordance with prespecified algorithms (the programs).

An artifact's *design* is an abstract description that encompasses both its structure and its functions. The difference between a design description and other descriptions is that, *with respect to the structure of the artifact,* the design description needs to be both precise and complete. A thing that is imagined by a designer cannot become a thing that is real until all its structural elements have been chosen. The choices can be made implicitly, as when a potter shapes a bowl on a wheel, or explicitly, as when a designer prepares a set of blueprints or a computer-assisted design (CAD) file. But a complete set of structural choices must be made and then implemented for an artifact to come into existence.

Fortunately, designers do not need a formal, detailed list of structural elements to complete a design. A basic set of structural elements is preselected when a designer chooses a "medium" in which to work.[3] A medium may be a physical material, like clay or plastic, or a logical environment, like a computer language or musical notation. In any medium, a core set of structural elements will be "implied by the medium," and thus will not have to be actively selected. If the medium is clay, for example, a ceramic mug can be left unglazed. It will then have the color and surface texture of its clay instead of the color and texture of a glaze. Knowing the consequences of "not selecting" some structural element is part of the detailed knowledge (technology) that grows up around the medium used for making a class of artifacts, like ceramic for mugs.[4]

In contrast to its structure, the functions of an artifact need not be completely specified in the design description. Functions are not neglected, however. Usually, a fundamental function is associated with the artifact class—mugs hold liquid, copiers

---

3 McCullough (1996).

4 Designers also use many "tricks of the trade" like prototyping, modeling, and simulation to explore different imagined designs and to delay the fixing of some of the structural elements, until the last possible moment. For descriptions of some of these methods, see, for example, Bucciarelli (1994); McCullough (1996); Iansiti (1997a, b); Thomke, von Hippel, and Franke (1998).

copy, and computers process information. More detailed functional objectives motivate and shape the design process for a particular artifact, and guide the process toward a final design. But the functions of an artifact are not fixed by its design in the way that its structure is fixed. Old functions for an artifact may be superseded and entirely new ones may be discovered after the fact. A disk drive may end up as a doorstop, and a microprocessor may become a piece of jewelry.

The descriptions of the two mugs given above are neither complete enough nor precise enough to qualify as design descriptions. The paragraphs are sufficiently detailed to distinguish between the ceramic and the plastic mug, but we could envision another plastic mug with different spatial dimensions (height, diameter, etc.), and the written description would not distinguish between the two plastic mugs. Hence what is written is only a description—it is not the actual design of a mug.

Structure: Design Parameters

Because they are fundamentally descriptions of artifacts, designs may be recorded and communicated in various ways: through natural language, scientific language, pictures, blueprints, prototypes, and, with the advent of CAD programs, through computer files. The CAD program representation is especially interesting because it gives us a way of visualizing the information content of a design and thereby comparing the designs of wildly different artifacts.

Suppose we had a powerful CAD program that we could use to describe any kind of mug. The design of any particular mug would then be a file readable by this program.[5] In its CAD file representation, a complete, makable mug design would consist of a string of 0s and 1s. Some of the string would be common to all mugs (descriptors of a cylindrical shape, handle, walls, cap, etc.). This portion of the file would constitute the global design of the artifact class.[6] Nested in this framework would be strings that varied within the class—these would be the design parameters for that class.

---

[5] Recipes for converting designs into real things are algorithms, hence we know (by the Church-Turing thesis) that such a file can be constructed for any design. See Penrose (1989), chapter 2.

[6] The concept of an artifact class is subtle. For a CAD program, the artifact class (of the program) may be formally defined as all forms that can be described by that program. That is a well-defined, though usually large, set. However, in the minds of human beings, artifact classes are subjective and informal, and their boundaries depend on both perception and natural language. As a result, the boundaries often seem fuzzy or arbitrary. For example, by changing the shape of a mug, one can "get to" the class of cups; by dropping the handle, one can

The design parameters are thus one part of the structure of a design. The other, complementary part is the "configuration"—the global framework in which the design parameters are embedded.[7] The entire CAD file in turn specifies (describes) the structure of the artifact.

The CAD program for mugs must have "spaces" for specific values of every design parameter. For example, in the string of 0s and 1s that constitutes the whole design of a mug, there must be a "block" that corresponds to the mug's diameter. The program would look for an input from the designer for that block. The designer might specify 8 cm (or some other number); and in the block corresponding to diameter (in centimeters), the string `1000` (binary 8) would appear.

The two mugs we placed on the table earlier have a long list of design parameters, which are given in table 2.1. This is not in fact a complete list of design parameters for the class of artifacts we call "mugs," but it is sufficient to allow us to make our points. From the rough count of structural elements found in table 2.1, in a complete mug design there would be something on the order of forty blocks of information corresponding to forty design parameters. Many of the parameters are bounded within certain ranges (see the rightmost column of the table for probable ranges of the mug parameters). Values for each parameter, within the appropriate ranges, must be selected, and put into the corresponding blocks in order to complete a mug design.

Notation: The "Space" of Designs

It is helpful to have a set of symbols for the individual design parameters of a particular class of artifacts. Therefore, let us label the design parameters $x_1, x_2, \ldots, x_N$. The set of design parameters needed to describe completely a particular artifact (one of

---

"get to" the class of glasses. These issues are of concern to a wide range of scientists, including design and artificial intelligence theorists (Alexander 1964; Simon 1969; Bell and Newell, 1971, chapters 1–3); neurobiologists (Edelman 1992; Deacon 1997); neuropsychologists (McClelland and Rumelhart, 1995); neurolinguists (Pinker, 1994); archaeologists (Clarke, 1978; Mithen, 1996); and marketing scientists (Zaltman, 1997). For our purposes, however, we may take the existence of artifact classes as given, and look at how designs vary within a class.

[7] We thank Karl Ulrich for this information. Designers develop expertise in particular configurations, thus for them configurations are "essential," whereas parameters are ephemeral. Note, however, that a configuration is at once an artifact and a means of creating artifacts. Hence, one can design something that assists in the creation of designs, which in turn assists in the creation of new artifacts. This recursive property is what makes the innate complexity of artifacts and their designs manageable for human beings.

Table 2.1 Design parameters for two mugs

Parameter	Symbol	Mug 1	Mug 2	Parameter range
Material	x_1	Ceramic	Plastic	All materials
Tolerance	x_2	0.5 cm	< 0.1 cm	Varies with material
Mfr. process	x_3	Shape/bake/ glaze	Injection mold	Varies with material
Height	x_4	9.5 cm	16 cm	6–25 cm
Vessel diameter	x_5	8.0 cm	8.1 cm	5–11 cm
Width of walls	x_6	0.3 cm	0.1 cm	0.01–0.5 cm
Type of walls	x_7	Single	Double	Single, double, complex
Weight	x_8	xx oz	yy oz	1–12 oz
Handle material	x_9	Ceramic	Plastic	All materials
Handle shape	$x_{10} \ldots x_{12}$	Complex	Complex	Varies within 1–25 cm
Handle attachment	x_{13}	Integral	Glued on	Various (glue, pins, etc.)
Cap/No cap	x_{14}	No cap	Cap	Yes/no
Type of cap	x_{15}	—	Fitted, internal	Fitted, hinged, etc.
Cap material	x_{16}	—	Plastic	All materials
Cap diameter	x_{17}	—	8.0 cm	4.9–10.9 cm
Cap shape	$x_{18} \ldots x_{20}$	—	Complex	Varies within 11 cm
Large hole shape	x_{21}	—	Square	All polygons
Large hole size	x_{22}	—	1 cm	0.5–1.5 cm
Small holc position	x_{23}	—	180° from LH	All degrees from LH
Small hole shape	x_{24}	—	Circle	All polygons
Small hole size	x_{25}	—	0.2 cm	0.1–0.3 cm
Vessel color	x_{26}	Blue	Black	All colors
Handle color	x_{27}	Blue	Black	All colors
Cap color	x_{28}	—	Black	All colors
Decoration	$x_{29} \ldots x_{40}$	Complex	Complex	Many

the two mugs) is then a vector $\boldsymbol{x}{:}\boldsymbol{x} = <x_1, x_2, \ldots, x_N>$. The categories in the set or list—material, height, diameter, and so on—in turn make up the dimensions of an imaginary "space"—the space of designs (of mugs).[8]

One way of characterizing the design of a particular artifact is as a "point" in the "space" of designs. This sounds mysterious, but it is not. A "point" in the space of designs is simply a combination of parameters that differs in at least one dimension from all other possible combinations. One can think of each point as a list (as in table 2.1) with all the entries filled in. If any entry were changed, then that would constitute a different point, and a different design.

In reduced form then, we can think of the process of designing a particular artifact (one of our mugs) as a set of actions that "fills in the blanks" in the list of design parameters.[9] Versions with identical entries all the way down the list would be structurally identical, and we would say they have the same design. Versions that differed on some (or all) dimensions would be structurally distinct, and we would say they have different designs.[10]

Hierarchical Design Parameters

Looking down the list of parameters for the two mugs, we see that there is a set of parameters, $<x_{15}, \ldots, x_{25}, x_{28}>$, that are left blank for the ceramic mug, but filled in for the plastic mug. These parameters describe dimensions, color, and so on, of the *cap* of the plastic mug. They are "called into being" by the decision to have a cap. The decision to have a cap (or not) is also a design parameter, but of a special kind. We call it a hierarchical design parameter.

Hierarchical design parameters are like ordinary parameters in that they correspond to "spaces" that need to be filled in to complete the design of an artifact. Thus within the global framework of the CAD program for mugs, a "slot" exists that corresponds to the "cap/no cap" decision. As with any design parameter, the program

---

8 Technically, the parameter categories $<x_1, x_2, \ldots, x_N>$ form the basis of an N-dimensional vector space.

9 This is not necessarily the way the process would be perceived or carried out by designers, but it is what the process does. Hence, this representation is an "abstraction" or a "reduced form" of the design process. On abstraction (equivalently, reduction) as a profitable strategy for investigating and representing complex phenomena, see Simon (1969); Alexander (1964); Bell and Newell (1971, chapter 1); and more recently, Bucciarelli (1996); and Wilson (1998).

10 If two mugs had identical entries, but were different in some important way not captured by the list, we would add the dimension on which they differed to the list of parameters.

would look to the designer to fill that slot. However, with a hierarchical design parameter, the designer's decision has the character of a logical switch.[11] If the decision is "yes, have a cap," then a new set of design parameters comes into existence. Decisions on these parameters must then be made to complete the design. But if the decision is "no, don't have a cap," the set of parameters related to caps will not be relevant to that particular design.

The design framework related to caps remains a possibility for other mug designs. It can be called into existence by simply changing the "setting" of the hierarchical parameter. The designer can decide that the next mug will have a cap, and a framework for designing caps becomes instantly available.

Hierarchical parameters are the way in which designers delimit and thereby bound the space of designs they plan to search. Without these "on/off' switches, the space of possible designs is impossibly immense, amorphous, and unstructured. However, early-stage, hierarchical design parameter choices (e.g., the choice to make a mug, not a cup or a glass) literally "create" some dimensions and exclude others.[12] This shows how a *recursion* (the choice of a set of choices) can be used to manage the innate complexity of artifacts and designs.

In effect, by setting a series of switches early on, designers can bound their immediate problem (the design at hand) to one that is manageable given their knowledge and resources.[13] Modular designs, which we formally describe in the next chapter, are the result of a purposeful, consistent, and rigorous application of this bounding technique.

Functions: Primary, Subsidiary, and Supplemental

A particular artifact can have several functions, although some will be more important (and more intrinsic to the artifact's definition) than others. For example, mugs hold liquid—that is their primary and defining function. In addition, mugs are often meant to hold hot liquid—hence their surfaces do not melt, as happens with some

---

[11] In programming, this is a "conditional branch."

[12] Brown (1969).

[13] Bounding of the problem at hand is always necessary because both human beings and machines have limited capacity to process information and make decisions. It is in fact easy to make a problem too large to be solved in some allotted time. On the importance of human beings' limited processing capacity (so-called bounded rationality) for the design of organizations and the economic system as a whole, see, for example, Hayek (1945); Simon (1969); and Nelson and Winter (1982).

waxed paper cups. The handle of a mug moreover ensures that the heat of the vessel does not burn the hand that holds it.

Artifacts can also have subsidiary and supplemental functions. Subsidiary functions elaborate and extend the primary function: they serve to differentiate artifacts within a given class, hence are very important in competitive settings. For example, the mug with the cap and the double thermos walls keeps liquid warm for several hours, and can be transported without spilling the liquid. Keeping liquid warm and making transport easy are subsidiary functions that extend the primary function of holding hot liquid.

Supplemental functions offer alternatives, or run in parallel to the primary function. For example, the ceramic mug holds pencils and looks good on a desk. Neither of these functions has anything to do with holding liquid—hence they supplement the primary function. Displaying emblems (like the Jayhawk and the corporate logo) is another supplemental function of many modern mugs.

In general, an artifact's value or worth in society is determined by its functions, and not by its structure.[14] If the functions are important and the artifact performs them well, the artifact's value in society will be high. If identical functions can be performed using a different structure, users will be indifferent, and will value the structures equally.[15] Also, other things being equal, users and designers almost always prefer structures that are less costly over those that are more costly.[16]

---

[14] Technically, by "value of the artifact" we mean its impact on general social welfare its contribution to the sum total of human happiness. The *price* of the artifact will generally depend not only on its value, defined this way, but on the particular characteristics of its market, e.g., how many competing firms are making it or can make it. We discuss how the organization of the larger economic system affects the value of artifacts and designs in chapter 4. We take up the issue of market structure and pricing in chapter 15 and in volume 2.

[15] The exceptions to this rule are informative. In some cases, users may value the "look and feel" of a particular structural element, like "real wood" or "natural fiber." In these cases, the substitutes do not exactly replicate the functions of the original item, since users can tell the difference between them. Even more subtle exceptions exist; for example, users may derive psychic benefit from their knowledge of how the artifact was made: "handcrafted" and "made in U.S.A." are two examples of this. As a practical matter, in the computer industry, users and designers of computers appear to be more sensitive to functionality than to structure. Evidence of this can be found in the fact that hardware and software implementations of the same function often compete with one another. (If structure mattered for its own sake, then artifacts with such different structures could not be substitutes.)

[16] Exceptions are again informative. They consist of goods consumed for the purpose of impressing or exploiting others (see Veblen 1899).

The Design, Production, and Use of an Artifact

Human beings interact with artifacts in three distinct ways: (1) they design them; (2) they produce them; and (3) they use them. Virtually all verbs that describe a human action on an artifact pertain to its design, production, or use. Thus, Ann conceives of a mug, or draws it (design); Bill molds the mug or glazes it (production); Jane fills the mug, drinks from it, or puts it away (use).

We know and accept that a tangible artifact must be produced before it can be used. A mug sketched in a designer's drawing or one molded from hot plastic or wet clay will not hold coffee well! As such, we understand that tangible artifacts pass from production to use in strict sequence. Therefore, it is easy to conceive of a "domain" of production, and a separate "domain" of use, and the idea that production precedes use fits easily into our mental frameworks.[17]

The "domains" of design and production are not so easily distinguished, however, because in practice the activities of design and production are often carried out simultaneously and by the same person; moreover, that person may begin a project not really knowing what will come out. The particular medium selected may suggest paths of expression that in turn determine the ultimate form of the artifact. If that is so, the activities of design and production become intertwined.[18]

Although the distinction does not come naturally, differentiating between design and production is a critical step in the construction of our theory. The essence of the distinction is as follows:

- The output of a design process is a design, that is, a *description* of a particular artifact.
- The output of a production process is the artifact itself, the thing that is used.
- When separated, design precedes production.

At its core, a design process is a search for something unknown. The ultimate form of the artifact is unspecified at the outset of the process. A search is mounted

[17] This logical sequence does not always hold for intangible artifacts. For example, processes (which are an important category of intangible artifacts) are produced (executed) and used (employed) at the same time. For further discussion of these domains of activity, see, Alexander (1964); Christensen and Rosenbloom (1995).

[18] McCullough (1996).

in hopes that a form (a particular set of parameters) may be discovered that jointly satisfies certain objectives and takes account of certain constraints (the thing holds liquid, conforms to the laws of physics, and costs less than $1.00). The result of this search is a description of the thing to be made, *including instructions about how to make it.*[19] Once what is to be made and how to make it are known, exactly and completely, the design process is over. What remains then is "pure" production.

For its part, a "pure" production process is an algorithm, a recipe that can be executed. There is a straight line, a sequence of steps from concept (the design) to realization (the thing itself). In a completely specified production process, no blanks or gaps exist: the process can proceed automatically. If there are deviations from the recipe—for example, experiments aimed at improving the process in some way—then those are by definition part of the search for an unknown solution, hence part of a design process, not part of the production process.

This distinction between design and production is important for two reasons. First, when design and production are carried out at the same time and in the same place, they will engender conflicting objectives. Searching for something better slows down production, and uses resources. The search for new designs also introduces uncertainty into any process, for if the "destination" were known, the search would be over. In contrast, the output of a "pure" production process is a certainty. The managers of many actual production processes strive for the ideal of certainty, but in so doing they must abandon the search for new and better designs.[20]

Second, as a logical necessity, when tangible goods are being produced the design process must stop at a point that leaves enough time for the existing production algorithm to be executed. Designers can and do work to postpone that "freeze point."[21] However, the existing production algorithm(s) must always be inserted between the concept (i.e., the design) and its realization.

Detailed representations of artifacts, that is, designs, increasingly have become intermediate products embedded in complex technologies. Thus, technologies that once would have been mapped this way:

---

[19] Instructions on how to make something are an important part of the overall design. However, there may be generic instructions (a production technology) that are applicable to all artifacts in a given class or category. In that case, the instructions pertaining to a specific design may simply be its unique parameter settings within the more general design of the production process.

[20] Abernathy (1978).

[21] Iansiti (1994).

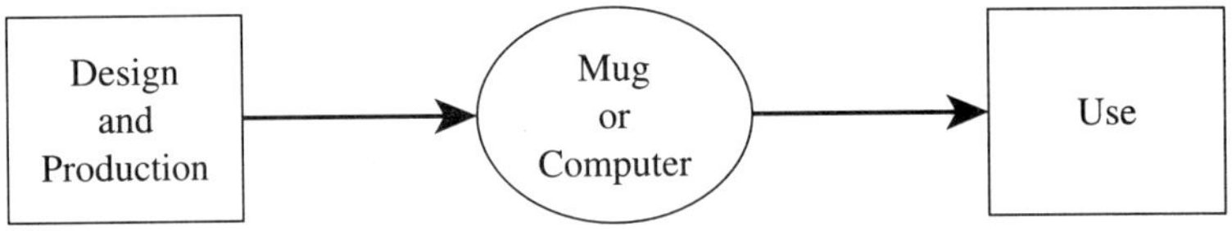

now have design and production broken down into two discrete steps as follows:

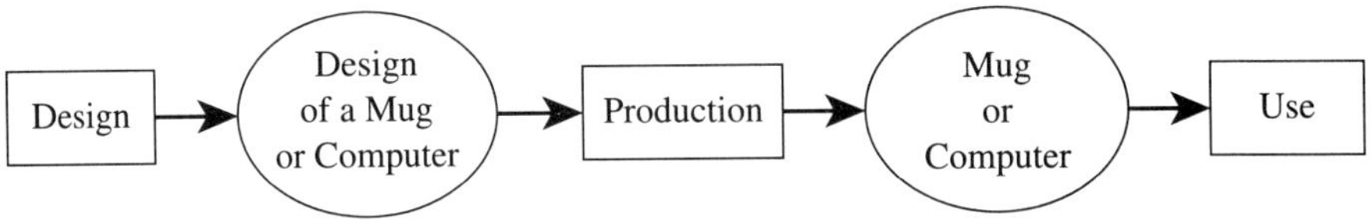

In the second map, a formal "design artifact" lies between, and links the design and production steps of the technology.[22]

The idea of developing a complete description of an object or process before attempting to produce the object or execute the process is relatively new. Indeed, formal designs (of both objects and processes) have been important forerunners of production in our economy only during the last hundred years. The spread of formal designs coincided with the diffusion of modern mass production techniques, and the two trends were related.

Mass production, generally speaking, relies on automatic processes that execute precise algorithms at high speeds.[23] Searches of any kind are highly disruptive and costly to such processes,[24] and thus the search for new designs came to be divorced from ongoing production.[25] At the same time, new scientific methods made possible

---

22 This splitting of design and production may occur even when the "thing being produced" is itself a production process. One first designs the process, representing it in some formal way (e.g., as a flow chart or a recipe). One then debugs the process (more design), and configures the resources needed to implement it. The process may then be executed, that is, simultaneously "produced and used." If the design of the process is complete (i.e., if there are no mistakes and no steps left out), then the process will be an algorithm and its execution can proceed automatically.

23 Abernathy (1978); Chandler (1977); Hounshell (1984).

24 This principle is captured in Henry Ford's famous aphorism, "In mass production, there are no fitters." Fitting was a step in which operators sought to match the internal and external parts of a piston by trial and error (a search process) (Hounshell 1984).

25 Abernathy (1978). Provisions for learning and systematic experimentation have since been reincorporated into the design of many high-speed production processes. (Japanese manufacturers led the way in this initiative.) However, great care must be taken to ensure that the search for better ways of doing things does not disrupt the flow of goods. For interesting

more systematic and efficient searches for better objects and processes.[26] However, many of these scientific methodologies were based on detailed and categorical descriptions of the objects of study. Thus, to reach their full potential, both high-speed mass production and scientific methods required more formal descriptions and stricter divisions between stages of work than older technologies had demanded.

Under the impetus of these two driving forces, the practice of preparing formal designs became widespread in the first half of the twentieth century, although it did not penetrate every nook and cranny of the economic system. In the second half of this century, the cost of preparing formal designs fell dramatically and continuously with advancing computer technology. Not surprisingly, then, there was no letup in the creation of new designs, nor in the invention of new design methodologies.

Indeed, following an economic logic that we describe in chapters 9 and 10, the trend has been to split designs even further to create options and thereby realize greater value from the effort of design. In the computer industry, moreover, the process of splitting designs has led to the formation of new markets and a growing number of ever more specialized firms. Thus, explaining how the splitting of designs occurs and what it implies for artifacts, firms, and markets is one of our main objectives in these two volumes.

The Role of Designers

Designers work with the design parameters of a class of artifacts to represent and then realize a real artifact. A real artifact is one that conforms to the laws of nature so that it "works," and can be made with an appropriate production technology. Anyone can imagine a desirable form of an artifact—the computer that knows everything, the mug that never runs dry. But such imaginary artifacts are the stuff of fairy tales. The true effort of design is the effort of imagining artifacts that can become real, and then making them real.

The designer, therefore, selects the design parameters of an artifact so that the artifact "works," is makable, and, if possible, pleases the user as well.

If a designer is working in an existing class of artifacts, then his or her goal is to create a new design that is better than its predecessors on one or more functional

discussions, see, for example, Jaikumar and Bohn (1986); Hayes, Wheelwright, and Clark (1988); Spear (1999).

26 See Landes (1998) on the "invention of invention."

dimensions. To imagine a better new design, the designer must know the relationships between structural elements, which are dictated by the laws of physics and logic. The designer must also know which functions he or she intends for the artifact to fulfill, and how these functions constrain the artifact's structure.[27]

For example, in the realm of mugs, the designer may intend for a mug to hold liquid—that will be its primary function, as we have said. He or she must then have "working knowledge" of how the law of gravity affects liquids, and of what materials and spatial forms can hold liquid. That structural knowledge is relevant to the mug's intended function. Based on this combination of functional intent and structural knowledge, the designer might conclude that a cube-shaped vessel would be a curious design, but would "work" (function) as a mug. He or she might also surmise that a cylindrical vessel with a hole in the bottom might look like a mug, but would not "work" as a mug.

A designer must also be able to predict changes in functionality that will arise from specific changes in structure (specific changes in the design parameters). He or she might reason as follows: If I make a larger vessel with double-sided walls, it will hold more liquid and keep it hot longer. If I put a cap on the vessel, the liquid can be carried around with fewer spills.

Finally, a designer must be able to weigh potential changes in structure and function, adjust for uncertainty (the fact that the structural changes may not have the desired effect), and decide whether a new design is worth trying or not. In other words, the designer must have mental concepts of value and of value changes. The designer's concept of value may be primitive (it works or it doesn't) or refined (market prices, option values), but some concept of value must be in the designer's head, to provide a guide for action (the new design is worth trying or isn't).

The labor of design is thus a complex mental effort that relies on both imagination and intent. The designer must first associate changes in the structure of the artifact (the design parameters) with changes in its functions (what it will do).[28] Then the predicted changes in function must be projected onto a change in "value." The change in value must in turn be compared to the effort of implementing the changes in the design (including the cost of producing and marketing the new artifact). If the result is positive, the designer will have a reason to go ahead and try the new design.

---

27 This knowledge forms what Bucciarelli (1996) calls the "object world" of the designer.

28 Even wholly new artifacts are generally imagined in relation to something that already exists (Clark 1985).

This mental effort, we believe, lies at the core of all human design processes. Our view is consistent with Gerald Edelman's characterization of the nature of higher-order (e.g., human) consciousness:

[T]he conceptual centers of the brain treat the symbols and their references and the imagery they evoke as an "independent" world to be further categorized. . . . [Such] higher-order consciousness leads to the construction of an imaginative domain. . . . It constructs artificial objects that are mental. In culture, these acts lead to studies of stable relations among things (science), of stable relations among stable mental objects (mathematics), and of stable relationships between sentences that are applicable to things and mental objects (logic).[29]

One of these imaginative domains, we believe, is what Leo Bucciarelli has called the designer's "object world." This mental arena is filled with symbol objects bound together in stable relationships—the "laws" of design in the object world of a particular artifact class.[30]

Our theory of design evolution thus rests on the idea that human beings are capable of visualizing new artifacts in an imaginative domain, and then acting purposefully to bring those artifacts into existence in the real world. We capture this idea in an axiom: *Designers see and seek value in new designs.*

Although the concepts of structure, function, and value must be present in designers' minds, the relationships between these concepts need not be framed with precision. The notions may be very fuzzy, perceived as intuitions and hunches, not articulated or voiced. Moreover, designers' beliefs about structure, function, and value are always speculations or hypotheses about the true state of the world. They are often proved wrong: in these cases, the newly-designed artifact will not work as expected, or it will not have the value expected. However, as the store of knowledge surrounding an artifact grows, designers' hypotheses will tend to become more accurate and precise.[31]

Design Parameter Interdependency: What Designers See

It would be relatively easy to design an artifact if all its design parameters were independent, and each contributed to the artifact's functions in its own way. Unfortunately, designers face a far more difficult set of problems. Design parameters depend

---

[29] Edelman (1992, pp. 150–152). Deacon (1997) offers a compatible theory of the mind.

[30] Bucciarelli (1996).

[31] Clark (1985).

on one another, often in very complex and convoluted ways. This in turn constrains the way designers go about their tasks.

There are many types of design parameter interdependency, and listing them is beyond the scope of this discussion. However, it is essential to have an intuitive grasp of the nature of these interdependencies in order to appreciate the complex constraints designers must grapple with, and how these constraints affect the benefits and costs of a design effort.

Design parameter interdependency can be illustrated by a simple but general example: "fit interdependency." This type of interdependency arises in every realm of design that involves tangible artifacts—in mugs, in computers, in automobile pistons, and in plugs that must fit into wall sockets.

Recall the plastic mug with a cap. The cap had several functions: it retained heat and minimized spills, allowing the user to sip liquid from the vessel. The cap in question was designed as a plastic "stopper" that fit snugly in the top of the vessel. A snug fit was essential to the functioning of the mug—if the cap was the slightest bit loose, hot liquid would leak out when the user tried to drink.[32]

Unfortunately, loose caps and tight caps are indistinguishable to the naked eye. The diameters of a loose cap and a tight cap differ by less than a millimeter, yet one "works" and the other doesn't. Note that if the mug did not have to "work" (if it was a piece of decorative art, for example), the cap would not have to fit tightly. But users for whom mugs must hold liquid unambiguously prefer those that do not leak to those that do. As a result, a mug with a snug-fitting cap is more valuable than one with a loose-fitting cap, by quite a large margin.

The snug-fitting cap requirement in turn creates a very tight interdependency between the mug's vessel and its cap. Whatever the diameter of the vessel, the diameter of the cap has to be just a tiny bit (e.g., 1 mm) smaller.[33]

Figure 2.1 illustrates this "fit interdependency." The x-axis shows the diameter of the vessel; the y-axis shows the diameter of the cap. The xy-plane represents different combinations of vessel and cap diameters. The figure shows how the value of the mug varies for different diameters of the vessel and the cap. In constructing this figure, we assumed that the "tolerance" of the design was about 1 mm.[34] Hence if the

---

[32] We know because we have a loose cap in our collection and have suffered spills as a result.

[33] More precisely, the internal diameter of the vessel must have this relation to the external diameter of the cap.

[34] "Tolerance" is determined by the material and the manufacturing process. Knowledge of tolerances that can be achieved with a given material and process is part of the structural knowledge (what Bucciarelli [1996] calls "the object world") that designers bring to their work.

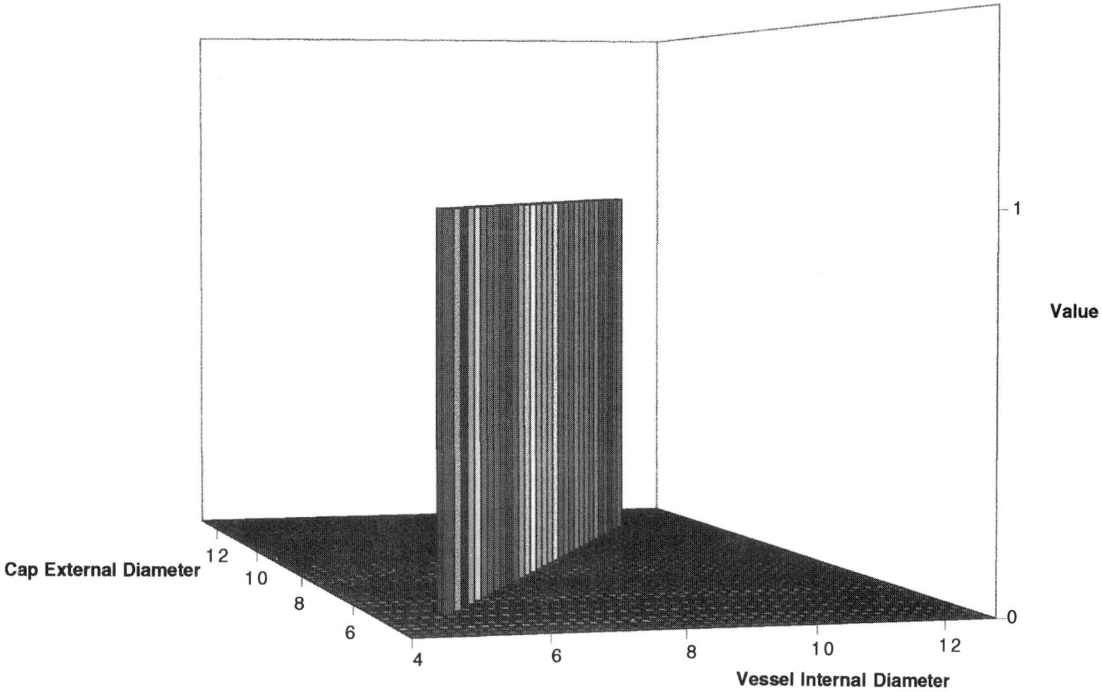

Figure 2.1 An illustration of a "fit interdependency" (cap diameter = vessel diameter − e; $e < 1$ mm).

diameter of the cap was less than the diameter of the vessel, but within 1 mm of it, the cap would "work." The value of the mug, we assumed, would then be one. Otherwise the cap would not "work," and the value of the mug would be zero.

Figure 2.1 shows a sharp wall, jutting out of the plane at the point (x = 5 cm, y = 4.9 cm), continuing along the 45 degree line, and then dropping back to the plane at the point (x = 11 cm, y = 10.9 cm). The value of a particular (x, y) combination is *one* if it lies on the wall and *zero* elsewhere on the plane. The vertical width of the wall is 1 mm, set by the tolerance of the design. The upper and lower bounds are determined by the ergonomics of mugs: if the mug is less than 5 cm or more than 11 cm wide, it becomes hard to drink from the vessel.[35]

[35] Note, however, that if the opening of the vessel shrinks to around 1 to 2 cm, it is again possible to drink, but the artifact becomes a bottle, not a mug.

The picture shows how important and how unforgiving design parameter interdependencies can be. If the designer or manufacturer fails to get the cap diameter within 1 mm of the vessel, the artifact will cease to perform one of its key subsidiary functions (prevent spills), and its value will drop significantly. This is a case where being close to the "right value" (say within 1.5 mm of it) is not good enough.

The "fit interdependency" shown in figure 2.1 is one of the simplest examples of a design parameter interdependency. Such interdependencies are what make the effort of design complicated, risky, and at times frustrating.

Seeing and Seeking Value in the "Space" of Designs

The imaginary effort of designing can be thought of as a mental process of "seeing and seeking" in the "space" of design parameters.[36] This "space" spans all possible combinations of parameters, and so is very large indeed. Many combinations don't work, hence have no value—these correspond to the points on the plane ("the sea") in figure 2.1. But other combinations do work—these jut out of the "plane" like islands in a sea. The designers' job is to find the islands in the sea of possible designs—in other words, the designs that "work." Then, within and across all the islands, designers must seek the highest peaks—the designs that work better than any others, and generate the highest values.

Where interdependencies exist, a seemingly small change in one design parameter can interact with other parameters and completely destroy the value of the specific artifact. Figuratively, where there are interdependencies, the islands of value in the sea of designs will be small, have complex shapes, and be riddled with steep cliffs and crevasses.

Fit interdependencies—where one parameter must equal another plus or minus a tolerance value—are pervasive in all types of artifacts. In electronics, every plug must fit into the corresponding socket. This single requirement generates a myriad of fit interdependencies that constrain the design of wall plugs, telephone jacks, connectors, chips (their pin count, diameter, and spacing), boards, disks and disk drives, and a host of other devices.

---

[36] This is a metaphor but a useful one. It is connected to the concept of a "fitness landscape" in biology and complexity theory. (See, for example, Waddington 1977; Dawkins 1996; Kauffman 1993.) It is also connected to fundamental concepts in economics and operations research, for example, the metaphor of "hill-climbing" in a parameter space that underlies the simplex method of linear programming, and the Kuhn-Tucker theorems of price-quantity optimization. For recent applications of the concept in economics and business strategy, see, for example, Levinthal (1994a) and Rivkin (1998a, b).

But there are other, more complex interdependencies, for example, weight interdependencies (every tangible component contributes to the weight of the whole) and volume interdependencies (every tangible component takes up a certain amount of space). In computer designs, one of the most important and complex interdependencies involves the *speed* at which the computer does its work. Speed, as perceived by users, is a function of many elements (design parameters) in the computer's hardware and software, which interact in a number of obvious and not-so-obvious ways.[37] Computer designers of all kinds endeavor to understand the determinants of speed, in order to design faster machines. Working in a very complicated design space, with millions upon millions of parameters, computer designers must "see" the structural interdependencies and their functional consequences, and "seek" those combinations that deliver the most speed for the least cost.

Design Structure

Hierarchical relationships and interdependencies among design parameters can be formally mapped using a tool called the design structure matrix (DSM). The mapping procedure was invented by Donald Steward, and has been extended and refined by Steven Eppinger.[38]

Mapping the structure of a design begins by listing parameters, for example, the list of mug parameters in table 2.1. In some cases, there will be a strict hierarchical relationship between two design parameters—thus, the parameter "to have a cap or not," x_{14}, precedes all the parameters that pertain to the detailed design of the cap, x_{15} through x_{25}. Only if a decision is made to have a cap will the other parameters be relevant.

In other cases, there may be interdependence of design parameters without a strict hierarchy. As we have seen, the decision to have a snug-fitting cap creates an interdependency between the diameter of the vessel, x_5, and the diameter of the cap, x_{17}.

---

37 Bell and Newell (1971, Chapters 1–3); Hennessy and Patterson (1990).

38 This section and the next rely heavily on the concepts of "design structure" and "design structure matrix" invented by Donald Steward, as well as the work of Steven Eppinger and his colleagues. Eppinger's theoretical and empirical work characterizing complex design processes has been particularly influential in the development of our own thinking about these problems (Steward 1981a, b; Eppinger 1991, 1997); Eppinger, Whitney, Smith, and Gebala (1994); Krishnan, Eppinger, and Whitney (1997); Smith and Eppinger (1997a, b); Novak and Eppinger (1998).

(a) Hierarchy

		x14	x17
Cap/No Cap	x14	•	
Cap Diameter	x17	**x**	•

(b) Interdependence w/out Hierarchy

		x5	x17
Vessel Diameter	x5	•	**x**
Cap Diameter	x17	**x**	•

Figure 2.2 Design structure: hierarchical and interdependent design parameters.

The two diameters must be chosen consistently, but either can be chosen *first,* as long as the correct functional relationship is reflected in the second choice.

The hierarchical relationships and interdependencies among design parameters can be summarized in matrix form, as shown in figure 2.2. In the case of strict hierarchy, an **x** appears in the column of the first parameter (the initiating parameter) and the row of the second (the resulting parameter).[39] Where interdependency exists, two **x**s appear in symmetric spots around the main diagonal of the matrix.

A *design structure matrix* (DSM) summarizes and displays all this information. In this sense it is a map of the design structure.[40]

To construct a DSM, one assigns the individual parameters of a design to the rows and columns of a square matrix. Then, going down the list, if parameter *a* is an input to parameter *b*, one puts a mark (an **x**) in the column of *a* and the row of *b*. One continues until all (known) hierarchical and interdependent relationships have been accounted for.[41]

Figure 2.3 displays a DSM corresponding to the first ten parameters of mugs listed in table 2.1. The point of this figure is not to characterize perfectly the design struc-

---

[39] We are risking confusion by using the letter "x" to denote both a design parameter (x_1, x_2, etc.) and a point of interaction between two parameters, as in figure 2.2. In the text and in our figures, design parameters will always be denoted in lower case and in a Roman font (x); points of interaction will always be denoted in bold (**x**). In Chapter 10 we introduce and formally define "the value of a module," which will always be denoted in uppercase roman (X).

[40] It is only a map, because one could display the information in other ways, for example, as a list, as a directed graph, and so on.

[41] We are defining DSMs constructively because analytically they are very complex mathematical objects. They are related to Hessian matrices in economics and to protein and genome spaces in complexity theory (see, for example, Kauffman 1993), but they are not exactly either of these things. However, children of elementary-school age can construct and explain simple DSMs. (We know this because we have taught the technique to a class of third graders.) We invite the reader to construct a DSM for an everyday artifact he or she knows well.

Design Parameter

		1	2	3	4	5	6	7	8	9	10
Material	1	•	x	x			x	x	x		x
Tolerance	2	x	•	x			x	x	x	x	x
Mfr. Process	3	x	x	•			x	x	x	x	x
Height	4			x	•	x			x		x
Vessel Diameter	5		x	x	x	•	x	x	x		
Width of Walls	6	x	x	x	x	x	•	x	x		
Type of Walls	7	x	x	x		x	x	•	x	x	
Weight	8	x		x	x	x	x	x	•	x	
Handle Material	9	x	x	x				x	x	•	x
Handle Shape	10	x	x	x	x					x	•

Figure 2.3 Design structure matrix: a map of a portion of the design structure of a mug.

ture of mugs, but to show a typical pattern of parameter dependencies. The matrix was constructed by asking, with respect to each parameter: If this parameter were changed for any reason, what other parameters might have to be changed too?

Suppose, for example, the designer wanted to change the material from which the mug was made. A change in material would clearly affect tolerances, and the manufacturing process. It might also affect the width of walls, the type of walls, the weight, the handle material, and the handle shape. However, within limits, a change in material would not necessarily affect the height of the mug or its diameter, because most candidate materials would be able to support the full range of mug heights and diameters.[42]

Overall, a DSM like that shown in figure 2.3 can be read as a kind of input-output table of design parameter choices.[43] The **x**s in the *row* of a particular parameter indicate which other parameters affect this one. The **x**s in the *column* of a parameter indicate the parameters that are affected by it.

We emphasize that most of the dependencies shown in figure 2.3 arise because of

---

[42] Readers who are sticklers for accuracy will maintain that a change in material *could affect* height and diameter as well as everything else. They might want to imagine very light **x**s in some of the boxes we left blank, denoting weak dependencies.

[43] The primary difference is that input-output tables typically summarize flows of material, whereas here the "flows" are of design information.

the laws of physics and logic as they apply to mugs. A mug handle "must" attach to the vertical surface of the vessel; that is what a "handle" does. This being the case, the height of a mug and its material will affect the shape of the handle and *its* material. Conversely, if a designer wanted a handle of a special shape or a special material (say, ivory), the height and material of the vessel would have be made consistent with that choice. There is no way around these facts: they are inherent in the physics and the logic of the artifact itself and how it can be made.

However, like any map, a DSM is only as good as the knowledge that goes into it. Just as shoals can exist hidden in the ocean, unknown interdependencies can exist, and emerge unexpectedly in a design space. Unknown interdependencies can be fatal or merely annoying, requiring extra work. Or they can be fortunate, revealing new "peaks" of value. Hence, unknown interdependencies are one of the primary sources of uncertainty in design processes. Their existence affects how design processes can be carried out—that is, they influence the organization of the *tasks* of design.

The Tasks of Design

A design is a complete description of the structural elements of a particular artifact. It is analogous to a CAD file with all the design parameter "slots" filled in. If there is an empty slot, the design description will be incomplete and ambiguous—it will correspond to more than one point in the space of designs.[44]

It is now time to consider how human beings arrive at a complete design. We will do this by looking first at the outputs of a design process, and then reasoning backward to the inputs. The *tasks* of design are what connect the two.

A design process has both a beginning and an end. It begins with a set of unchosen parameters ("empty slots" in a CAD file), which correspond to uncompleted design tasks. The structural elements of the final artifact are not fixed—they exist as possibilities, known and unknown, in the space of designs. As the design process

[44] For tangible artifacts, ambiguities must be resolved before the artifact can be made, and the thing produced will have a unique design corresponding to its actual structure. However, in practice, ambiguities can be resolved randomly or by default—for example, if the design does not specify color, the producer can default to the natural color of the material. Intangible artifacts may have persistently ambiguous designs—think of a system of law or a set of policies, for example.

moves forward in time, the parameters are selected; the "empty slots" are filled in, and the process converges on a single, final, complete design. The expectation, which justifies the effort, is that the value of the final artifact will be high, relative to other known "points" in the space of designs, that is, other potential versions of the artifact.

Thus a design process starts with a definition of the artifact to be created,[45] and then "selects" the design parameters needed to arrive at a final design. The choices are constrained by parameter dependencies and interdependencies, which arise from the laws of physics and logic. The microscopic "units" of the design process are the "acts" of selecting individual design parameters. To come up with a design that "works," the actors must put effort into their choices. The act of selecting a design parameter, coupled with the effort, is a "task of design."[46]

Put succinctly, the design process begins and ends with a list of design parameters that are also design tasks. At the beginning, the list is a set of decisions that will be made, that is, tasks that *will be* performed. At the end of the process, the list tells of the decisions that have been made, that is, tasks that *were* performed. The value of the design is determined by the quality of the artifact that the final list describes.

The Task Structure of a Design Process

The identification of design tasks with the selection of design parameters was pioneered by Donald Steward and continued by Steven Eppinger and his colleagues. It is a powerful analytic device because by using it we can see *with clarity* how the physical and logical structure of an artifact gets transmitted to its design process, and from there to the organization of individuals who will carry the process forward.

---

[45] This initial definition may be very narrow (a travelling mug with a cap), or very broad (something to drink out of). The initial definition of the artifact to be created determines the scope of the designers' search in the "space" of designs. Narrow definitions call for focused searches; broad definitions call for wide searches.

[46] We are following Steward (1981a, b); von Hippel (1990); Ulrich (1995a, b); and Eppinger (1991) in this naming convention. The definitive characteristic of a task is that it require some mental or physical effort. The tasks involved in production are often physical tasks: a pallet is moved from point A to point B; a chemical reaction takes place; a wire is wrapped around a knob. In contrast, the key tasks of design are not physical but mental and symbolic. However, not all mental and symbolic acts qualify as design tasks. Daydreaming is not designing, for example.

The *task structure* of a design process consists of

- a list of tasks, corresponding to the design parameters to be selected;
- a list of all precedence relationships between tasks;
- an assignment of each task to one or more "doers" of the task.

A "precedence relationship" means that the output of task A is an input to task B. In physical processes, "precedence" means that material must flow from the location of task A to the location of task B. Design processes are not physical processes, however, and so instead of material flows, we must track the flow of information. In this context, a precedence relationship exists if the choice of parameter A influences the choice of parameter B.

A task structure is not completely specified until we know who is going to perform the tasks. Every task must be assigned to an actor—a human being, a machine, or a human-machine combination. We can represent these task assignments by matching agents (e.g., Ann, Bill, Clare, Dave, etc.) with actions (e.g., "selects") and objects (e.g., design parameter x_1, design parameter x_2, etc.). For example, we might say Ann will select parameter x_1, Bill will select x_2, and so on down the list.

In this mapping of actors to tasks, there is room for multiple task assignments (Ann will select x_1 and x_2); and joint responsibility (Ann and Bill will select x_1). It is even possible to specify delayed, conditional task assignments (whoever selects the material, x_1, will select the manufacturing process, x_3). What is problematic, however, is an unassigned task, for, by definition, the design process will not be complete, until all necessary tasks have been accomplished.[47] (See box 2.1.)

Donald Steward was the first to investigate the precedence relationships among the tasks of engineering design, and to summarize them in matrix form. In this fashion, he invented both the DSM and the task structure matrix (TSM). His instructions for constructing a matrix of task relationships were as follows[48]:

---

[47] Sometimes the failure to specify a doer for a particular task can be fatal to the job. For example, if no one selects a material for a mug, the mug cannot be made at all. In other instances, however, failure to perform a task results in a default setting for the design (and the resulting artifact). For example, if no one explicitly makes a cap/no cap decision, the default design is a mug with no cap. Learning to perceive defaults as explicit choices (i.e., there exists the possibility of creating a mug with a cap) is part of what designers do as they "explore" a complex design space. A catalog of these choices then becomes part of the designers' working knowledge—their object world (Bucciarelli 1996).

[48] Steward (1981b).

Box 2.1 The Fractal Nature of Tasks

A very deep problem arises when specifying the "basic unit" of a task. For example, if selecting a design parameter involves running experiments and collecting data, is each parameter that is selected a task? Is each experiment a task? Is gathering each data point a task?

Ultimately, tasks are defined by the instructions given to those who will do them. For the same task (e.g., "Fetch and return the ball"), different instructions are given to a human being, an animal (a dog), or a machine (a computer or a robot). A crucial difference between human beings and animals or machines is that we have an innate ability to assemble very basic tasks into *meaningful groups,* and to conceive of a job in terms of larger task units, not smaller ones.[1] For example, "play right field" is a large task, which subsumes catching or fetching and throwing a ball to different points. "Play right field" is an adequate instruction for a human being who knows the game of baseball. Most dogs and all machines need more detailed instructions to perform this task.

Hence, human task structures have a fractal property—within each large task lies a set of smaller tasks, and within each of those there may lie a set of still smaller tasks. (This subdivision continues until eventually one reaches the level of neural programming.) In writing a book, for example, each chapter may be defined as a task. Alternatively, each paragraph, sentence, word, or even keystroke might be construed as a task. Each of these levels of task definition is valid for some purposes, and this makes it difficult to conceive of a single, unique, and complete list of tasks.

Human beings do not seem to be stymied by this inherent ambiguity, however. We "chunk" and "unchunk" tasks quite easily, moving from one level of task definition to another, according to what is most convenient for the job of the moment. When learning to type, we may focus on keystrokes. When learning to write, we may focus on sentences and paragraphs. Later, we may group sentence and paragraph tasks together, to construct essays, articles, and books.

---

[1] For theories of how humans do this, see Edelman (1992), Pinker (1994), or Deacon (1997).

1. First, list and name all pertinent tasks (for convenience we will assign tasks to the letters A, B, C, etc.).
2. Identify the tasks, A, B, C, and so on with the rows, columns, and diagonal elements of a matrix.
3. For each task, identify all predecessor tasks.
4. If task *i* precedes task *j*, put a mark (**x**) in column *i*, row *j* of the matrix.

In Steward's view, the appropriate sources of data on tasks and precedence relationships were the engineers themselves:

Each engineer responsible for a task should be able to tell you which of the set of other tasks are needed to do the work. If the engineer cannot give you this information, there is good reason to doubt whether he can perform the task at all.[49]

A *task structure matrix* is a map of the tasks of design and the precedence relations among tasks. Whereas a DSM begins with a list of design parameters (e.g., the elements in table 2.1), a TSM begins with a list of tasks: select x_1, select x_2, and so on. The list of tasks becomes the rows, columns, and diagonal elements of the TSM. The off-diagonal elements represent the necessary connections between the tasks—they indicate that the doers of the column task and the row task will want to confer with one another before arriving at their respective final decisions.

The Fundamental Isomorphism between Design Structure and Task Structure

Although the DSM and TSM are closely related, they are not identical. The DSM characterizes the "topography" of a design "space." It indicates both hierarchical dependencies among parameters (parameter x_1 calls parameters $x_2 \ldots x_i$ into being); and interdependencies among parameters (a change in x_j makes a change in x_k desirable and vice versa). For its part, the TSM characterizes a process—a set of actions (choices) and coordination links.

For the final design to have a high value, it must be possible to recognize and deal with the hierarchical relationships and interdependencies within the structure of the artifact to be designed. In a design process, the way to address a dependency between two parameters is to *link* the corresponding tasks. This is most easily done if these tasks can be performed by one person who understands the nature of the dependencies very well.[50]

When the artifact's complexity exceeds one mind's capacity, however, assigning all the tasks to one person will not be feasible: the design tasks must then be divided up among different people,[51] meaning that the design process must allow for *connections* between them. In practice, designers whose parameters are dependent must be aware of and must communicate with one another. They must transfer and share

49 Steward (1981b, p.161).

50 Brooks (1975).

51 Von Hippel (1990).

relevant knowledge, and coordinate their final parameter selections.[52] Should they not do so, the design effort as a whole could well fail.

Returning to our basic mug example: in designing a mug with a fitted cap, if the cap diameter and the vessel diameter decisions were to fall to different people, those people would need to talk with one another, and each would need to make a decision in light of what the other planned to do. Only if they coordinated their selections would the design stand a chance of being "on the ridgeline" of high value in the space of designs (see figure 2.1). In the absence of such coordination, the realized value of the mug would almost certainly be zero.

In sum, every hierarchical relationship and interdependency in the creation of a design requires a corresponding connection among tasks. *Thus every off-diagonal element in the design structure matrix needs to appear in the task structure matrix, too.*[53] In other words, the DSM and the TSM should be made to "look" identical.

Hence, for the design process to work effectively within the organization producing the designed artifact, the TSM of a design process should be the "image" of the corresponding DSM. We call this principle the "fundamental isomorphism" of design structure and task structure. It is a milestone in the development of our theory for it relates the microstructure of the thing being designed (its parameters and their physical and logical dependencies) to the microstructure of a set of tasks (the tasks and their coordinating links).

The Layers of Structure in the Design of an Artifact

At this point we have three "levels" or "layers" of structure implicit in the design of an artifact: (1) the structure of the artifact, (2) its design structure, and (3) its task structure. Each of these layers is conceptually distinct, but related to one another. Figure 2.4 shows how they relate to one another. Effectively, we are looking at a "complex adaptive system."

---

52 Iansiti (1997b).

53 A complexity arises because the designers of task structures may not know the actual design structure of their artifact when they design the task structure of the process. However, they will have knowledge about the design structure of previously created artifacts in the same class. On the basis of knowledge gleaned from prior designs, they must create a task structure from which good parameter choices for the next design can emerge. We discuss this issue further in chapter 5.

Name	Elements	Structure	Function
Design Process (Activity)	Tasks	**Task Structure =** Arrangement of Tasks	To create a design
		↕ **Fundamental Isomorphism**	
Design (Description)	Design Parameters	**Design Structure =** Arrangement of Design Parameters	To describe an **Artifact** especially its **Structure**
Artifact	Molecules Spatial Forms Colors Words Musical Tones	**Artifact Structure =** Arrangement of Molecules, Spatial Forms, Colors, Words, Musical Tones	To perform specific functions—to be seen, heard, touched, and used

Figure 2.4 The "layers of structure" in an artifact's design.

A layering of structures is typical of such systems—it is what makes them fascinating but hard to study. "Images" of one layer in another are common, too. For example, we are all familiar with a wide variety of maps—topographical maps, road maps, political maps, weather maps, and so on. Taken as a group, these maps show different layers of information (structure) about the region they depict. A topographical map indicates features such as altitude, rivers, coastlines, and so on; road maps show routes; political maps show borders; weather maps show weather patterns. One can often see "shadows" of one set of features in another—roads and political borders follow rivers; climate reflects altitude, and so on.

It is the same with artifacts and their designs. Complete, detailed designs provide a set of images of the artifact. The relationships among the design elements constitute the "design structure." Preparing the design involves tasks, and the relationships

Box 2.2 Deductive vs. Inductive Mapping of Design Structures and Task Structures

The fundamental isomorphism suggests that there are two equally valid ways to map a design structure and its corresponding task structure. The first is to look at an artifact, list its design parameters, and categorize the hierarchical relationships and interdependencies. This is basically what we did for a subset of mug parameters to arrive at figure 2.3. The second is Steward's procedure: seek out the people involved in the design process, and ask them "What parameters do you select?" and "Whose decisions do you need to know in order to make your decision?"

The first is a deductive approach; the second is an inductive approach. Both are valid and important. The inductive approach can tap the specific, tacit knowledge of designers about dependencies. The deductive approach, in contrast, requires there to be a generally accepted body of theory that describes what the artifact is and how it works. Hence the inductive approach requires less in the way of codified, formal knowledge about the artifact and its designs than the deductive approach.

among tasks are what we call the "task structure." And finally, the artifact—the thing that we see, hear, touch, or use—has its own structure, which *only* comes into existence when the design is converted into a real thing (see box 2.2).

Interconnected Design and Task Structures

Figure 2.5 shows a design and task structure matrix for the design process of a laptop computer, constructed by Kent McCord and Steven Eppinger.[54] This matrix has the "block-interconnected" structure characterizing many designs of an intermediate level of complexity.[55] Along the main diagonal of the matrix are four densely interconnected task blocks, corresponding to the four main components of the laptop computer system: the drive system, the main board, the LCD screen, and the packaging. Other points of connection (**x**) are scattered outside the main task blocks. These are less densely distributed than the connections within the blocks, but there are still quite a few of them. Significantly, there are as many out-of-block points of connection *above* the main diagonal as there are *below* the main diagonal.

[54] McCord and Eppinger (1993).

[55] Simple designs, like our mug, may be totally interconnected.

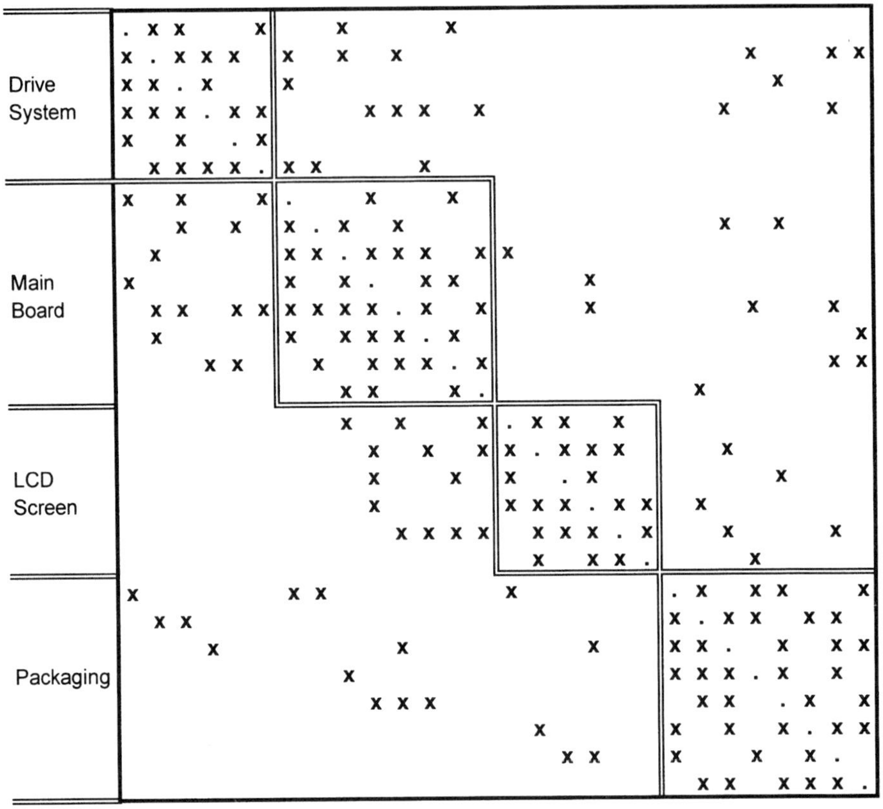

Figure 2.5 An interconnected, ordered task structure matrix: design of a laptop computer after reordering tasks.

Figure 2.5 begs an important question: How does the work of design proceed with a design and task structure such as this? The tasks do not fall into a natural sequence. If one starts at the top of the matrix and tries to work through the tasks in fixed order, cyclical interdependencies, denoted by **x**s above the main diagonal, quickly drive one "back" to reconsider previous decisions.

In fact, there is no way to divide the overall design into independent subdesigns. If by some chance the drive system, say, were designed independently of the main board, the matrix indicates that there are twenty-seven ways that the resulting designs might be in conflict (27 out-of-block **x**s). Any one of these dependencies, if unaddressed, could lead to the catastrophic failure of the whole design, just as an unaddressed vessel-cap dependency leads to mugs whose liquid cools too quickly or falls out. Unintended, possibly dire consequences arise when innate interdependencies in the design are not addressed in the task structure of the design process.

Cycling through Design Parameters

In cases like this designers must cycle and iterate through the tasks, searching for a combination of parameters that "works." Such cycles are extremely common in design processes. Christopher Alexander describes them in his classic work on design theory, *Notes on the Synthesis of Form:*

It is a common experience that attempts to solve just one piece of a problem first, then others, and so on, lead to endless involutions. You no sooner solve one aspect of a thing, than another point is out of joint. And when you correct that one, something else goes wrong. You go round and round in circles, unable to produce a form that is thoroughly right.[56]

Donald Steward, inventor of the design and task structure mapping tool, gives the following example:

[I]n the design of a heat exchanger, the temperature depends on the heat flux, the heat flux depends on the heat transfer coefficient, and the heat transfer coefficient depends on the temperature. This is a circuit. [In our language, this is an interdependency, involving three parameters.] *The engineer [resolves this circuit] . . . by iteration. He guesses the temperature to determine the heat transfer coefficient, which he then uses to determine the heat flux; the heat flux is used to re-estimate the temperature.*[57]

Steward goes on to explain that interdependencies like this are usually cheaper to resolve in the design phase than later, when the artifact is being fabricated or used:

Once a $100,000 heat exchanger is installed, it [is] very expensive to find it was too small, or worse yet, to find there was no room to install a larger one.[58] *. . . It is the primary role of engineering to resolve . . . circuits, usually by iterating through trial designs with pencil and paper or the computer, so as to minimize the more costly iteration which might otherwise occur during fabrication [or use].*[59]

---

[56] Alexander (1964, pp. 116–117).

[57] Steward (1981b, p. 20).

[58] Ibid., p. 21

[59] Ibid., pp. 62–63. Conflicting design constraints leading to cycles is not something that engineers and professional designers alone experience. Anyone who has tried to schedule a meeting with three or more busy people has had direct experience of both conflicting constraints and cycles.

We can depict the phenomenon of cycling in a TSM by drawing arrows from one design parameter choice to another through the off-diagonal points of connection. The arrows in figure 2.6 show one such path through a subset of the laptop computer's design parameters.

In this hypothetical example, a drive decision triggers an issue in the main board, which triggers a question about the screen. The provisional screen parameter choice then raises a different issue on the main board, which in turn feeds back to the drive system. Fixing the drive system then brings up a packaging issue, and the process continues from that point. It all becomes very complex.

Problem-solving sequences such as this are not at all unusual in design processes: it might take weeks, or days, or hours to work through this chain of interdependencies.[60] Moreover, the path drawn in figure 2.6 is a tiny subset of the problem-solving paths that will be initiated as designers attempt to arrive at a functional laptop computer design. At any given moment, many parallel investigations might be occurring. A designer working on the screen might at one time be concerned about the screen's interface with the main board, a drive system constraint that affects the resolution of the screen, as well as six or seven interdependencies within the screen unit itself.

At the same time, this process of searching, problem solving, and cycling is one of the most creative known. Through it, new parts of the design space can be explored, and new "peaks" of value can be discovered. At the same time, as is clear from the diagram, this process can be very inefficient. Provisional decisions get made only to be revisited over and over. If the number of interconnections is large, it may take a very long time to converge on a solution that enables a complete design to emerge.

There is, therefore, a delicate balance between "creative cycling" that helps determine the "best" design, and the demand for a solution to "any" design. With many artifacts whose designs are inherently complex, this balance is struck by modularizing the design and its corresponding task structure—but within a set of "design rules" that characterize the core interdependencies of the artifact class.

Modularization is described in detail in the next chapter; it is a procedure that uses knowledge of design structure and design parameter interdependencies to create design rules. The design rules support an efficient and flexible task structure, in which "parts" of the design are worked on independently and in parallel with one another. Independent, parallel efforts are possible because the design rules address all the implicit interdependencies of the artifact. Thus, when the parts are brought together, they will function seamlessly as a system.

---

60 Clark and Fujimoto (1989); Wheelwright and Clark (1992); Iansiti (1997b).

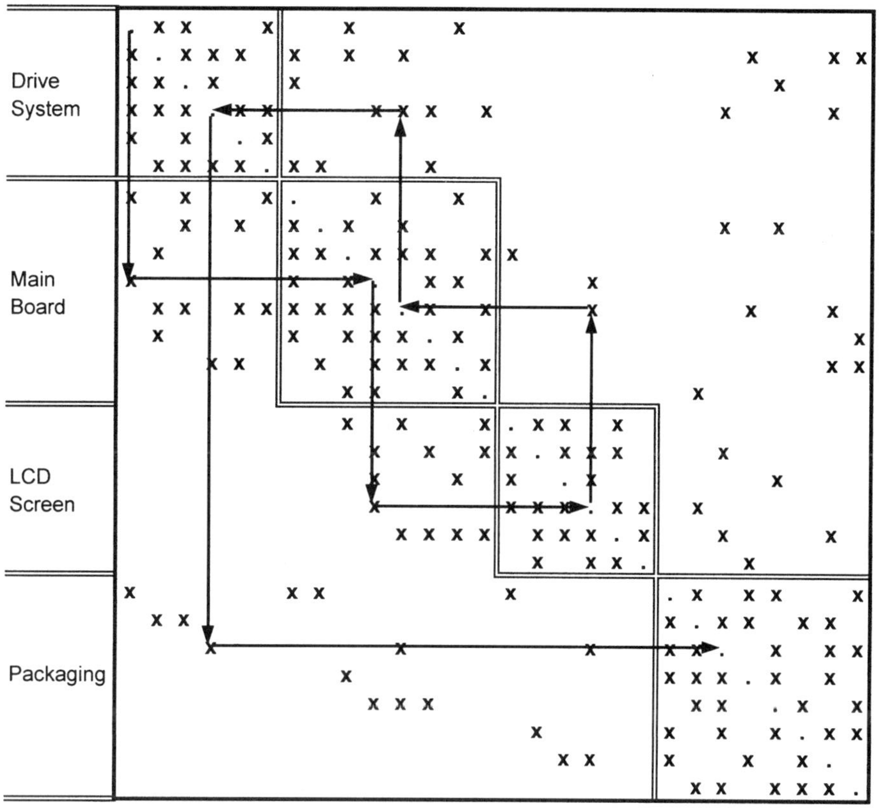

Figure 2.6 One path through an interconnected task structure shows how cycling occurs to resolve interdependencies.

The Task Structure and the Internal Organization of an Enterprise

As we have said, the task structure associated with the design of an artifact has a direct influence on what work gets done and how it will be organized. For example, to get the job of designing a laptop computer done, the first requirement is to find people to perform the tasks. For simplicity, let us assume that each row and column of the TSM corresponds to a set of tasks that can be performed by a single person. Thus in figure 2.7, we have put people's names (or initials) in the rows and columns of the matrix.

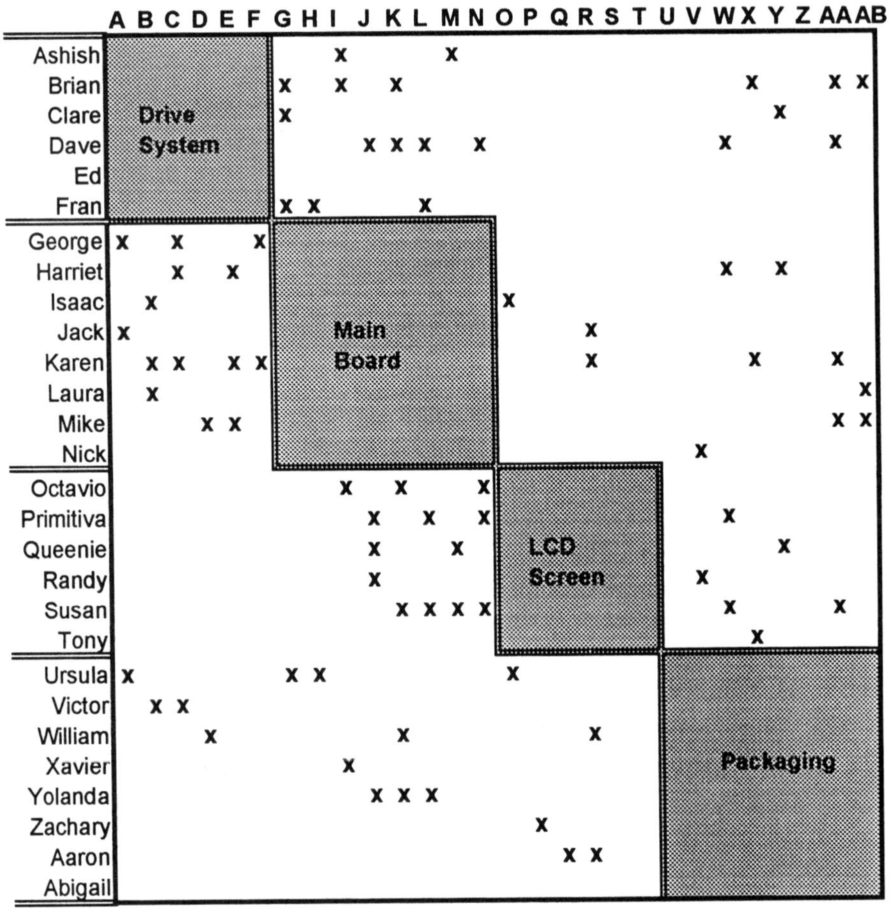

Figure 2.7 Mapping people to tasks: the organizational reflection of a complex task structure.

In their paper on the relationship between product design and organizational structure, Rebecca Henderson and Kim Clark broke the internal organization of an enterprise into four elements[61]:

1. *workgroups*—the groups of people in day-to-day contact that are charged with performing certain subsets of tasks;

[61] Henderson and Clark (1990).

2. *communication channels*—the mechanisms by which information moves across and within workgroups;
3. *information filters*—mechanisms that focus workgroups on certain types of information and block off other sources;
4. *a repertory of problem solving strategies*—ways of searching out solutions to recurring classes of problems.

Three of these four elements can be inferred from the TSM—the fourth can be observed by watching people work within a particular TSM.[62]

Starting at the top of the list, workgroups will correspond to the highly interactive "knots of structure" in the matrix. By definition, people with tasks in the same block need to be in close, constant contact with one another to get their jobs done. That is the definition of a workgroup. To recognize the existence of these workgroups, we have "grayed out" the pattern of **x**s within each of the task blocks and given the block or group the name of the component that it must design. Thus Ashish, Brian, Clare, Dave, Ed, and Fran are in the "drive system" group, George et al. are in the "main board" group, and so forth.

The communication channels across groups are indicated by the **x**s outside the main blocks. For example, Isaac, a designer in the main board workgroup, may know that Octavio, a screen designer, cares about the configuration and size of the memory cache. This dependency is indicated by the **x** in Isaac's column and Octavio's row. There is also an **x** in Octavio's column and Isaac's row, indicating that their design decisions are *interdependent.* Thus even though they are not in the same workgroup, Isaac and Octavio need to have ways of communicating with one another and coordinating their actions.

The information filters of the organization are indicated by the white spaces outside the main blocks. These are the interconnections that this organization has decided—implicitly or explicitly—are irrelevant to its design goals.

The problem-solving strategies cannot be read from a TSM. In effect, problem-solving strategies are *paths* through the matrix—ways in which members of the organization approach a complex, interconnected set of tasks. (See figure 2.6.) People working in organizations that implement complex task structures will tend to develop

---

[62] Implicitly, we are assuming that the TSM was derived inductively, that is, by asking designers what they "had to know" to do their jobs. When TSMs are obtained deductively, they reveal a normative organizational structure, which may or may not be the actual organizational structure (see Morelli, Eppinger, and Gulati 1995).

problem-solving strategies based on their own past experiences. These strategies may be documented, but more often they will be part of the tacit knowledge that resides in individuals and groups in the enterprise.[63] For example, in a laptop computer company, Marco in sales may know that he should call Alice in production if customer A calls and needs a rush order delivered. Alice and Marco have a shared tacit understanding that A deserves priority in the scheduling of jobs. In the language of task structure, Marco and Alice have developed an expeditious way of dealing with an interdependency between their two functions—an off-diagonal **x** that links order fulfillment to production scheduling.

Thus there is a close and powerful connection between the task structure and the internal organization of an enterprise. Tasks are mapped to people and the way they work together. This mapping, once established, may be both deep and irreversible, because it is often both tacit and complex.[64] The people who do the tasks will know, with varying degrees of clarity, how things "really" get done where they work, but much of their knowledge will be specialized, local, and incomplete.[65]

Indeed the connection between the task structure and the four elements of internal organization goes deeper still. A particular task structure creates a specific pattern of knowledge in the organization. Interactions within workgroups are, by definition, frequent, and thus people will tend to build and renew knowledge within their own groups. In contrast, interactions across workgroups are less frequent and may seem less important. Hence those same persons will "not perceive" or "forget" knowledge that crosses over workgroup boundaries.

As the number of tasks and the number of unrecorded linkages increase, any single change may have a host of ramifications with unforeseen, possibly costly consequences. Then only those changes that are isolated from the core of interdependencies can be tolerated. In all other respects, the organizational image of the task structure will become quite inflexible: the workgroup structures, communication channels, information filters, and problem-solving strategies will all become relatively fixed.[66] This set of circumstances leads to a particular type of "complexity

---

63 See Abernathy (1978); Nelson and Winter (1982); Bohn and Jaikumar (1986); Leonard-Barton (1995).

64 They will be complex in the precise sense of having many dimensions.

65 In many instances, people may be unaware of their own task structures (Eppinger et al., 1994).

66 Henderson and Clark (1990), following Nelson and Winter (1982) and Abernathy (1978), argue that stability in routines is informationally efficient as long as the external environment

catastrophe," as we describe below. Modularity, in turn, is a potential solution to this catastrophe.

Dominant Designs, Frozen Organizations, and Complexity Catastrophes

In the 1970s and early 1980s, William Abernathy and James Utterback, and then Michael Tushman and Philip Anderson, worked out a theory of product and process evolution that was a precursor of our theory.[67] Their central thesis was that product and process technologies usually evolve along problem-solving trajectories defined by an early major innovation—the so-called dominant design.

Abernathy went on to suggest that, as designers worked on the agenda of problems established by the dominant design, the locus of competition between firms would shift from product improvement to cost reduction. A "productivity dilemma" would then emerge, as the search for lower cost drove out, first, large innovations, and later, all but the most minor innovations. Thus, Abernathy argued, as an enterprise becomes more efficient, it also "necessarily" becomes more rigid and inflexible. Metaphorically, it gets trapped in its own problem-solving trajectory, and then evolves toward a "frozen state" in which no substantive innovations are possible.[68]

This outcome can be understood as the end result of the development of an interconnected task structure in which actions taken to resolve dependencies are not recorded, but instead become part of the tacit knowledge of an enterprise. In this setting, the mechanisms for making decisions, sharing information, and resolving conflicts will come to be embedded in unmapped organizational routines. This in turn leads to a classic "complexity catastrophe," in which as the number of tasks and the number of unrecorded connections increase, only very small, isolated changes can be made to the overall system.[69]

---

does not vary too much. This is theoretically plausible; however, the trick is to find the amount and type of experimentation that is just right for the environment. See Bohn and Jaikumar (1986), Iansiti (1997a, b) and Thomke et al. (1998) on "optimal" amounts of costly experimentation.

67 Abernathy and Utterback (1978); Abernathy (1978); Tushman and Anderson (1986).

68 We are using the language of complexity theory here. See, for example, Kauffman (1993).

69 "Complexity catastrophe" is a term Stuart Kauffman (1993) used to describe the trapping of a search process in a suboptimal equilibrium. The trapping arises because of interdependencies in the underlying parameters of the value ("fitness") function that are not known to

For example, in automobiles, it was possible in the mid-1970s to switch from nylon-ply to radial tires because tires had only a few, well-defined dependencies with the rest of the vehicle. However, it was virtually impossible to change a parameter like the weight of a vehicle, which affected the engine, drive train, and body all at once.[70] One of us observed and described what happened when such design changes were attempted. For example:

> *The introduction of [high-strength] steels during the 1970s allowed body panels to be made much thinner and lighter. [However, in one company] testing revealed that the new hoods resonated with engine vibration and even oscillated. . . . The problem [might have been] with the thinness of the steel, the shape of the hood, the metallurgical properties of the steel or the stamping process. On further investigation, it became clear that the body and stamping engineers . . . had no idea why their traditional methods produced a part that did not resonate or oscillate. Established rules of thumb,* probably derived from some long-forgotten process of trial and error, sufficed to meet traditional requirements. *[Emphasis added.]*[71]

In our language system, the established rules of thumb were undocumented hierarchical design parameters that "fixed" certain implicit interdependencies without recording how or why the fixes worked.

Organizational and design rigidities similar to those Abernathy and Clark observed in the U.S. auto industry have been documented in other settings, by (for example) Michael Tushman and Philip Anderson, Rebecca Henderson, Clay Christensen and Richard Rosenbloom, and Karen Wruck.[72] In some cases, an organization's failure to change was caused by a pure lack of perception by those involved. In other cases, however, people within the organization perceived the need for change, but foundered in implementing it. All of these failures, we believe, were fundamentally manifestations of the "complexity catastrophe," which develops in large interconnected task structures when the organizations in question fail to invest in knowledge about embedded interdependencies.

The essence of the problem lies in the fact that the complex systems needed to create artifacts and their designs do not necessarily evolve toward flexibility. First,

---

the searching entities. Complexity refers to the degree of interdependency among parameters; catastrophe refers to the fact that outcome is very inferior relative to the global optimum in the space. We develop the notion of a value function in a complex space in greater detail in Chapter 9.

70 Sull, Tedlow, and Rosenbloom (1997).

71 Clark (1988b, pp. 61–62).

72 Tushman and Anderson (1986); Henderson (1993); Christensen and Rosenbloom (1995); Wruck (1998).

the bodies of scientific knowledge that naturally grow up around a class of artifacts (like automobiles) by themselves do not give rise to flexible technologies. Scientific knowledge may give designers powerful and flexible tools for thought, but flexibility in imagination is a far cry from flexibility in implementation. It is always much easier to change a design in one's head than "on the ground."

Compounding the problem, the organizations that implement task structures have no predisposition to be flexible, either. Social practice tends to focus individuals on interactions and improvement within local blocks of tasks. There is a corresponding tendency to "freeze and forget" interactions across blocks, just as the body and stamping engineers "froze and forgot" the reasoning behind their own rules of thumb.

Modularity offers a way to avoid this complexity catastrophe, and to preserve flexibility within a complex system. The driving force behind a quest for modularity is always a desire to achieve the right balance between fruitful uncertainty and paralyzing complexity. The architects of a modular design want to admit enough uncertainty and interdependence into the design process to allow new things to happen, but do not want to admit so much that the process cycles endlessly without converging, or settles into a frozen state.

Independent, Sequential, Hierarchical, and Hybrid Designs and Task Structures

To end this chapter, and as a prelude to our exploration of modularity and design rules, we will provide a taxonomy of design and task structures. Figure 2.8 shows the most common configurations which we will discuss in the next four sections.[73]

Independent Block Structure

The first panel in the figure shows two blocks that are individually interconnected but are wholly separate from one another. There are many connections within each block, but no connections across the blocks. There is also no information common to both blocks: actors in one are wholly ignorant of what those in the other might be doing.

---

[73] In addition to design processes there are processes that involve the production and use of artifacts. Although design processes are the principal focus of this book, we have not limited the examples in figure 2.8 to design task structures alone. Examples of all three types of task are given in figure 2.8.

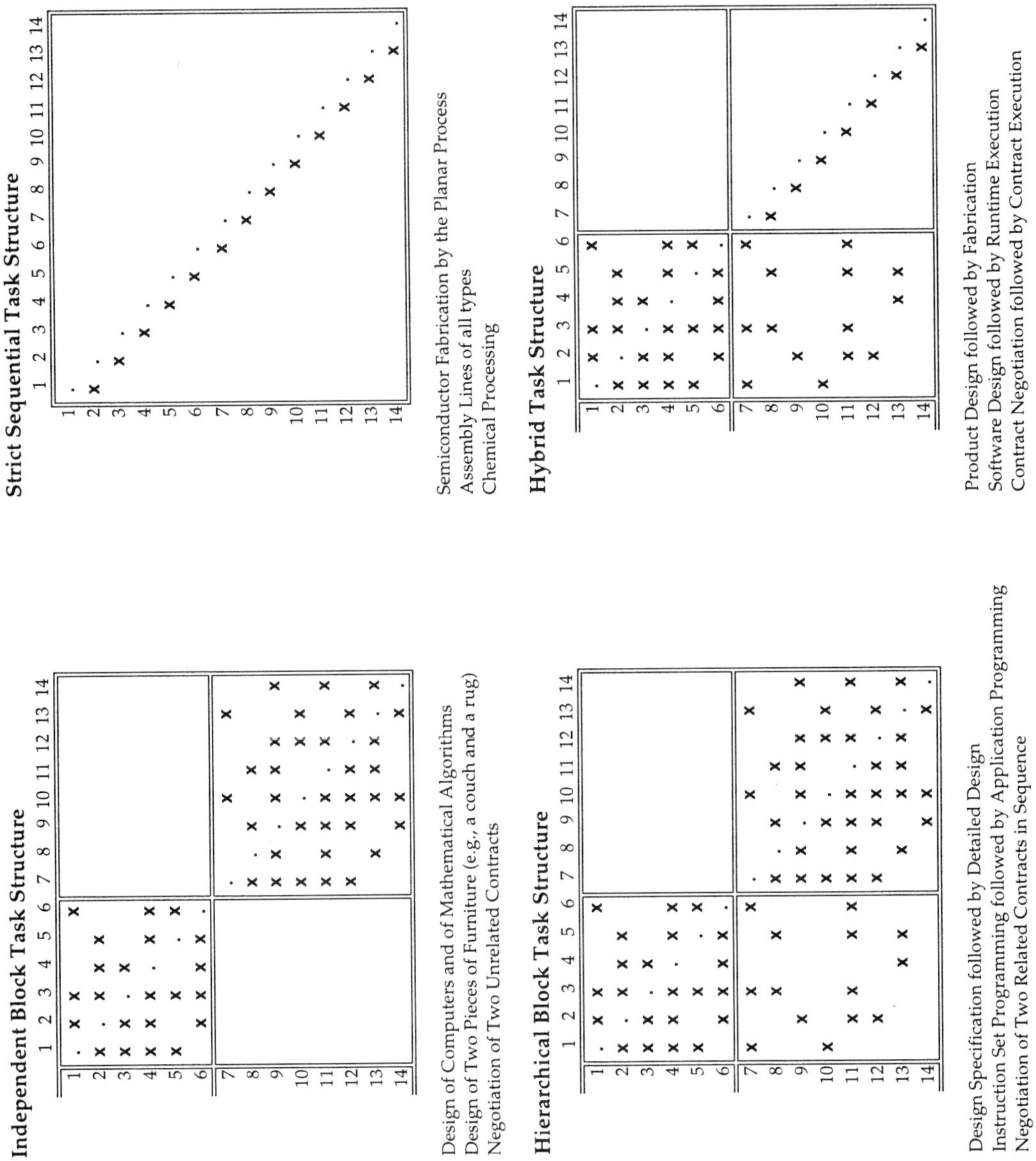

Figure 2.8 Paradigmatic task structures.

In modern economies, strictly independent block task structures are quite hard to find. However, this pattern can be an important element in more complex structures. For example, we will see that this block-independent structure lies at the core of every modular design.

Strictly Hierarchical or Sequential Structure

In the second panel, the parameters have a strict hierarchy, implying that tasks must be performed in a strict order. Moreover the only input to a particular task is the task immediately before it. This structure is representative of the flow line processes underlying modern mass production: Henry Ford's assembly line is probably the best-known example of this type of task structure.[74]

The economies obtained by balancing and synchronizing sequential task structures can be quite large.[75] However, strictly sequential task structures are generally not appropriate for design processes. The reason is that the strict sequencing of work "locks in" the process, making change, improvement, and learning itself difficult. By definition, in a design process, the "one best design" for the artifact is not known—if it were, the design process would be over. Thus a sequential design process, in which each step follows predictably upon the one before, is also a sterile design process.

Hierarchical Block Structure

The third panel shows two densely interconnected blocks, arranged in a strict order. The outputs of the first block are inputs to the second block, but the reverse is not true. This is a very common pattern in design processes: decisions made in phase 1 affect parameters chosen in phase 2, but phase 1 decisions may not be revisited.

---

[74] Hounshell (1984); Chandler (1977). Frederick Taylor, a management expert active at the turn of the century, used data from time-and-motion studies to create sequential task structures in many production facilities of his time. The sequential approach is also evident in the modern practice of "reengineering."

[75] We also have much to learn about such processes. For example the "kanban" feature of the Toyota production system has the effect of inserting a backward feedback loop between successive tasks. This corresponds to inserting a set of **x**s just above the main diagonal of the TSM. Instituting this form of local control on the sequential process can in some cases dramatically improve its efficiency. See Spear (1999).

In modular structures, design rules cannot be changed after the fact without drastically compromising the process. Thus the creation of a modular design requires a hierarchical decomposition of the design parameters, like that shown in this figure. We will revisit the issue of hierarchy when we discuss "visible" and "hidden" design parameters in the next chapter.

Hybrid Structure

Different types of structures may be combined to form many hybrid types. The figure shows one common pattern: an interconnected process provides inputs to a sequential process. Many design-and-manufacturing processes exhibit this pattern. The reason is that (as Donald Steward pointed out) it is often easier to go through a process of exploration and cyclical trial and error with a pencil and paper or on a computer, than in a factory or in the field.

However, when the relative magnitudes of the fabrication and design costs are reversed, it is sensible to fabricate a very large number of trial designs (including many that do not "work") and then test them. Such "inverted" processes are common in agricultural and pharmaceutical research. Here researchers may first grow an experimental field of hybrid corn, or generate a large number of chemical compounds via sequential processes. When the experimental results are in, researchers may then analyze the crop or the compounds to see which were most "successful." Analysis of the results may uncover interdependencies among different parameters, and, attempting to understand these, the researcher may cycle through the results many times.

In this case, an interconnected set of tasks *follows* a sequential set of tasks. Patterns like this are becoming quite common in fields that enjoy a very low cost of experimentation.[76]

Now that we have laid out the concepts of design structure and task structure, described their relationship to organizations, and provided a taxonomy of common structures, we are ready to focus on what it means to create modular designs and task structures. The next chapter, therefore, addresses the question, What is modularity in the domain of designs?

---

[76] Thomke et al. (1998); Iansiti (1997b); Holland 1995.

3 What Is Modularity?

The Core Concept

Modularity is a concept that has proved useful in a large number of fields that deal with complex systems.[1] Two subsidiary ideas are subsumed in the general concept. The first is the idea of *interdependence within and independence across modules.*

> *A module is a unit whose structural elements are powerfully connected among themselves and relatively weakly connected to elements in other units. Clearly there are degrees of connection, thus there are gradations of modularity.*[2]

In other words, modules are units in a larger system that are structurally independent of one another, but work together. The system as a whole must therefore provide a framework—an architecture—that allows for both independence of structure and integration of function.

The second idea is captured by three terms: *abstraction, information hiding,* and *interface:*

---

[1] The fields range from brain science and psychology, to robotics, artificial intelligence and industrial engineering.

[2] This definition is adapted from McClelland and Rumelhart (1995). Similar definitions are sprinkled throughout much of the literature. In contrast to this definition, Ulrich (1995b, p. 422) defines a modular (product) architecture as one that "includes a one-to-one mapping from functional elements in the function structure to the physical components of the product, and specifies de-coupled interfaces between components." After some analysis, we concluded that it is difficult to base a definition of modularity on functions, which are inherently manifold and nonstationary. Hence our definition of modularity is based on relationships among structures, not functions. Although we depart from his definition, we are very indebted to Karl Ulrich for his prior thinking on this issue.

A complex system can be managed by dividing it up into smaller pieces and looking at each one separately. When the complexity of one of the elements crosses a certain threshold, that complexity can be isolated by defining a separate abstraction *that has a simple* interface. *The abstraction* hides *the complexity of the element; the* interface *indicates how the element interacts with the larger system.*[3]

For human beings, the only way to manage a complex system or solve a complex problem is to break it up.[4] In the breaking apart, it is best to look for points of natural division, dividing "the Idea . . . at the joints, as nature directs, not breaking any limb in half as a bad carver might."[5]

Once the system or problem has been divided in this fashion, one can hide the complexity of each part behind an abstraction and an interface. The time to begin the process of breaking apart the system is when the complexity of the whole threatens to overwhelm progress toward the goal. (See box 3.1.)

But how is it possible to "break apart" a complex system, without destroying it? How does one find the tightly connected modules in a welter of interdependencies? And how can the modules be separated from one another? Our goal in this chapter is to explain how individuals with knowledge can split apart a large design with many innate interdependencies, and thereby create a modular design and task structure. The economic consequences of such modularizations will then be our central focus throughout the rest of this work.

Design Rules

Let us look again at the Eppinger-McCord task structure matrix (TSM) for the design of a laptop computer. (Readers may refer to figure 3.1, which reproduces the matrix in figures 2.5 and 2.7). Imagine what it would be like to participate in this type of design process. To the participants, the process would be very freeform and problem-driven, but at times it would also seem chaotic and maddeningly inefficient. In particular, each of the off-diagonal **x**s would require at least one conference, and some

---

3 This is a synthesis of many definitions sprinkled throughout the design literature.

4 Breaking up a problem helps human beings deal with complex issues. However, it is not especially helpful to machines, like computers. When instructions are translated from human to machine form, they are often demodularized. For example, compilers take modular programs and translate them into sequential code.

5 Plato, *Phaedrus,* 265D: This is the epigraph of Christopher Alexander's classic work on design theory, *Notes on the Synthesis of Form* (Alexander 1964).

Box 3.1 Metaphors for Modularity

In the 1960s Christopher Alexander and Herbert Simon put forward early theories about how to design complex systems. Although neither used the word "modularity" in his writing, the concept was central to their thinking. Each explained it using an extended metaphor: Simon tells a parable of two watchmakers, while Alexander describes an array of interconnected lights. Their metaphors are still powerful evocations of the power of modularity, and so we offer them here.

Herbert Simon's parable of the two watchmakers appeared in his famous essay "The Architecture of Complexity:"

There once were two watchmakers, named Tempus and Hora, who manufactured very fine watches. . . . The watches the men made consisted of about 1000 parts each. Tempus had so constructed his that if he put it down—to answer the phone, say—it immediately fell to pieces and had to be reassembled from the elements. The better the customers liked his watches, the more they phoned him and the more difficult it became for him to find enough uninterrupted time to finish a watch.

The watches that Hora made were no less complex than those of Tempus. But he had designed them so that he could put together stable subassemblies of about ten elements each. . . . Hence when Hora had to put down a partly assembled watch to answer the phone, he lost only a small part of his work, and he assembled his watches in only a fraction of the man-hours it took Tempus. . . . [If] there is one chance in ten, that either watchmaker will be interrupted while adding any one part to an assembly—then . . . it will take Tempus on the average about four thousand times as long to make a watch as Hora.[1]

In his book *Notes on the Synthesis of Form,* Christopher Alexander likened a design to an array of lights. The state of the lights represents the quality of the design:

Imagine a system of a hundred lights. The lights are so constructed that any light which is on always has a 50-50 chance of going off in the next second. . . . Connections between lights are constructed so that any light which is off has a 50-50 chance of going on again in the next second, provided at least one of the lights it is connected to is on. . . .

The off state corresponds to fit: the on state corresponds to misfit. . . . The state of equilibrium, when all the lights are off, corresponds to perfect fit or adaptation. . . . Sooner or later the system of lights will always reach this equilibrium. The only question that remains is, how long will it take for this to happen?[2]

How long it takes the array to reach equilibrium depends on the pattern of interconnections between the lights. If there are no connections, there is nothing to prevent a

[1] Simon (1969, p. 200ff.).

[2] Alexander (1964, p. 39 ff). Alexander adapted this example from W. Ross Ashby's *Design for a Brain* (1952).

(continued)

light from staying off for good, and the system will reach its ideal state in a few seconds. But if every light is connected to every other light, then the only way for the system to reach equilibrium is if by chance every light happens to go off at once. That takes on average 2^{100} seconds, or 10^{22} years.

Alexander then constructs a third possibility: ten principal systems of ten lights each:

> *The lights within each subsystem are so strongly connected to one another that again all must go off simultaneously before they will stay off; yet at the same time the subsystems themselves are independent of one another as wholes, so that the lights in one subsystem can be switched off without being reactivated by others flashing in other subsystems.*[3]

This system will reach equilibrium "in about the same amount of time it takes for one system to go off. . . about a quarter of an hour." Thus, says Alexander, dividing a large design into its principal components, and then *severing connections* between the components turns an impossible design problem into a manageable one.[4]

[3] Ibid.

[4] Alexander fudged a bit here: Equilibrium arrives only after the last system switches off, and thus the average time lapse is the expectation of the *maximum* of ten trials. The distinction is potentially important as one adds components to the system (as 100 lights goes to 200 divided into twenty subsystems), the time needed for the whole system to reach equilibrium increases. Even with perfectly independent components, there are limits to what modularization can accomplish!

would rcquirc many. In many instanccs, thcrc would bc tradcoffs and conflicts, some of which would require a costly and time-consuming dispute resolution process.

To the participants in this process, the off-diagonal **x**s are not simply placeholders. Corresponding to each is a specific real issue, involving at least two design parameters. Thus the matrix understates the depth of the designers' knowledge—for them, each of the **x**s stands for a specific problem or question.

For example, let us suppose that the first of the two **x**s circled in figure 3.1 corresponds to the issue, Will the central processing unit (CPU) manage the screen, or will there be a separate graphics controller on the main board?[6] The decision to have a graphics controller, in turn, would have consequences for the screen design, which

[6] In the 1960s, Myer and Sutherland observed that specialized display processors tended to evolve according to a "wheel of reincarnation." In each incarnation, display processors began as simple point-plotting controllers under the command of the CPU, and moved through stages to eventually become general purpose processors in their own right. Hence graphics control is a recurring issue in computer architecture. See Myer and Sutherland (1968); Bell, Mudge, and McNamara (1978, p. 201 ff).

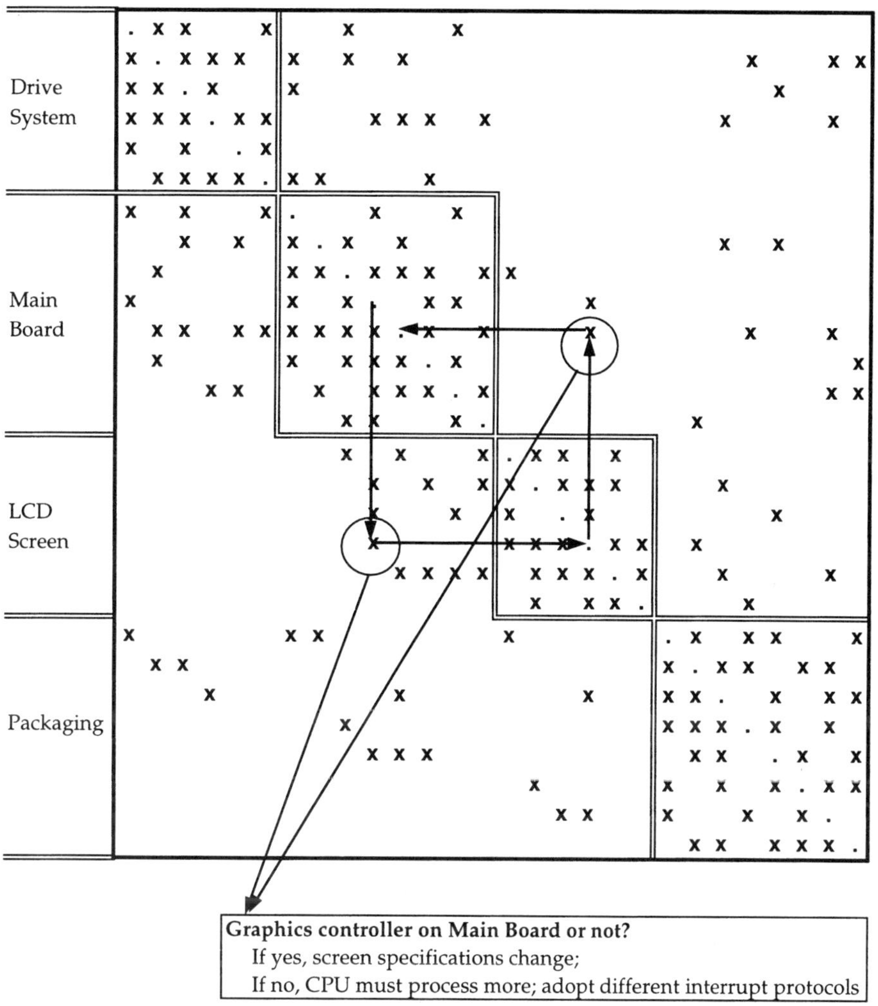

Figure 3.1 Design rationalization: cycling around the graphics controller issue.

would then feed back to other issues affecting the main board design. A fragment of the cycling that might ensue is depicted in figure 3.1.

If they have experience with previous designs of laptop computers, the designers will know that the question of whether to have a graphics controller is certain to come up. They will also have knowledge about what was done in previous designs, what the ramifications of the choices were, and how the resulting artifacts performed.[7] This knowledge gives the designers a set of preformed strategies, which can make their search for solutions more efficient. In particular, many of the routes to failure will be known, and presumably do not need to be tried again.

However, even before the design process gets underway, a special kind of designer (or team)—one called an *architect*—may foresee that the graphics controller issue will come up (because "it always does . . ."). Moreover, he, she, or they may conclude that cycling around this particular issue will not make much difference to the ultimate performance or cost of the computer. If this person or team also has the authority to make design decisions and enforce them, then he, she, or they can take this decision out of the interconnected set of tasks. Architects do this by making the choice before the rest of the tasks get underway and making the decision binding on all subsequent choices. In this fashion, the graphics controller decision becomes a *design rule.*

Figure 3.2 shows what happens to the design and task structure when a particular parameter choice becomes a design rule. First, one or more points of interconnection disappear—we have left circles in the figure to indicate where those points were. Potential cycles through those points of interconnection disappear as well. Instead of the original interdependency, a hierarchical pattern emerges in which certain parameters are "privileged"—they affect other parameter choices but they themselves cannot be changed.[8] These privileged parameters are the design rules.

Converting an ordinary design parameter into a design rule entails both benefits and costs. On the plus side, there will be a gain in efficiency through the elimination of cycles in the design process. On the minus side, designers will lose the ability to explore some parts of the space of designs—in effect, the architects will restrict the

---

7 In this fashion, there is a transfer of knowledge from one design to another via human intermediaries. Hence a design can be "descended" from another (or several others). Within organizations, effective "technology integration" can make transfers of knowledge from one design to another more efficient. See Iansiti (1997b).

8 Technically, the graphics controller "switch" is a hierarchical design parameter, whose setting would change some (but not all) of the subsequent parameters and tasks. We have suppressed this complexity in the figure, however.

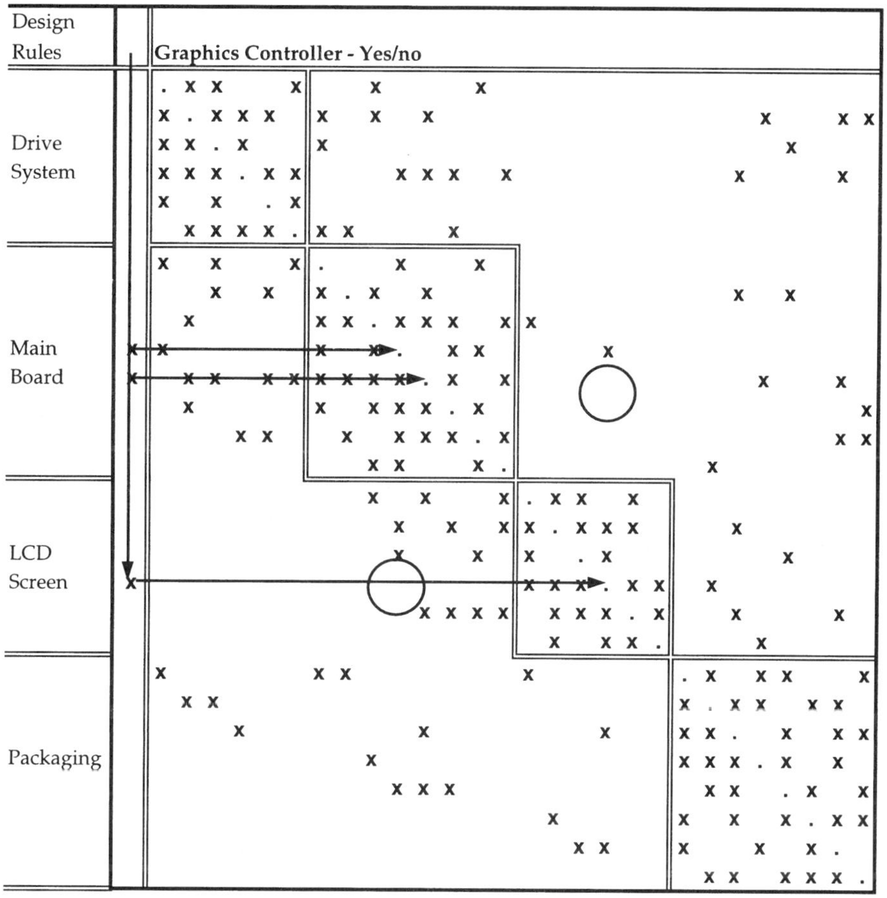

Figure 3.2 Design rationalization: creating a new design rule.

search, declaring some parts of the design space to be out of bounds. If those areas have already been explored and are known to have no "high peaks of value," then the cost of limiting the designers' search in this way will be low. But if the areas have not been explored and do contain "high peaks," then those superior designs will not be discovered, and the cost of imposing the design rule will be high.

This kind of rationalization of a design and its corresponding task structure is the first step toward modularization. Rationalization can go on for a long time—indeed, some designs never get out of this phase. The process takes time, because every move needs to be based on knowledge gleaned from experience with previous designs or scientific understanding of the underlying parameter interdependencies.

Imposing a design rule when one is ignorant of the true underlying interdependencies can lead to design failure. For example, in the case of a mug with a cap, suppose the mug designer declared that the cap and vessel diameters would both be 8 cm, not knowing that the molds used to make the vessels were 8.2 cm in diameter. Re-forming molds is very costly, but producing 8.2 cm vessels with 8.0 cm caps is foolish. The best course of action, if possible, is to rescind the original design rule, which was based on insufficient knowledge of the critical interdependencies.[9]

The result of the process of design rationalization will be a block hierarchical design and task structure, but not necessarily a modular one. Figure 3.3 is a typical example of a rationalized design and task structure. It is a map of Intel's chip design process as observed by Sean Osborne in 1993. This structure is clearly a hybrid form. First, there are a number of interconnected blocks along the main diagonal. Parameter choices are highly interdependent within these blocks, and less interdependent outside the blocks. The blocks also have a clear sequence, revealed by the dense scattering of **x**s below the diagonal. However, the ordering of blocks is not perfect: there are also points where the process cycles backward, indicated by the smaller, but significant number of **x**s above the diagonal. Because there are out-of-block **x**s both above and below the main diagonal, no block is truly independent of any other.

In the right circumstance, the blocks in this design and task structure could become modules. But this is not a modular design.

Design Modularization

In order to modularize a design and its corresponding task structure, the architects must have as their goal the creation of a set of independent blocks at the core of the design process. They must then set about systematically to sever all dependencies known to exist across the protomodules. As we have seen, parameter interdependencies can be severed by promulgating design rules early in the process.

However, at the time the architects want to modularize the process, the complete set of interdependencies may not be known. To address unforeseen interdependencies, tie up loose ends, and ensure that the end product is a functioning system, the

[9] This example may seem trivial, but costly mistakes like this are at the root of the "design for manufacture" (DFM) movement. The thrust of DFM is to change the manner in which product design parameters are chosen so that the costs of manufacturing may be taken into account. See Nevins and Whitney (1989); Eppinger, Whitney, Smith, and Gebala (1994).

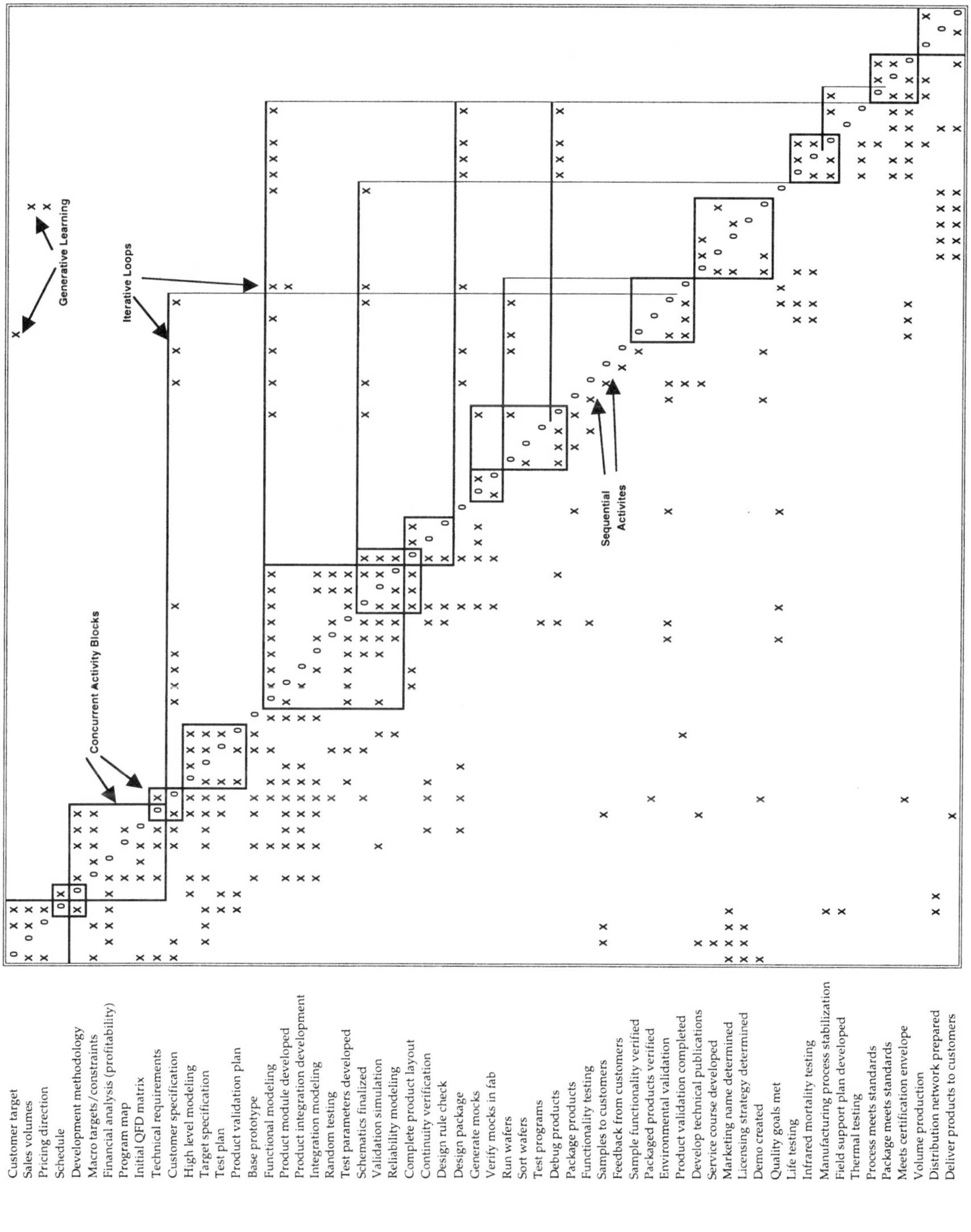

Figure 3.3 A task structure matrix for the design of a semiconductor chip (compiled by Sean Osborne, 1993).

architects must provide for integration and testing of the system at the end of the design process.

Figure 3.4 shows the design and task structure that might be derived from a modularization of the Eppinger-McCord laptop computer design. This is also the paradigmatic form of any modular design, and so we will discuss it in detail. It is a hybrid structure containing interconnected, independent, and hierarchical elements.[10] The matrix depicts a three-stage design process. There is (1) a *design rules stage,* followed by (2) an *independent parallel activity stage,* followed by (3) a *system integration and testing stage.* The design rules are inputs to all subsequent stages; the independent module parameters are inputs to the system integration and testing stage.[11]

At the core of this design are four independent blocks of parameters and tasks, corresponding to the four major components of the laptop computer. Internally these blocks were densely interconnected to begin with, and still are. One of the goals of modularization is to preserve the most fruitful interdependencies and tasks connections: in this design there is still ample room for innovative cycling and creativity, but it is contained within the modules.

Outside the four blocks, all task connections have been severed. There are no longer opportunities for cycling between and among modules. In addition, no decision in one module block is an input to any decision in another module block. This design structure imposes more stringent restrictions on the designers than does the Intel chip design structure (see figure 3.3), in which earlier blocks provided informational inputs into later blocks. In the modular case, only the design rules affect what goes on in any of the modules.

Information Hiding and Design Hierarchies

For a modularization to work in practice, the architects must partition the design parameters into two categories: *visible information* and *hidden information.* This partition specifies which parameters will interact outside of their module, and how potential interactions across modules will be handled.

---

[10] Modular task structures may contain strictly sequential blocks, too.

[11] For efficiency's sake, the integrators and testers should not have to know *everything* about the individual modules. In practice, the amount that integrators and testers do have to know varies across systems and over time. We return to this issue in chapters 5 and 10.

The principle of *information hiding* was first put forward in the context of software engineering by David Parnas.[12] However, the principle is perfectly general, and can be usefully applied to any complex system. With respect to software programs, Parnas reasoned that if the details of a particular block of code were consciously "hidden" from other blocks, changes to the block could be made without changing the rest of the system. The goal was then to enumerate and as far as possible restrict the points of interaction between any two modules of a program. The fewer the points of interaction, the easier it would be for subsequent designers to come in and change parts of the code, without having to rewrite the whole program from scratch.

"Information hiding" is closely related to the notion of abstraction defined above:

When the complexity of one of the elements crosses a certain threshold, that complexity can be isolated by defining a separate "abstraction" with a simple interface. The abstraction "hides" the complexity of the element. . . .

Abstraction is a technique for managing complexity that is deeply ingrained in human beings. As information processors we are always suppressing details we take to be unimportant; as problem solvers, we instinctively focus on a few manageable parts of a problem at one time. If our abstractions match the true underlying structure of the problem—if we are good carvers, not bad ones—then the analysis of abstractions will lead to a good solution to the problem. In any case, given our limited mental capacities, we have no choice but to work with simplified representations of complicated problems.

Information hiding begins as an abstraction. But to achieve true information hiding, the initial segregation of ideas must be maintained throughout the whole problem-solving process. This means that as points of interaction across problem boundaries arise, they cannot be dealt with in an ad hoc way. Instead, the interactions must be catalogued and a set of interfaces specified.

An *interface* is a preestablished way to resolve potential conflicts between interacting parts of a design. It is like a treaty between two or more subelements. To minimize conflict, the terms of these treaties—the detailed interface specifications—need to be set in advance and known to the affected parties. Thus interfaces are part of a common information set that those working on the design need to assimilate. Interfaces are visible information.

---

[12] Parnas (1972 a, b).

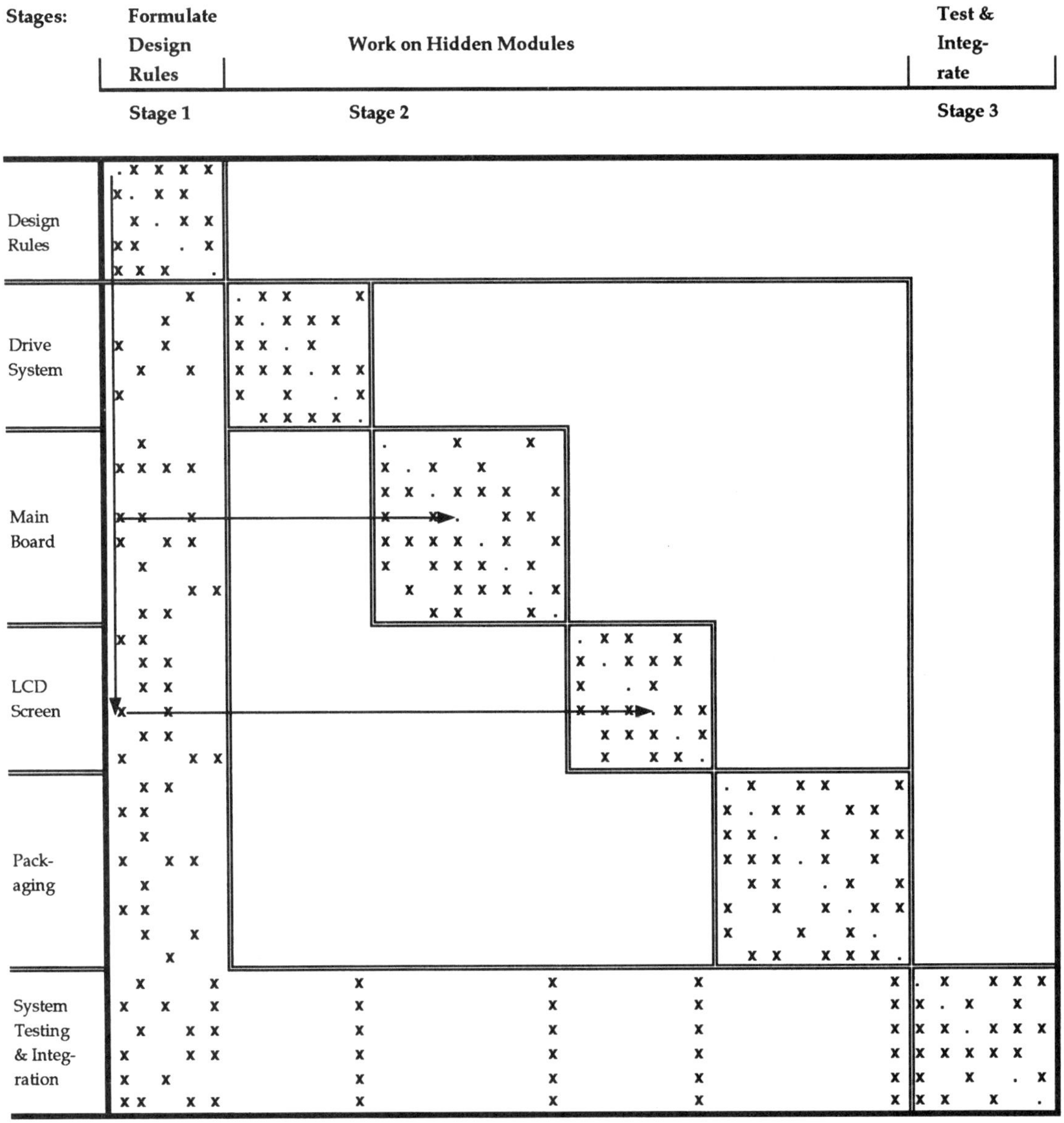

Figure 3.4 Hypothetical modularization of the laptop computer design.

Design parameters that are not in the visible set constitute the hidden information in the design. Hidden design parameters only affect their own piece of the system, hence they can be changed without triggering any changes in distant parts of the system. They invite tinkering, which can lead to improvements in performance.

Note that if we manage to hide a substantial amount of the information relevant to subsystems of the design, we will have modularized the system. Thus going from a set of conceptual abstractions to a complete and robust partition of visible and hidden information is the same process as modularizing a design. The design rules are the visible information; the parameters left to the discretion of the module designers are the hidden information.

The relationship between hidden and visible information can also be represented in the form of a design hierarchy, as shown in figure 3.5. We stress that a design hierarchy diagram and a task structure matrix are really two maps of the same thing—they are complementary representations of the same underlying reality.[13] The advantage of the design hierarchy representation is that it draws attention to features of the design that are hard to see in the matrix format. Specifically, it shows

- who has to know what;
- the temporal order of decisions;
- the reversibility of different choices.

The design hierarchy in figure 3.5 is made up of three levels. At the top are the global design rules that are visible to all subparts of the design. At the next level are design rules that are visible to some, but not all, parts of the system. And at the lowest level are design parameters that are within modules and thus hidden from the rest of the system. (Throughout this work, we will call the lowest level of modules in a design hierarchy the "hidden modules" of the system. By definition, these are modules which contain no visible information, hence their design parameters are hidden from the rest of the system.)

As the figure shows, a design rule can be visible to the modules below it in two different ways. In the diagram, a line drawn from a higher-level box to a lower-level box indicates that the lower-level designers have direct knowledge of the rules set at the higher level. But rules do not have to be known to be constraints. For example, hidden-module designers may not know about design-rule decisions taken two levels

[13] Marples (1961); Clark (1985); Henderson and Clark (1990); Tushman and Rosenkopf (1992).

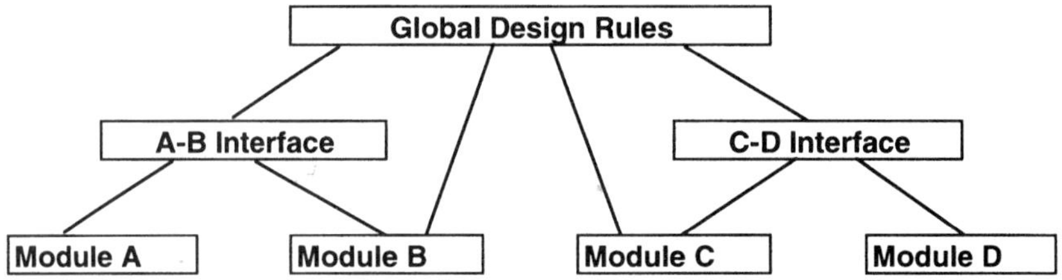

Figure 3.5 A design hierarchy with four hidden modules and three levels of visibility. There are three levels of visibility/invisibility. The global design rules are directly visible to modules B and C and indirectly visible to modules A and D. This means that designers of hidden modules B and C must know the global design rules as well as their local interface specifications, while designers of modules A and D need only know their local interface. However, the interfaces are based on the global design rules; if these change, the interfaces must change as well.

up. However, those decisions are visible to the intermediate-level designers, and their choices in turn are visible to the hidden-module designers. In this way, choices made at the topmost level are indirectly visible to all. (The two outer modules in figure 3.5 depict indirect visibility; the two inner ones depict direct visibility.)

Because they constrain later actions in every other part of the system, the parameters in the top level of the diagram need to be established first. Thus higher levels both "are visible to" and "come before" lower levels.

Finally, by definition, changing visible information requires changing parts of all designs that "see" that information either directly or indirectly. Therefore, changes at the top of the diagram will have far-reaching consequences, and are bound to be difficult and expensive. In other words, visible design choices are relatively irreversible. Conversely, changes at the bottom of the diagram are limited in scope, hence cheaper to implement, at least in comparison to visible changes. Hidden design choices are reversible.

A Complete Set of Design Rules

Perfectly modular designs do not spring fully formed from the minds of architects. When a design is first modularized, the designers' knowledge of interdependencies is usually imperfect, and as a result the initial set of design rules will be incomplete. Incomplete design rules give rise to unforeseen interdependencies, which in turn will require consultations and iterations between the hidden-module designers and the architects of the system. The integration and testing of these designs will be fraught with difficulty and the designs themselves will have high risks of failure.

As the properties of the system and the modules become better understood, the design rules will tend to become more complete. Then, as more of the innate interdependencies come to be addressed in the design rules, integration and testing of the system will become more cut-and-dried. Eventually, this part of the process may become so standardized and so simple that users themselves can take over the tasks of integrating and testing their own systems.

A complete set of design rules fully addresses the following categories of design information:

- *architecture,* in other words, what modules will be part of the system, and what their roles will be;
- *interfaces,* that is, detailed descriptions of how the different modules will interact, including how they will fit together, connect, communicate, and so forth;
- *integration protocols and testing standards,* that is, procedures that will allow designers to assemble the system and determine how well it works, whether a particular module conforms to the design rules, and how one version of a module performs relative to another.

In the time path of a modularization, the architecture of the system is usually defined first, followed by the interfaces and related conformance tests. System-level tests of performance are developed as people gain experience using the system.[14] Last to emerge usually are the comparative tests of performance of individual modules.

An Actual Modularization: The Mead and Conway Proposals

We have argued that interconnected designs and their corresponding task structures can be modularized, consciously and intentionally, by designers known as architects. Modularization involves promulgating appropriate design rules, and severing connections between task blocks. Successful modularizations must be based on detailed knowledge of the underlying parameter interdependencies. However, the architects' knowledge does not have to be complete: minor incompatibilities and loose ends may be dealt with in an integration and testing stage at the end of the design process.

[14] Clark (1985); von Hippel (1988).

Box 3.2 Modularity in Production and Use

In chapter 2 we said that humans interact with artifacts in three basic ways: they design them, produce them, and use them. There are, as a result, three basic types of tasks (or actions) and related task structures. In this book, we are primarily interested in designs and design task structures. However, one can observe modularity in the tasks of production and use as well.

Manufacturers have used *modularity in production* to simplify complex processes for a century or more. Generally, complicated products are made by dividing the manufacturing process into various process modules or "cells." Carmakers, for example, routinely arrange to manufacture the components of an automobile at different sites and bring them together for final assembly. They can do so because they have completely and precisely specified the designs of those parts. The engineering design of a component (its dimensions and tolerances) constitutes a set of design rules for the factories that supply the parts. Those suppliers may experiment with their production processes and logistical systems, but historically they have had little input into the design of the components themselves.[1]

Modularity in use allows consumers to mix and match elements to come up with a final product that suits their taste and needs.[2] For example, consumers often buy bed frames, mattresses, pillows, linens, and covers made by different manufacturers and distributed through different retailers. The parts all fit together because different manufacturers make the goods in standard sizes. These standard dimensions constitute design rules that are binding on manufacturers, wholesalers, retailers, and users. Modularity in use can spur innovation: the manufacturers can independently experiment with new products and concepts, such as futon mattresses or fabric blends, and find ready consumer acceptance as long as their products conform to the standard dimensions.

[1] On modularity in complex production systems, see, for example, Nevins and Whitney (1989); Uzumeri and Sanderson (1995); Fine (1998); Spear (1998).

[2] Christensen and Rosenbloom (1995); Baldwin and Clark (1997a).

Up to this point, we have based our argument on logic, without empirical evidence in support of our contentions. That is, we have not yet shown that design modularizations actually occur in the way we have proposed. The laptop computer design, shown in figure 3.1, might be modularized, but then again it might not. (In 1997, laptop computer makers disagreed on this issue, and were laying different architectural bets as a result.[15]) Hence, to make the case that modularizations do arise and

[15] Private communications in June 1997 from Harvard Business School alumni who were employed at different laptop computer firms.

have important consequences, we need to supply a real example. Then we can say, not only that the process might work as we have described but that it actually does work that way sometimes.

The example we offer is Carver Mead and Lynn Conway's approach to very-large-scale integrated (VLSI) chip designs.[16] In choosing this example, we are admittedly guilty of overkill. Theirs was not a simple modularization, as might happen with laptop computers. Mead and Conway did not pursue modularity piecemeal; instead they totally reconceptualized chip designs in terms of nested, regular, modular structures. Their rethinking of the process of creating chips encompassed all levels of the design, from the location of single transistors and the layering of material, to the economic arrangements that linked chip designers, maskmakers, and fabricators.

Let us begin where Mead and Conway began—with the planar process of integrated circuit fabrication, which was invented by Robert Noyce at Fairchild Semiconductor in 1958.[17] The process allowed electrical circuits to be constructed by depositing alternating thin layers of silicon dioxide and metal on a substrate. The appropriate patterns were etched into each layer by shining ionizing radiation through the transparent areas of an otherwise opaque mask.[18]

The planar process for making integrated circuits is inherently sequential. In contrast, for many years, the process of designing integrated circuits was unstructured, interconnected, and ad hoc.[19] This was acceptable because initially the circuits themselves were simple. However, as more transistors came to be packed on a chip, integrated circuit designs became inexorably more complex, and the unstructured design processes began to break down.

Mead and Conway saw a way to address this burgeoning complexity by taking advantage of unexploited potential modularity. In their eyes, "the most powerful attribute" of the planar fabrication process was the fact that it was *pattern independent.* By this they meant, "There is a clean separation between the processing done during wafer fabrication, and the design effort that created the patterns to be implemented."[20]

---

[16] We thank Sam Wood for pointing us to this example.

[17] Noyce, with Gordon Moore and Andrew Grove, later left Fairchild and founded Intel Corporation.

[18] Many books describe the planar process in detail. An especially good description is at the beginning of Chapter 2 of Mead and Conway's classic book, *Introduction to VLSI Systems.* Published in 1980, this book is at once a textbook for new engineers and an impassioned argument in favor of Mead and Conway's then-novel approach to chip design and fabrication.

[19] For an excellent description of the unstructured design process, see Gilder (1989).

[20] Mead and Conway (1980, p. 47).

In our language system, chip design and chip fabrication were intrinsically modular, that is, they constituted almost independent units embedded in a larger coordinating framework. A planar process could fabricate essentially any pattern as long as the pattern conformed to the basic design rules of the process.

The question Mead and Conway focused on was how to use the pattern independence of the planar process to get the most out of increasingly inadequate design resources. They attacked this question on many levels. We will not be able to do justice to their whole theory in this limited space, but we can describe some of the techniques they espoused, and show how these techniques allowed chip designs to be split into modular subsystems. The specific techniques we will describe are dimensionless, scalable design rules, and multiproject chips.

Dimensionless, Scalable Design Rules

In 1978 and 1979, when Mead and Conway were writing their book, Moore's law—the observable fact that the density of transistors on chips doubles every eighteen months—had been in effect for twenty years and showed no signs of abating. Each doubling of density required totally new processing lines to fabricate the chips. Mead and Conway fully expected processing lines to keep changing at this rate far into the future.

It was already common practice in the industry for chip designers to work within a set of geometrical constraints that reflected both the physics of transistors and the tolerances of a specific processing line. These constraints took the form of minimum allowable values for "widths, separations, extensions, and overlaps," in the geometry of the circuits. These constraints, called "design rules," were essentially a set of hierarchical design parameter choices that resolved the most critical interdependencies between the processing line and the designs of chips made on the line.[21]

Thus a typical chip-to-fab design and task structure in the 1970s would have had the form shown in figure 3.6. In this structure there are two sub-blocks—the design tasks and the fabrication (including maskmaking) tasks. These sub-blocks were united by a set of design rules, which were specific to the pairing.[22] Prototype chips

---

[21] Our terminology is derived from Mead and Conway. Their explanation of how design rules enabled the creation of modular chip designs allowed us to understand how design rules work in general.

[22] The task blocks were not necessarily cleanly separated: they were usually "protomodules" instead of actual modules. Intel's TSM, shown in figure 3.3, is a 1990s version of a chip-and-dedicated-fab block task structure. See Iansiti (1997b); Osborne (1993).

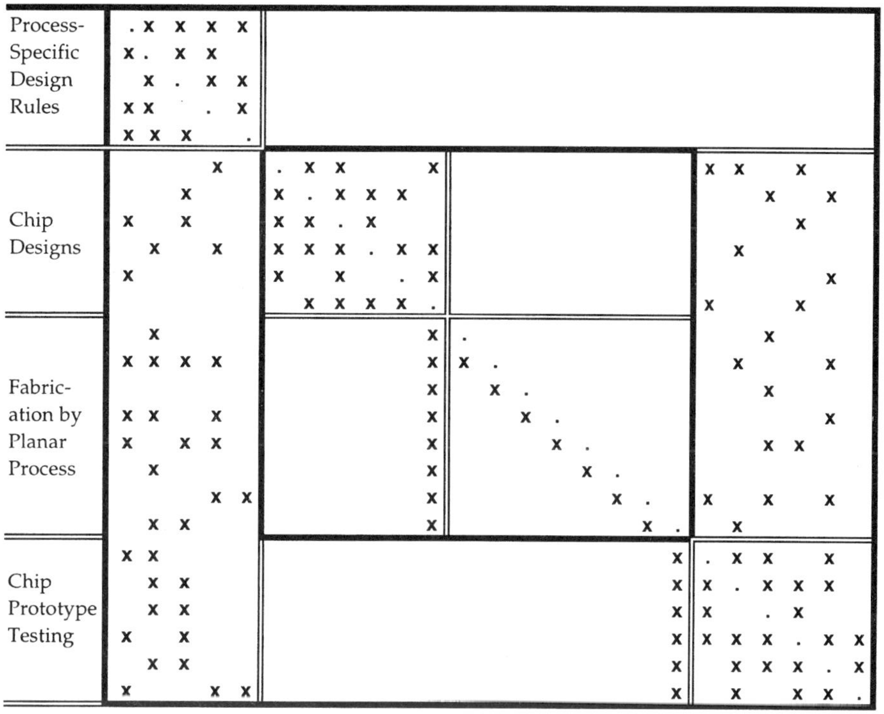

Figure 3.6 A schematic task structure for integrated circuit designs with process-specific design rules. The process is a hybrid type of task structure, made up of hierarchical, interconnected, and sequential elements. Initially a set of process-specific design rules is communicated to all participants. The chip designers then design a prototype chip conforming to the design rules, and the prototype chip is fabricated on the line. Testing of the prototype may reveal flaws in its design or in the fabrication process or both. These flaws are indicated by the **x**s in the last six columns above the main diagonal. They would cause the designers to cycle back to earlier stages, but (hopefully) not as far back as the design rules. There are many variations on this type of design process. The critical thing to note is that the chip and fabrication process is a dyad linked by process-specific design rules. The dimensions (e.g., line widths) of the process were embedded in these rules. Hence each time a new generation of equipment was introduced, a new set of design rules, new design practices, and new chip designs would all have to be created from scratch. This was the situation that Mead and Conway set out to change.

would then be designed, fabricated, tested, redesigned, refabricated, and retested many times prior to the ramp-up of the fab line to full production.

Knowing that the processing lines were continually changing, and wanting to carry over circuit designs from one generation to the next, Mead and Conway sought to express the fundamental geometrical constraints of the design rules as "allowable ratios of certain distances to a basic length unit." The basic length unit in turn would be a function of the achievable tolerances of a specific fab line. Lynn Conway later described their thinking as follows:

We . . . decided that we could normalize the design rule to the resolution of the process. Later on, we could ask about the value of the minimum line width. . . . We wanted to come up with something that students would learn once and retain [i.e., rules that would carry over from one generation to the next].[23]

In other words, as processes got better, tolerances would become finer, the basic length unit would decrease, but the geometrical relationships between line widths, separations, extensions, and overlaps would not change. As chip densities increased, circuit geometries could simply be rescaled. Practically speaking, this meant that circuits would not have to be redesigned each time a new generation of fabrication equipment came on line. Instead, the circuit designs could simply be shrunk to suit the tolerances of the new equipment.

Figure 3.7 shows the design and task structure implied by scalable design rules and reusable circuit designs. The new structure has *one* task block for design and *many* blocks corresponding to the different fab lines on which a rescaled circuit would work. The primary *interface* between the designs and the fab lines is the basic length unit, λ.

In their text, Mead and Conway provided thirteen fundamental design rules, scaled by this basic length unit.[24] These rules are hard to describe in words, but fairly easy to understand when expressed in terms of color-coded blocks. Although more needed to be done to achieve true continuity of designs across generations, these dimensionless, scalable design rules created system and structure at the most basic level of the artifact—the level of transistors and gates.

---

[23] Lynn Conway, quoted in *Electronics,* 20 October 1981.

[24] Mead and Conway (1980, pp. 48–50 and Plates 2 and 3).

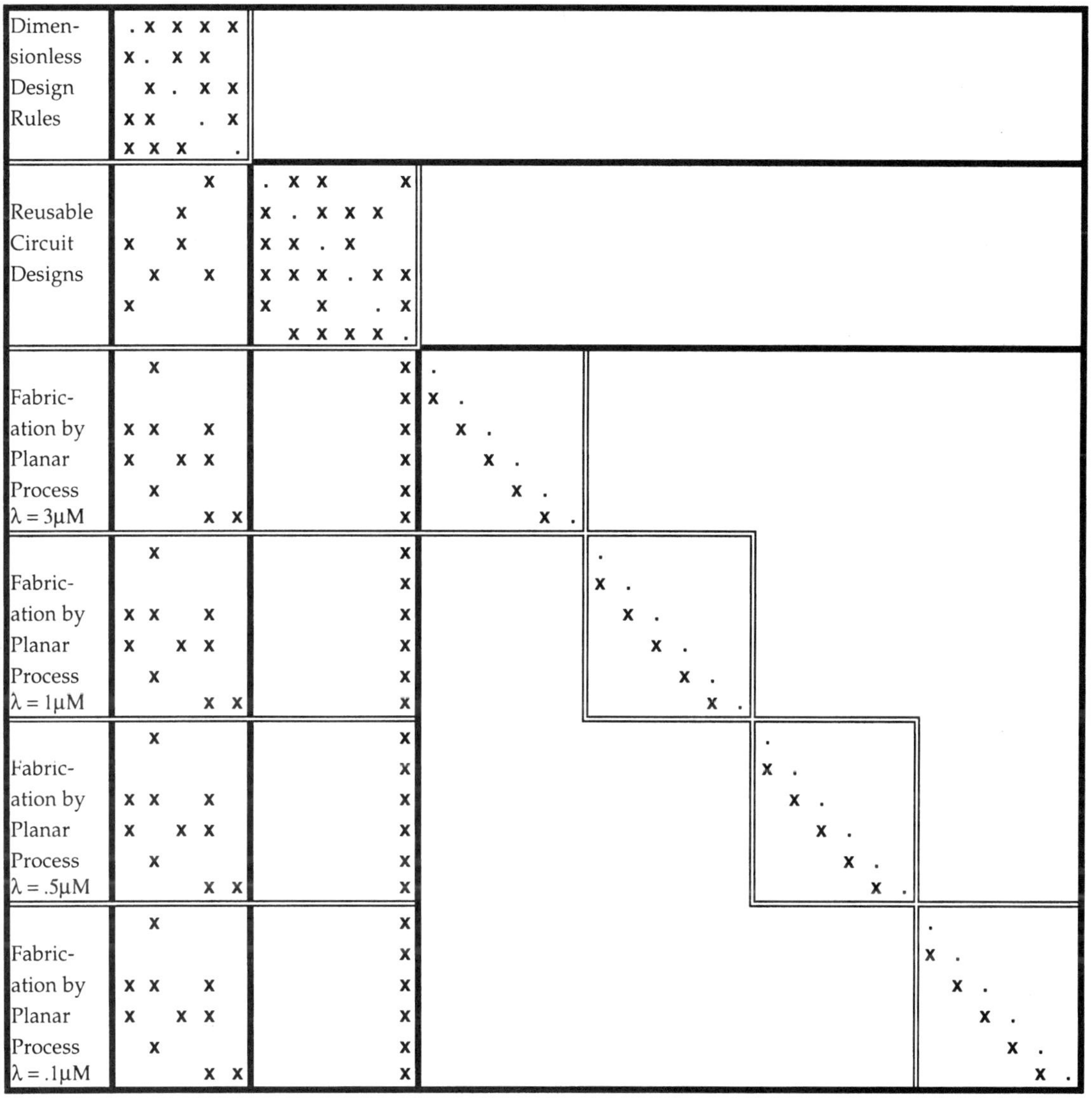

Figure 3.7 The impact of dimensionless design rules: a schematic task structure for reusable circuit designs over several generations of equipment. Initially a set of dimensionless design rules is communicated to all participants. Circuits are then designed to conform to those rules. The circuit designs are subsequently implemented on processing lines with line widths of 3, 1, .5, .1 μM. The critical thing to note is that, with the advent of dimensionless, scalable design rules, circuit designs and processor lines no longer need to be in a dyadic relationship. A scalable circuit geometry can be reused over several generations of equipment with very different densities and line widths.

The Multiproject Chip

Dimensionless, scalable design rules opened up new possibilities for chip designs in the future, but in 1980 these were only possibilities. There was no proof then that reusable circuit modules and subsystems would in fact become a reality. In the short run, at a totally different level of the VLSI system, Mead and Conway set out to show that in practice the tasks of design could be divided up among different teams, and that parallel independent efforts could be used to accelerate the design process. Thus in Chapter 4 of their book, they described the creation and management of a modular, decentralized design process in some detail.

They began with a sense of impatience and frustration with the slow pace at which complex chip design processes moved toward conclusion. They seized on the idea of dividing the system up, allowing teams of designers to work independently, and then assembling the parts:

A large, complex VLSI system could be quickly and successfully developed by designers able to easily implement and test prototypes of its subsystems. The separate subsystems can be implemented, tested, debugged, and then merged together to produce the overall system layout. . . . However, such activities are practical only if a scheme exists for carrying out implementation with a minimum turnaround time and low procedural overhead.[25]

To make this vision a reality, Mead and Conway created an organizational framework in which teams of students would design, implement, and test modular subsystems, all within a single academic semester. This was a great challenge because in 1978 maskmaking typically took three to five weeks and fabrication three to four weeks, while academic semesters were only twelve weeks long.[26]

Mead and Conway wanted to give each student team a small subsystem to design. The individual projects would then "be merged together onto one moderate sized chip layout, and implemented simultaneously as one chip type."[27] To make this possible, Mead and Conway, as the architects, had to

- document various parameters and specifications to be used during maskmaking and fabrication;

[25] Ibid., p. 128.

[26] Waits were longer if the mask or fab firms had large backlogs of work. Mead and Conway noted that "almost all the implementation time involves waiting in various queues. . . . Were an implementation facility to optimize its structure and procedures for fast turnaround, implementation times of less than one week should be easily achievable" (ibid., p. 130).

[27] Ibid., p. 130.

- create the layout of a *starting frame,* into which the projects would be merged;
- gather, relocate, and merge the project layouts into the starting frame to create one design file.[28]

The paramount goal, of course, was to preserve the freedom of the module designers. Mead and Conway did not want to predetermine or overly constrain what their students could do—they wanted to give them as much latitude as possible. But, for the success of the joint enterprise, they had to establish a set of design rules, including provisional boundaries on circuits, so that one module would not interfere with another.[29]

Their design rules established the basic structure of the system—its "architecture." The rules also governed the interaction of the modules with each other and with the larger system—the "interfaces." In the multiproject chip, for example, module designers had to know about shared resources, line widths, density limitations, and power usage. Finally, the design rules told the module designers how to put their individual designs into a common format that could be mapped onto the pattern generator for a single chip—the "integration protocols."

The multiproject chip design rules were added to the dimensionless design rules governing the geometry of the circuits. The first set of design rules made circuit designs scalable, thus able to span different generations of equipment; the second set partitioned the area of a particular chip and established common parameters applicable to just that chip. Mead and Conway also recommended other "rule-type" practices, including the creation of function cells, programmable logic arrays, and finite-state machines, all of which contributed to the regularity and modularity of intermediate structures on the chip. They summed up their approach as follows:

> *The beginnings of a structured design methodology for VLSI systems can be produced by merging together in a hierarchy the concepts presented [here]. . . . When viewed in its entirety, a system designed in this manner is seen as a hierarchy of building blocks, from the very lowest level device and circuit constructs up to and including the high-level system software and application programs. . . . [A] system of nested modules emerges if we view every module at every level as a finite state machine or data path controlled by a finite state machine. . . . The entire system may thus be viewed as a giant hierarchy of nested machines, each level containing and controlling those below it.*[30]

---

[28] Ibid., pp. 128–137 (actions placed in temporal order).

[29] In a comment that bespeaks painful experience, they say "it is a good idea . . . to have some bounds checking [of the pattern generator file] to prevent stray items of one project from clobbering another" (ibid., p. 130).

[30] Ibid., p. 89.

The evolution of designs within this "giant hierarchy" was not part of their vision, at least as it is described here. However, in later chapters we show that decentralized design evolution becomes possible following the creation of "systems of nested modules" such as those Mead and Conway taught their students to create.

The Reality of Design Rules

It its important to stress that the rules used to modularize a design are never arbitrary, and they do not arise costlessly. Modularizations are not free. A modularization requires the reorganization of fundamental design parameters based on knowledge gleaned from previous designs. Mead and Conway's modularization was based on very deep and detailed knowledge of both chip design principles and the operational constraints of actual maskmaking and fab houses.[31]

Setting design rules also adds a new group of tasks to a project, providing an overhead burden. Another overhead burden arises in the system integration and testing phase. Moreover, the costs associated with system integration and testing very much depend on the quality of the design rules. In a newly modularized design, when the modules are first brought together, problematic interactions usually appear and workarounds to those problems must be devised.[32] In the worst cases, the workarounds will feed back into the module designs, and the modularization itself will fail.

The Impact of Mead and Conway's Proposals on the Semiconductor Industry

Mead and Conway's insights into how to organize chip design processes came at a critical time in the semiconductor industry's history. In the late 1970s the increasing complexity of the various devices was creating serious problems for many firms. Mead and Conway's proposals attacked the looming complexity catastrophe head-on.

Following the three-part structure of any modular design process, we can divide Mead and Conway's overall contribution into three segments: (1) their design rules, (2) their partitioning of tasks, and (3) their activities with respect to system integra-

---

[31] As Lynn Conway said of the dimensionless design rules, "We chose the ratios based on Carver [Mead]'s knowledge of where the processes were headed." *Electronics,* 20 October 1981.

[32] Iansiti (1997b).

tion and testing. (In this last category, in addition to figuring out how to automate the creation of pattern generation files, Mead and Conway offered many criticisms of mask and fab practices and suggestions as to how these functions might be better performed.)

Twenty years later, Mead and Conway's *dimensionless, scalable design rules* have survived. The patterns were initially chosen to be independent of the line width of the fabrication process, and thus as line widths got smaller, the patterns were simply rescaled.[33] The basic task structure they posited—many small groups designing circuit modules—also became a highly competitive mode of chip design for VLSI chips. Hence their approach not only affected the design of large general-purpose chips (like Intel's microprocessors), it spawned a whole industry making application-specific integrated circuits (ASICs).[34]

Many of Mead and Conway's hoped-for changes in maskmaking and fabrication practices also came to pass. Specialized silicon foundries, committed to fast turnaround and reliable service, emerged to service the needs of the specialized design houses. With the arrival of these complementary specialists, the vertical integration of chip design and chip fabrication "under one roof" was no longer an economic necessity for chipmakers.

This in turn signaled a major change in direction for the semiconductor industry. In the late 1970s and early 1980s many observers were predicting that the industry would consolidate into a small number of vertically integrated suppliers.[35] Instead, the industry broke apart at a critical modular boundary—the design-to-fab interface. The potential for this split was inherent in the pattern independence of the planar process, but Mead and Conway's insight and design rules were required to make the modular structure a reality. Subsequently, many small U.S. companies entered the design stage, and a few large companies based in Singapore, Korea, and Taiwan entered the fabrication stage of chipmaking.

Mead and Conway's projects (there were a series of them, culminating in a multi-university project coordinated over ARPANET[36]) were extremely important in

---

[33] Chip designers are quick to point out that rescaling is not as simple as it sounds, especially now that line widths are coming up against fundamental physical limitations (the wave length of natural light).

[34] Gilder (1989).

[35] See, for example, Levin (1982); Alic, Harris, and Miller (1983); Ferguson (1985); Miller and Alic (1986).

[36] The ARPANET was the predecessor of the Internet.

demonstrating the feasibility of their modular, decentralized approach to chip designs. But this particular task structure did not in the end prove dominant. Instead of adopting the many-projects-to-one-chip format, designers attacked the chip vertically, partitioning the design elements into those common to most circuits (transistors and gates) and those specific to particular applications. The result was products with built-in design flexibility, such as programmable logic arrays (PLAs) and floating point gate arrays (FPGAs). These chips have a common and regular structure at the bottom levels, and then use different masks (and other techniques) to customize the top levels.

The last important change took place in the realm of system integration and testing. Within a modular system, efficient testing and verification *at the module level* becomes critical. (We will explain why this is so in Chapter 10.) Mead and Conway knew this. In order to streamline the testing of modules, they initially called for faster turnaround at wafer fabs. But one could look for a solution to this problem in software as well as hardware. Carver Mead and some of his students soon discovered that the software solution was both feasible and extremely cost-effective.[37]

Thus today, more often than not, a designer will not have to wait for a fab run to test his or her circuit design. If the circuit is not too big or complicated, the designer can test it using the software on a desktop computer system. The integration and testing of circuits on chips are performed by programs: verifiers, emulators, and simulators. Moreover, there is a healthy subindustry of firms that make and sell these "design tools" to other firms whose principal business is designing chips or parts of chips.

In the end, then, the new modular, decentralized approach to circuit design, which Mead and Conway pioneered, gave rise to new kinds of products, firms, and markets within the semiconductor industry. These changes in industry structure mirrored the underlying structural changes and the inherent logic of the modular designs. That is the power of modularity: it makes new designs and new patterns of design evolution possible, and the new designs in turn may give rise to new products, firms, and markets.

What Modularity Is: A Reprise

We began this chapter by asking, What is modularity? The answer seems deceptively simple: Modularity is a particular design structure, in which parameters and tasks are interdependent within units (modules) and independent across them. For any

[37] Gilder (1989).

design, the independence and interdependence of units can be determined by preparing a design or task structure matrix. Modular structures are different from interconnected or sequential structures: each generic structure has its own easily recognizable matrix pattern. (To recall these patterns, refer to figures 2.8 and 3.4.)

Modular task structures can be created through a process of modularization. This process requires architects to rigorously partition design parameters into visible and hidden subsets. Visible design parameters—also known as *design rules*—must be established before the module tasks begin; once fixed, they are relatively costly and difficult to change. In contrast, hidden design parameters affect only one module: they can be left to the discretion of module designers and are relatively inexpensive and easy to change after the fact.

In a complex design, there are often many levels of visible and hidden information. Some design rules will apply to all modules, while others will govern only a few. In addition, some "high-level" parameters may not be known to module designers two or more levels down, but through their effect on the intermediate levels, they still limit design possibilities. Such parameters are "indirectly visible" to the lower-level modules.[38]

A design hierarchy diagram summarizes the relationships between visible and hidden information in the system. It is a complement to the design or task structure matrix map. Some things that are difficult to see in the matrix representation are readily apparent in the hierarchical format.

Finally, there are degrees of modularity and the process of modularization generally proceeds by increments. Early in the process, the design rules may be wrong or incomplete. At that point, the integration of modules will be an arduous process, and risky in the sense that the system may not function as planned. However, as the architects gain experience, they will be able to foresee (and forestall) more of the negative interactions. In the end, the tasks of integration may become trivial: this is the ideal of "perfect" modularity with full "plug and play" flexibility.

What Modularity Does: A Preview

The concept of modularity spans an important set of principles in design theory: design rules, independent task blocks, clean interfaces, nested hierarchies, and the separation of hidden and visible information. Taken as a whole, these principles pro-

[38] Indirectly visible modules can exercise great power in the design, but they are also vulnerable to certain types of competitive attack. We will have more to say about this in volume 2.

vide the means for human beings to divide up the knowledge and the specific tasks involved in completing a complex design or constructing a complex artifact.

However, the requirement to divide knowledge and tasks may arise for different reasons in different settings. Therefore modularity is a versatile concept, which in practice may be used to solve a variety of design problems. For example, in the case of integrated circuits Mead and Conway felt that the complexity of the designs of VLSI chips threatened to overwhelm the resources available. Thus their key goal was to simplify the design process. They also felt that fast turnaround and rapid prototyping would make designers more productive, and they saw modular designs as a means to that end.

In other circumstances, the purpose of a modularization may be to obtain scale and scope advantages in production. For example, if all items in a product line use common parts ("all screws the same size"), economies in parts sourcing may substantially reduce the total cost of production. The same type of modularity can support "mass customization," wherein high levels of variety are achieved at a cost comparable to mass production. The trick in these production systems is to partition the product design and production process into (1) platforms common to the whole family, and (2) features specific to one or a small group of customers.[39]

In all of these examples, and others that we could cite, modularity does three basic things. If any one of them is desirable, then designers are likely to want to move down a path toward greater degrees of modularity:

1. Modularity increases the range of "manageable" complexity. It does this by limiting the scope of interaction between elements or tasks, thereby reducing the amount and range of cycling that occurs in a design or production process. As the number of steps in an interconnected process increases, the process becomes increasingly difficult to bring to successful completion. The time spent goes up, the probability of success goes down, and the quality of the output may go down as well. By reducing the set of allowed interactions, modularity reduces the range and scope of potential cycles. Fewer cycles means less time spent on the process. In addition, if cycling is unproductive, modularity will increase the project's probability of success and the quality of the final output.[40]

---

39 Nevins and Whitney (1989); Shirley (1990); Uzumeri and Sanderson (1995); Fine (1998); Spear (1999).

40 Sequential processes also limit the scope for interaction among tasks, and may be able to encompass even more steps than modular processes. However, sequential processes work best when the final output is predetermined.

2. Modularity allows different parts of a large design to be worked on concurrently. The independent blocks in a modular task structure can all be worked on simultaneously. Thus as long as the "design rules" and the "integration-and-testing" stages do not eat up too much time (a crucial assumption!), a modular division of tasks will shorten the end-to-end time needed to complete a given process. When the benefits of concurrent processing are added to the reduction in time spent in cycling, the time savings can be dramatic.[41]

3. Modularity accommodates uncertainty. The defining characteristic of a modular design is that it partitions design parameters into those that are visible and those that are hidden. Hidden parameters are isolated from other parts of the design, and are allowed to vary. (Sometimes their range is limited, but they can vary within the range.) Thus from the perspective of the architects and the designers of other hidden modules, hidden parameter values are uncertain. They lie inside a black box known as "the module."[42]

However, this structure, which permits uncertainty in the first round, can also accommodate uncertainty in later rounds. Therefore, modular designs and task structures are flexible. If new knowledge later yields a better solution for one of the hidden module designs, then it will be relatively simple to incorporate the new solution without changing much in the rest of the system.[43]

This ability to accommodate uncertainty in subsets of the overall system is a property that is unique to modular designs. Interconnected design processes also have uncertain outcomes—in fact, in some ways, their results are even more uncertain, because they allow interactions that are excluded from modular structures. However, interconnected design processes take an all-or-nothing approach to uncertainty: One either sticks with the present design (with all its flaws) or starts over, practically from scratch. Because everything is related to everything else, there is no way to attack problems piecemeal.

Speaking technically, modular mix-and-match flexibility creates *options* within the design and task structure. The value of those options in turn depends very much

---

[41] Sequential processes also provide the benefit of concurrent processing, although it is implemented in a different way (through synchronization and line balancing, instead of by severing connections).

[42] Parnas (1972a, b).

[43] Langlois and Robertson (1992); Garud and Kumaraswamy (1994); Sanchez and Mahoney (1996a, b); Baldwin and Clark (1994, 1997a, b); all have made the connection between modularity and flexibility.

on the context. Are customers interested in and willing to pay for improvements in the design? Do they value variety or want customized solutions? If so, the option value of a modular design will be high.

However, high option values imply more than simply financial profit. Pursuing option value in the context of a modular design requires a different organizational approach from managing either a sequential process or an iterative, interconnected process. To take advantage of the value of embedded options, designers and the companies that employ them must be ready to experiment and equally ready to discard solutions that turn out to be inferior. They must be prepared for rapid change in the designs of hidden modules, and capable of rapidly redeploying human and capital resources as new opportunities appear.[44]

Moreover, if this dynamic of experimentation takes hold, then as new solutions are tried out and fitted into the larger system of artifacts, the system as a whole will begin to change and to evolve. Individual modules will deliver higher performance and exhibit greater variety. The whole system will become more complex as old modules are split and new ones hooked up to the existing interfaces. This is the process we have labeled "design evolution," which we will describe in detail in part III.

Evolution in the designs of artifacts can have profound implications for the organization of the surrounding economic system. The firms that make and sell modules and systems; the markets in which those goods are exchanged; the contracts entered into and agreed upon; and the conduits that convey resources and information to and from firms and markets—all of these constituent parts of the economic system may need to change in response to a newly modular artifact design. Thus in Chapter 4, we turn to a consideration of the organization of the economic system that surrounds artifacts and designs.

---

[44] Christensen (1997); Tushman and O'Reilly (1996); McGrath (1997).

4 The Economic System Surrounding Artifacts and Designs

A search for higher value is the primary motivating force in the theory of design evolution presented in this book. *Designers see and seek value*—that is the first and only axiom in our analysis. This axiom recognizes that human beings have an innate ability to improve artifacts by changing their designs. Value-seeing and -seeking are what ultimately cause designs—hence artifacts—to improve and to become more complex. We accept this as a fact, and will not attempt to explain it.

In this book, we are seeking to understand the evolution of a set of designs—computer designs—of great economic significance. The aggregate market value of these designs today is in excess of $1 trillion.[1] Because of their value, these designs cannot be separated from their economic surroundings. Directly or indirectly, the greater economic system licenses every computer investment and evaluates every new computer design. It provides both the motivating force and the organizing framework for collective efforts aimed at generating new artifacts in this class.

In Chapter 1, we said that methods of managing complexity and dividing effort and knowledge arise at two levels in human society—the level of engineering design and the level of economic organization. In this chapter we will describe a complex multilevel system of economic organization in which artifacts and designs play an important role. This system is an "advanced market economy." A schematic overview of the system, which extends from the humblest physical artifacts to social institutions and governments, is shown in figure 4.1.

---

[1] Our data on the market values of companies whose stock is traded in the United States shows their aggregate value to have been $850 billion at the end of 1996. To this number, we must add the value of privately-held companies, foreign companies, and the net value *to users* of all computer hardware and software purchased in the past that is still in use. The sum gets us easily over $1 trillion.

Level	Elements	Structure
	Knowledge of How to Design Property Rights; Valuation Technologies	
Governments	Mechanisms to Create and Enforce Contracts; Systems of Adjudication and Dispute Resolution	**Contract Structure**
Capital Markets	Accounting and Payment Systems based on Money (a Contractual Claim on a Government); Practices that standardize the Exchange of Money and other Contractual Claims	**Contract Structure**
Financial Institutions	Firms (= Enterprises) that Design and Supply Monetary and other Contractual Claims	**Contract Structure**
	Contracting Technologies	
Goods and Labor Markets	Marketplaces for Goods and Services; Practices that standardize the Exchange of Goods, Services and Employment	**Contract Structure**
Organizations	Firms (= Enterprises) that Design and Supply Goods and Services	**Contract Structure**
	Scientific and Technical Knowledge	
Design Processes	Tasks	**Task Structure**
Designs	Design Parameters	**Design Structure**
Artifacts	Molecules, Spatial Forms, Programs, Words, Colors, Sounds	**Artifact Structure**

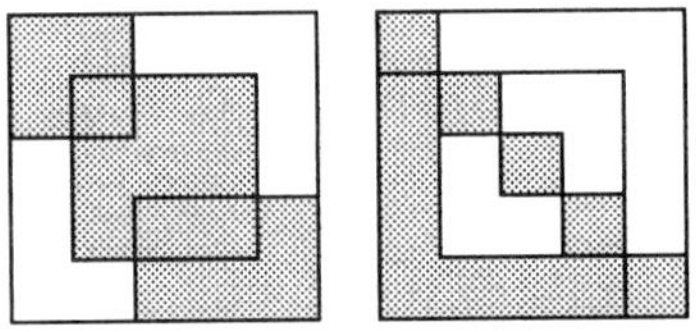

Figure 4.1 The economic system surrounding artifacts and designers.

Artifacts, designs, and tasks occupy the lowest level of this complex system; firms and markets, the next level; and financial institutions, capital markets, and governmental institutions, the next level. Because we are interested precisely in how the different levels interact, we do not maintain a strict stratification among them. Higher or lower has no special significance in the diagram, but each level does have its own body of relevant knowledge and accompanying technologies—for instance, physics is used to analyze the structure of a computer, while finance is used to analyze the value of a firm that makes computers.

Our theory of design evolution holds that a "force of value" originates in the minds of designers, and is then magnified by product markets (depicted in the middle panel of the figure). In advanced market economies, this "force" is further intensified by the workings of the capital markets. The force of value operates on the perceptions and intentions of designers and "pulls" them in certain directions.

The force of value is not a physical force, like the force of gravity. It is an economic force that is created and sustained by human conventions and social interactions. However, the fact that value is a social artifact does not make it any less real. The perception that "value lies here" is what causes designers to invest time and effort in the creation of new things, and those new things in aggregate are what give rise to change and progress in the economic system as a whole.

The economic "force of value" operates through a set of *valuation technologies*—these are the methods human beings use to define, transfer, and measure wealth. These technologies have developed dramatically and fundamentally over the last fifty years, in parallel with computer technologies. The ongoing development of valuation technologies is an important external factor, which has affected both the evolution of computer designs and the structure of the computer industry.

In addition to valuation technologies, in advanced market economies designers have access to *contracting* and *guidance* technologies. Contracting technologies make it possible to form and sustain collective enterprises efficiently and at low cost. Guidance technologies make it possible to coordinate large and small groups of people and "point" them in particular directions.[2] Contracting and guidance technologies have also progressed over the past fifty years, although not as dramatically as valuation technologies.

The pull of value, mediated by valuation, contracting, and guidance technologies, influences the paths of development of designs and artifacts. The promised reward to those seeking value will be greatest along the paths that correspond to high perceived market value. For their part, the technologies of contracting and guidance will

---

[2] Barnard (1938).

affect both what designers can see and how they seek that value. As a result, the artifacts that arise in an advanced market economy will be different from those realized in primitive or planned economies. That is why the organization of the economic system is important to us as we attempt to understand the evolution of complex designs of artifacts like computers.

In this chapter, we will look at the economic system as a whole, using the three technologies—valuation, contracting, and guidance—as our lens. The major force in this system is *value,* and so it is appropriate that we begin there.

Valuation Technologies

What makes a design worthy of being tried out? What makes it likely to survive? How can we represent the worth of a design to its creator? What is its worth to society? The answer to all these questions is the same: *value.*

Value is a measure of an artifact's worth in a particular social context. Mugs have value because they hold liquid and perhaps pencils. In Inuit society, harpoons have value because they are a key part of a seal-hunting system that serves the function of getting food. (See box 4.1.) Computers and computer programs have value because they process, store, and transfer information.

The formal definition of value begins with a comparison of alternatives. Consider two artifacts—for example, the two mugs described in Chapter 2. Although they might be used for the same purpose, they differ on some dimensions that could matter to the user (e.g., one has a cap, while the other doesn't).

Suppose the user would willingly exchange V_1 units of some good for the first mug, and V_2 units for the second. V_1 and V_2 are respectively measures of the value of the first and second mug. If the user prefers the second one to the first, then (under weak rationality conditions[3]) $V_1 < V_2$. The value to that user of switching from design 1 to design 2 (ignoring switching costs, if any) is simply the difference in their values: $\Delta V = V_2 - V_1$. If ΔV is greater than zero, as we have assumed, then the switch is worth making.

[3] The user's preferences depend on many contextual variables and may change over time. If this happens, the value the user assigns to each artifact may change, even to the extent of reversing the choice. All we require in terms of "rationality" is that at any point in time, if the user is willing to exchange V_1 units of the numeraire for artifact 1 and V_2 units for artifact 2, that user should be indifferent between artifact 1 plus $V_2 - V_1$ units of the numeraire and artifact 2.

Box 4.1 Complex Designs in Human Culture[1]

Today we take it for granted that designs change, usually for the better. Our computers operate faster and run new software. Our automobiles become more luxurious yet use less fuel. Airlines offer more departures and lower fares. All of this can happen only if there are ongoing, value-increasing changes in the designs of computers, automobiles, and airline schedules.

This state of affairs is relatively new in human history. For approximately two million years, humans have been toolmakers,[2] but until 60,000 years ago, the most complex artifacts were made up of at most two or three components, and were based on a narrow set of stone-working technologies. A typical artifact was a hand axe—a shaped piece of flint that could be used in myriad different ways. For fifty thousand generations, hand axes and other stone tools hardly changed at all. The archaeologist Glynn Isaac said of this period that "toolkits tended to involve the same essential ingredients seemingly being shuffled in restless, minor, directionless changes."[3]

Then, quite suddenly, about 60,000 years ago,[4] tools improved dramatically, becoming more specialized and complex in the process. Components were made out of different materials, and were assembled into composite wholes—like a spear with a serrated point and a detachable foreshaft.[5] An Inuit harpoon is representative of these complex tools. A typical harpoon was made up of twenty-six separate components, individually made of wood, bone, ivory, leather, sealskin, and stone.[6] Different technologies, each with its own tools, were needed to shape the raw materials. Making the harpoon thus required many different types of knowledge, and the artifact was not always the product of one mind and one set of hands.

In addition, harpoons were used with kayaks to hunt seals in waters off Greenland. The kayaks and the harpoons, plus the clothes that the hunters wore, were all artifacts that contributed in various ways to making this type of seal hunting feasible. Together, the artifacts and humans made up an effective seal-hunting system.

The system of designing, making, and using harpoons, kayaks, and clothes to hunt seals was itself a complex artifact, like the engineering design systems and the economic systems that surround computers today. In its context, the seal-hunting system's ability to achieve its purpose (getting food) was impressive. It is impossible to conceive of naked human beings hunting seals in frigid waters with hand axes alone.

---

[1] This material draws heavily on Steven Mithen's arguments as well as his excellent review of theory and evidence from the archaeological record (Mithen 1996).

[2] "Humans" includes the species *Homo habilis, Homo erectus, Homo neanderthalis,* and *Homo sapiens.*

[3] Isaac (1982), quoted in Mithen (1996, pp. 21 and 123). The nature of artifacts and technology, as well as the debates among archaeologists as to the role of tools in early human society, are summarized in Mithen's chapters 6 and 7, pp. 95–146.

[4] The shift is called the "Middle/Upper Paleolithic transition."

[5] There are even examples of modularity in the archaeological record: toolmakers made standardized "blanks," which could then be refined into a variety of different "blades" (Mithen 1996, pp. 167–171).

[6] Mithen (1996, pp. 125–127; drawing by Margaret Mathews, p. 127).

In the language of mathematics, the value of an artifact (measured by a particular consumer's willingness to pay for it) is a mapping of the vector of the artifact's functions onto the real line. This means that in the end, all the many ways the artifact may vary in the eyes of a user get reduced to one number, which we call its value.[4] Value is measured in units of some other good—in more advanced economies there is a standard "numeraire" good called "money."[5]

To complicate matters, however, the nature of value will change depending on the economic system in which users and designers operate. Because value has a chameleon-like quality, we want to be very clear about (1) how it can be measured, and (2) how different types of value come into existence as an economy moves from a primitive state to more advanced states. In particular we want to distinguish between the following different types of value:

- ex ante predicted value and ex post realized value;
- personal value and market value;
- value measured in product markets and value measured in the capital markets.

In an advanced market economy all of these different types of value are present at the same time. Because these different concepts are "in the air," they may sometimes pull designers in different directions.[6] For example, ex ante predictions often differ from ex post rewards. Personal values may not accord with market values. However, contracts and guidance mechanisms can be brought to bear to align different values so that they direct designers toward the same goal.[7] The different types of value are compared in the next three sections.

---

[4] For our purposes, the value of an artifact to a user/purchaser is a single real number (see Hicks 1946). With many purchasers and no capital market, however, the value realized by a seller of artifacts can be a vector of real numbers. See the discussion of capital markets below.

[5] A "numeraire good" is a good generally valued in a society that is relatively divisible and transferable, and thus can be used to balance unequal exchanges. Modern fiat money is an abstract numeraire good whose supply is controlled by the state. Tangible commodities such as gold, sheep, and grain can also serve as numeraire goods.

[6] Exchange markets are the ne plus ultra of social mechanisms for aligning values among heterogeneous populations. However, even the most advanced market economies are a very long way from having complete and perfect markets and thus different types of value may still pull people in different directions.

[7] Skill in contracting, negotiation, and organizational design and management often amounts to skill in understanding subtle value differences across individuals (e.g., "Andy's family comes from San Francisco; Mary values good colleagues; John is looking for a challenge"), and acting on them.

Ex Ante Predicted Value vs. Ex Post Realized Value

How do variations in designs arise and how do the better variations survive? One possibility is the process of *blind selection.* Variations in artifacts might arise through accidental modifications, followed by some form of reward, and the retention of favorable innovations.

Biologists believe, for example, that in living organisms, genes blindly recombine to create new forms. The environment equally blindly "selects" the best variants—those that perform well under local conditions. Neither genes nor the environment has the ability to create an organism based on what the other is likely to do, but *evolution by natural selection* occurs nonetheless. This was Darwin's essential thesis, which has been accepted, refined, and expanded by subsequent generations of biologists.

Humanity might originate and retain new artifacts by a similar process of blind natural selection.[8] Humans could, without forethought, generate accidental variants of existing artifacts, and then be rewarded for the most favorable alterations. Those *unanticipated ex post rewards* (remember the process works without foresight) would then confer advantage on the originators, who would then (again blindly) perpetuate the successful design. Indeed this may have been the way in which the earliest tools evolved; however, the evidence from the archaeological record (see box 4.1) indicates that in early human history, design improvements came very slowly, and artifacts had a tendency to "get stuck" in simple forms.

Modern human beings have foresight, however. That is, they have the ability to "look ahead," and to imagine both alternative worlds and the consequences of their actions.[9] Thus human design evolution does not have to wait for the arrival of accidental variations and unanticipated rewards. Human designers have the capacity to imagine, to analyze, and to act with intent. Because human beings have these abilities, it is the expectation of gain, not blind chance, that draws human effort into design processes. Ex post realizations are important, but mostly because they serve to confirm (or refute) what designers have already foreseen in their imaginations.

---

[8] Beginning with Darwin, exponents of the theory of evolution by natural selection have used designs and designing as a contrasting archetype to explain what biological evolution is not. Here we want to turn the tables, and use natural selection to explain what design evolution is not. In particular, design evolution is not blind. (See, for example, Darwin 1859; Dawkins 1989; Plotkin 1993; Dyson 1997.)

[9] Edelman (1992); Kelly (1994); Bucciarelli (1994).

The fact that expectations drive actions will affect how, in the chapters that follow, we model design evolution. The first and most obvious effect is that design changes must be characterized in terms of their expected or market values. (Expectations are subsumed in market values.) Our theory states that designers (or the firms that employ them) will make design changes that have high expected or high market values.

An equally important, but less obvious effect of human foresight is that designers can select not only the designs of artifacts but also the designs of the processes that generate the designs of artifacts. In other words, foresight makes possible the design of the design process itself. This property is called "recursiveness" and it will play a very important role in our theory of design evolution.[10]

Personal Value vs. Market Value

Designers try new designs because they believe they will gain some advantage—some value—from a new artifact. The ability to visualize alternative states of the world and to seek more advantageous ones is a universal human characteristic. But the nature of the advantage sought depends on the type of economy in which the designer dwells.

Complex artifacts and systems, such as the Inuit harpoon and seal-hunting system, may be found even in nontechnological societies. Such artifacts and systems emerged in human culture long before the first market economies came into existence.[11] However, we do not think that this is inconsistent with our core idea that designers see value and seek value.

In a society without markets, a designer might "see" a better way to make an artifact, and predict a gain from making it along the new lines. If he or she attempted to make the artifact in the new way, we would call that "seeking value." If the new way turned out to be better than the old, he or she would have "created value" by improving the design.

In this process, the designer does not have to frame his or her projections of gain in terms of a precise numerical accounting. Using intuitive reckoning alone, with a fuzzy concept of advantage, the designer could still "see value"—project a positive

---

[10] Ashby (1952); Kelly (1994).

[11] Although complex artifacts appeared in human culture around 60,000 years ago, the first cities only arose around 10,000 years ago. They may have had markets for some goods, but we do not know if any were in fact market-based economies. Many, like the cities of ancient Egypt, appear to have had a central authority that exercised coercive control over the allocation of goods and labor.

ΔV—and "seek it"—attempt to capture the positive ΔV by turning the imagined design into a real artifact.

The magnitude of the projected advantage, however, would be limited by the scale and the scope of the surrounding economy. For example, in a tribal society, the advantage to be gained—and the resources that could be used to realize the new design—would be restricted by the size and wealth of the tribe. As the society grew and developed links to other societies, however, the potential gain from a new design would grow too, roughly in line with the number of people who could adopt it. Nevertheless, unless the society developed markets, concepts of value in the society would remain personal and idiosyncratic.

In economies that lack markets, individual calculations of value are necessarily imprecise, and exchanges of goods and services are generally inefficient.[12] Markets offer two important benefits. First, they centralize and simplify trade, thereby reducing the costs of exchange, and increasing potential rewards to the designers of new artifacts. Second, they give designers a valuation technology, that is, an easier way to calculate the values that others will place on a new design for an artifact.

To appreciate the importance of a superior valuation technology, consider the calculations a designer would need to make in a world where all transactions were idiosyncratic and unstandardized. If John made pots, for example, customer A might want to trade wheat for a pot; customer B, a pig; customer C, some days of labor; and so forth. In this kind of primitive economy it is hard to know what a pot is worth, much less what a new design for a pot would be worth.

However, if the economy became a little more advanced, so that pots were sold for money in an organized market (a town or a fair), the value of a new pot design could be represented as the product of a price per pot times a quantity sold, less the cost of making a set of pots: $V = pQ - C$. This value could be thought of—quite informally—in terms of "money," an abstract numeraire. An anticipated gain in value could then be conceived of—again informally—as an expectation over probable realized values.[13]

---

[12] The effort of design is costly and consumes resources, and thus those who improve designs without the prospect of gain stand to lose time, energy, and wealth. At a personal level, these losses may be offset by the pleasure of the activity itself. Many people do appear to be willing to spend time and money on designing and improving artifacts for their own consumption. Think of art produced for personal enjoyment, gardens, living rooms, parties—all of these require some effort of design, as well as an expenditure of resources to implement the design.

[13] Bernstein (1996); Crosby (1997). Markets reduce the cost of searching for buyers of an artifact and for suppliers of inputs to its production. In this way, markets increase the total

New calculations and value-seeking behavior become possible as a result of this valuation technology. (New definitions of property arise at this juncture, too; see Box 4.2.) Thus, quite apart from the transactions themselves, markets provide new information that can enhance the precision of a designer's imaginative foresight.[14]

The precise definitions of gain and loss that arise when product markets emerge also give us a way to measure the gain that occurs when a new design is introduced into a society. Consider the following thought experiment, well known in welfare economics: When an artifact with a new design is introduced, consumers will be polled on their willingness to pay for it. The sum of the amounts revealed at that time less production costs we define to be the "value" of the new design.

Throughout the rest of this book, we will denote this measure of value "V." In general, V is denominated in units of "money." In the absence of a capital market, V may be a vector of risky inflows and outflows of money arising at different points in time.

The Role of Capital Markets

New designs demand effort and are therefore costly. The cost of achieving the new design is an investment, which we denote as "I." I is always positive, and always precedes the realization of value from the design. This can be a problem in an economy that lacks a capital market, for expected values in the future must then be filtered through a designer's preferences for delayed gratification and risk. A designer who must eat and feed a family may be unwilling or unable to invest today for an uncertain reward in the future.

Before the development of advanced capital markets,[15] the search for valuable new designs was sharply circumscribed. The pursuit of any particular opportunity was limited by the resources, time preference, and risk preference of the designer. Only

---

reward that can be garnered by the innovator, as long as he or she has a monopoly over the artifact in question. However, markets also accelerate flows of information, especially between buyers and sellers; hence the innovator may find that his or her monopoly is short-lived.

[14] This is exactly the "information-generating" function of a market described by Bodie and Merton (1995).

[15] By "advanced" capital markets, we mean capital markets that can efficiently create, price, and transfer debt, equity, futures, and option claims on commodities and securities. The names given to the claims are less important than their substantive definitions. (See Modigliani and Miller 1958; Merton and Bodie 1995.)

Box 4.2 Property Rights and the Rewards to Design

To the extent that a designer can capture a portion of the value created by a new design, he or she will be rewarded for introducing the innovation. But there are many reasons why some of the value may escape the grasp of the innovator. The most salient reason has to do with the presence (or absence) of property rights.

In general, rewards can be realized only if a superior design is implemented in a real artifact. Thus there is always a split between the value realized by the designer (who has the idea) and the producer (who produces the artifact). How the value will be split in turn depends on what counts as property in the economic system at large.

In their early stages, markets are generally organized around property rights in goods, not property rights in ideas. The more tangible the good, roughly speaking, the more clear and enforceable will be the rights to dispose of it and pocket the proceeds.[1] For example, in 1450, pots were property bought and sold in markets, but pot designs were not. To realize the gain in market value from a better pot design, the designer needed to make the actual pot and sell it.[2]

Property rights in intangible things—like pot designs and the profits of pot-making enterprises, for example—are a newer set of social concepts than property rights in tangible goods.[3] In fact, the development of property rights in ideas goes hand in hand with the development of the contracts that are needed to sustain collective enterprises.[4] The recognition of property rights that inhere in ideas and in contracts is in turn a prerequisite of the emergence of advanced capital markets.

---

[1] This is the concept of "alienability" put forward by Jensen and Meckling (1992).

[2] Pirenne (1917); Braudel (1979a); Landes (1998).

[3] On the economics of property rights, see Demsetz (1967); Alchian (1969); Alchian and Demsetz (1972, 1973); Jensen and Meckling (1976); Williamson (1985). There is a subtle chicken-and-egg problem in the definition of property rights. Until an intangible thing comes to be perceived as valuable, a society has no reason to call it "property" and delineate the rights over it. Thus new types of intangible things—like design rules, interfaces, software, data, or genetic recipes—will have very imperfectly described and weakly enforced rights attached to them, even in advanced market economies that have embraced the notion of intangible property in general. An argument that software is ill-protected under current U.S. definitions of intellectual property is put forward in Davis, Samuelson, Kapor, and Reichman (1996). For an opposing view, see Dam (1995). We return to this issue in volume 2.

[4] For example, articles of apprenticeship gave a master craftsman rights over the apprentice for a period of time (an intangible good), in return for food and housing (tangible goods) and education in the craft (an intangible good). This contract formed the basis of a craftsman's establishment, which was an early form of collective enterprise.

designs that furthered the interests of a state or ruler (like machines of war) could access investment pools greater than the wealth of one family.

In contrast, in an advanced market economy of the late twentieth century, a computer designer can convert his or her daily effort into food, a home, a vacation, and medical care. The benefits of the computer design will be reaped in the future, while food and shelter are consumed in the present. The vacation will take the designer to a distant place, and the medical care will be available to the designer in the event that he or she falls sick. This conversion of goods back and forth in time and space, and across probability states, occurs automatically because all these goods are exchanged in product markets that are themselves integrated by a large and efficient capital market.

Therefore, in an economy with advanced capital markets, it does not matter if value, *V*, and investment, *I*, occur at different points in time, or if *V* is uncertain. It does not even matter if both *V* and *I* depend in complex ways on future events and contingencies. Whatever the structure of payoffs, the workings of an advanced capital market will make them measurable in the present, independent of preferences, and divisible by contract.[16]

Advanced capital markets thus provide incentives to entrepreneurs to form collective enterprises for the purpose of designing and producing complex artifacts. By tapping into the capital markets, entrepreneurs can stand as intermediaries between the preferences of designers (and those involved in the production of artifacts) and the aggregate preferences of the society.[17]

In addition to channeling resources into long-term, risky ventures, capital markets provide an even more powerful valuation technology than product markets do. The valuation technologies of the capital markets can take complex and risky payments occurring at different points in time, apply appropriate prices, and come up with a single number called the "present value" of the new design.[18] Subtracting the investment cost, *I*, from the present value yields the "net present value" (NPV) of the new design opportunity: $NPV = V - I$.

---

[16] According to Merton and Bodie (1995, p.13), one of the six principal functions of the financial system is "to provide a mechanism for the pooling of funds to undertake large-scale indivisible enterprises or for the subdividing of shares in enterprises to facilitate diversification."

[17] Fama (1978).

[18] The architects of the modern theory of asset pricing are Tobin (1958); Debreu (1959); Markowitz (1952, 1959); Sharpe (1964); Arrow (1964); Lintner (1965a, b); Samuelson (1973); Black and Scholes (1973); Merton (1973a, b); Ross (1976); Breeden (1979); and Cox, Inger-

The present value (or net present value) that the capital market attaches to an opportunity is both like and unlike the "values" designers see and seek in premarket and product-market economies. It is like these values in that it is based on an expectation of benefit in the future that justifies effort in the present. It is unlike these other values, however, in that it prices out *all* inputs against all (appropriable) outputs. All inflows and outflows therefore are reduced to a single number.

This number summarizes the society's preferences with respect to time and risk, but it is independent of any one individual's preferences in those dimensions. Thus if the designer needs to eat today, or prefers the certainty of a secure wage, entrepreneurs operating in the capital markets can bridge the gap between the designer's present needs and the design's future rewards.[19] The designer's current needs and desires can be priced out in the valuation of I; future rewards (net of production costs) can be priced out in V. The net present value, $V - I$, if positive, is a gain in the present, which is available to any person (or group) that can assemble a credible set of contracts with which to pursue the opportunity.

Hence in an economy with advanced capital markets, value-seeking is almost a tautology.[20] Valuable designs, by definition, are those for which everyone involved can be paid their price, and the organizers of the enterprise—the entrepreneurs—still have something left over. Valuable designs pass the acid test of collective action—everyone needed to make the design into a real artifact will be able to agree that the investment should be made.[21]

Valuation technologies have developed greatly in the last fifty years, fueling in turn an explosion in capital markets and financial institutions. In the computer industry in the United States, for example, beginning around 1970, burgeoning financial innovation interacted with design innovation to create a highly dynamic pattern of development. A new type of financial firm (the venture capital firm) appeared whose purpose was to invest in new firms making new artifacts. Venture capitalists joined forces with computer designers and entrepreneurs to organize new firms and new markets to supply and exchange new computer products.

By this time, most new computer products were based on the principles of modularity. Many of these products had never been seen before—previously they existed

soll, and Ross (1985a, b). For a discussion of capital market equilibrium and the aggregative properties of asset prices, see Bodie and Merton (1999, Chapters 11–16).

19 Mises (1949).

20 It is a tautology under the assumption of "no arbitrage" in the larger economy.

21 Fama and Miller (1972).

as "knots of structure" (protomodules) within interdependent design structures. However, as we shall describe in volume 2, new contract structures and financial arrangements arose, quickly and efficiently, to allow companies to capture the market values implicit in the new designs. Eventually, through the combined efforts of artifact designers, entrepreneurs, contract designers, and financiers, the old computer industry broke apart, and a new industry structure emerged.

Contracting Technologies

We have said that a search for value leads designers to improve designs. If the artifact in question is sufficiently simple, a single designer will be able to accomplish the tasks required to create the new design. But designing a complex artifact like a computer requires the involvement of many people. Where artifacts are complex, therefore, the process of design must be a collective enterprise.

In advanced economies, collective enterprises can be formed efficiently and can be sustained for long periods of time via voluntary agreements between and among individuals.[22] Following what is now common practice in economics, we will call these agreements "contracts." Voluntary contracts are the basis of most economic undertakings in advanced market economies. They do not have to be formal, explicit, or legally enforceable to be effective: many are informal and implicit, and either "self-enforcing," or enforced by personal standards of honor or deeply-ingrained cultural norms.[23]

As Ronald Coase observed in 1937, there are advantages to using standardized contracts to define and sustain collective enterprises.[24] The main advantages are that with standardized contracts, many of the organization's "rules of operation" will be familiar to most participants,[25] will not have to be invented from scratch, and will not have to be renegotiated each day. These three characteristics translate into an enor-

---

[22] Barnard (1938); Williamson (1985).

[23] A self-enforcing agreement is one in which all parties to the agreement always find it in their best interest to take the actions they have promised to take. For examples, see Telser (1980); Baldwin (1983); Baker, Gibbons, and Murphy (1997).

[24] Coase (1937).

[25] "Participants" in this sense includes both employees and the parties with whom the enterprise transacts—customers, suppliers, investors, and so on. The totality of interested parties are sometimes called "stakeholders."

mous savings in the effort, time, and cost associated with forming and maintaining a collective enterprise.

We shall use the term *contract structure* to describe the set of contracts used to form a collective enterprise, and to organize its dealings with the greater economic environment.[26] In general, the contract structure of an enterprise will include agreements with suppliers of labor, material, and capital; with customers, owners, and managers; and with governments. Some contracts will be implicit, some will be explicit, and some will be a combination of the two.

Every firm that makes, buys, or sells something has a contract structure, that is, a set of implicit and explicit agreements that encase its activities and define its boundaries. Similarly, every marketplace has a contract structure, that is, a set of rules that regularize and standardize transactions in that particular forum.

A contract structure secures the resources for and captures the value of a task structure, but it is not itself a task structure. It is distinct from a task structure in the same way that the wrapping of a package is distinct from its contents. (See figure 4.2.) To get the tasks done and to capture value, there must be people to carry out the tasks, raw material, supplies and capital equipment for them to work with, and customers ready to buy the output and willing to pay for it. The purpose of a contract structure is to bring those resources together to allow the internal organization to implement the task structure as efficiently as possible.

A great deal of knowledge is involved in the design of contract structures. This technology encompasses knowledge of human behavior, economics, accounting, and the law. In advanced economies, individuals like lawyers, and enterprises like investment banks, specialize in the design of contracts. Often these individuals or firms will experiment with "new and improved" contract structures, just as the designers of physical artifacts experiment with new and improved artifact designs. Through this process (which is itself a design process), problematic contract structures tend to improve over time, and innovations that solve frequently encountered problems may become widely adopted throughout a society.[27]

---

[26] Alchian and Demsetz (1972) call the contracts that define a firm the "contractual structure" of the enterprise. We have adopted this phrase, but have left off the suffix to make the term parallel to "artifact structure," "design structure," and "task structure."

[27] For examples of "problem-solving" contract structures, see Tufano (1989, 1995, 1997); Williamson (1985, pp. 274–297); Chandler (1977); Jensen and Meckling (1976). One difficulty with innovation in the contractual domain is that contract *designs* are not considered "property" in our society. Therefore, a new form of contract can be imitated almost as soon as it is introduced (Tufano, 1989).

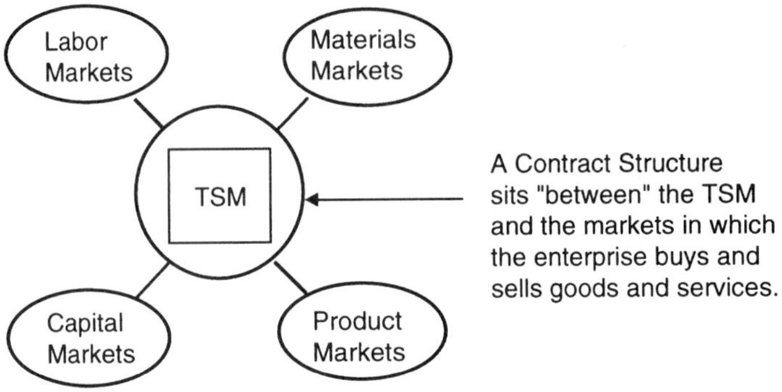

Figure 4.2 The contract structure surrounding a simple task structure.

Hence *contracting technologies*—knowledge about how to craft contracts and design enterprises—are an important part of an economic system that supports the creation of new designs and new artifacts. By facilitating the formation of stable, long-lasting collective enterprises, like firms and markets, contracting technologies break the "one mind, one set of hands" constraint on artifacts and designs. They support the creation of complex artifacts as an ordinary, day-to-day occurrence.

Just as scientific and technical knowledge determines what designs are feasible at a given time and place, contracting knowledge determines what kinds of firms and markets are feasible at a given time and place. Contracting knowledge grows, and thus what is possible changes over time. For example, in the late Middle Ages, preexisting markets in towns and a predesigned set of employment contracts (e.g., the "articles of apprenticeship") allowed a journeyman potter to settle down and become a master potter with a fixed establishment and apprentices. Later, more advanced versions of these markets and employment contracts allowed Josiah Wedgwood to set up a pottery works in Etruria, Staffordshire, England, which turned out high-quality porcelain artifacts, and employed hundreds of workers.[28] The firms and markets surrounding us today in turn are more complex and have capabilities that were unthinkable in Wedgwood's time.[29]

---

[28] Koehn (1995).

[29] Obviously, by providing ever more powerful and cheaper methods of storing, processing, and transferring information, computers have contributed greatly to the development of contract structures, including new types of firms and new markets for good and services.

Contractual Design Rules

In both firms and markets, the contract structure surrounding a task structure has two purposes (1) to protect internal activities from disruption, and (2) to set up efficient avenues of access and communication with the outside world. The parameters of the contract structure then become an integral part of the *design rules* that govern the performance of tasks in that enterprise. (See figure 4.3.)

Contractual design rules arise for the same reasons as technical design rules: they establish a framework for getting a particular job done efficiently. The architects of enterprises use contractual design rules to rationalize and standardize decisions when (they believe) there are few benefits to be gained by exploring other ways of solving a particular problem.

The practice of establishing contractual design rules for collective enterprises is ubiquitous in all advanced market economies. To see how universal are these rules, it is interesting to think of collections of people that do not have predefined contract structures. There are very few. Affinity groups (as long as they are unorganized) are one example; also mobs, audiences, and passengers on a bus or a plane. In the last two cases, however, there are social codes of behavior that operate like contract structures while the group remains in existence.[30]

A good contract structure defines the boundaries of a particular enterprise, and "shelters" some of those "inside" it from minute-to-minute changes in the outside world. For example, the contract structure of a firm usually takes the issues of asset ownership, terms of employment, and the division of gains "off the table" for most of the time that work is being done. Similarly, the contract structure of a marketplace takes questions of membership, ability to pay, and mechanisms of settlement out of the flow of normal exchange. In their respective spheres, these contractual design rules simplify the day-to-day activities of a firm or a marketplace, making each form of enterprise more efficient at carrying out its particular set of tasks.[31]

This sheltering is cost-effective in two ways. First, as we already indicated, it allows people to get a job done without constantly having to respond to changing external conditions. This in turn makes the enterprise efficient with respect to a particular set of tasks.

Second, and equally important, contract structures package transactions and tasks into meaningful units, making valuation technologies easier to apply, hence more

---

[30] Hansmann (1996).

[31] Williamson (1985); Merton and Bodie (1995).

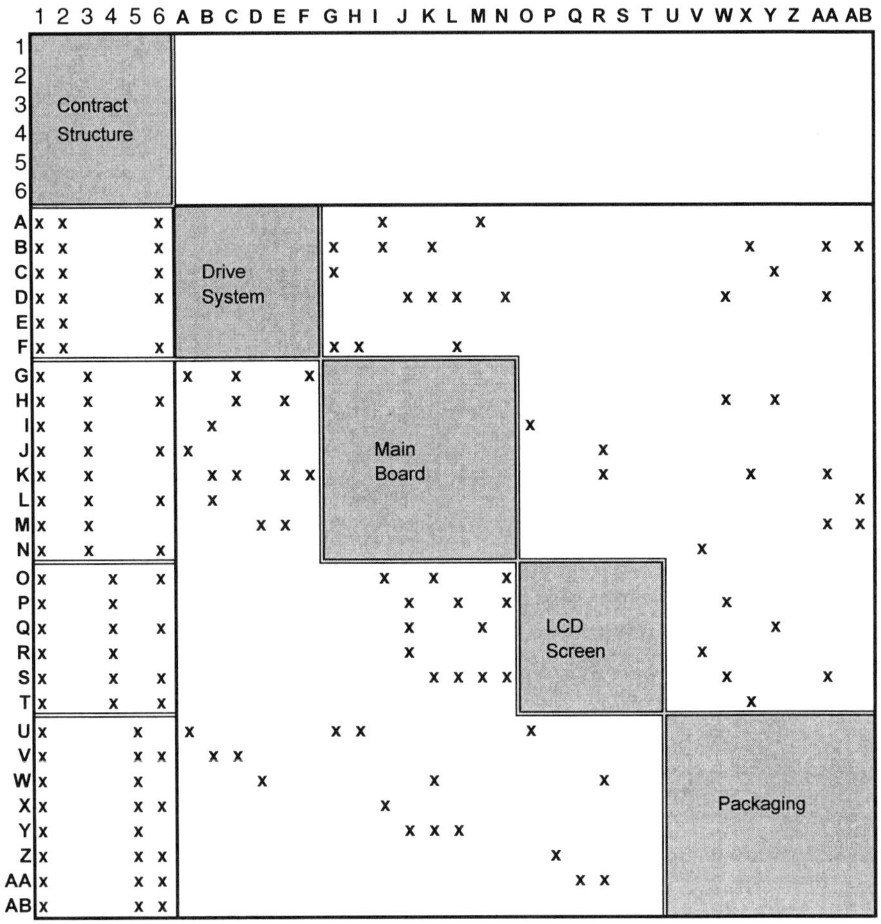

Figure 4.3 A contract structure sets design rules for a collective enterprise.

cost-effective. For example, in a world with no stable contract structures, corporations and long-term employment contracts would not exist. The "market value" of an individual might then need to be calculated and new terms of employment negotiated every day.[32]

[32] Spot markets for day labor work this way. They were common in the past, but today are rare in advanced market economies.

By way of contrast, in our world, though the market value of a company may be calculated each day (or each minute), the value of an individual normally is calculated only when he or she changes jobs. By providing a meaningful aggregate unit of valuation (the company), contract structure design rules economize on the use of market valuation technologies.[33] We look more closely at the question of economical valuation units in volume 2.

Corporations

When a particular task structure requires a very stable, long-lasting contract structure to be efficient, it makes sense to form the kind of enterprise known as a "firm."[34] An especially useful contractual design rule is then *The firm shall be incorporated.* We define a *corporation* to be a collective enterprise that is legally able to

- hold title to property;
- enter into contracts and participate in transactions as an enterprise;
- issue mathematically defined, divisible claims against the surplus generated by its activities.

This is a functional definition, which does not necessarily correspond to the legal definition of corporation in any jurisdiction.[35] (The precise legal definitions of corporations generally differ from one place to another.)

---

33 This implies that appropriate "units of valuation" are a function of the valuation technologies in existence, hence endogenous elements of the larger economic system. This fact is not surprising to financial economists, steeped as they are in valuation methodologies and the costs of using them. However, it is a fact sometimes overlooked in other branches of economics, wherein "the firm" is assumed to be an exogenously determined unit of analysis.

34 This was Ronald Coase's thesis (Coase, 1937). One of Alfred Chandler's great contributions was to tie the need for stability to specific modern technologies—the "coordinated high-throughput" technologies also known as the technologies of mass production (Chandler, 1962, 1977).

35 A contract structure that is not legally a "corporation" may embody these attributes, in which case we would consider it to be a corporation for purposes of this discussion. Viewing the decision to incorporate as a design rule (i.e., a hierarchical design parameter) allows us to view "incorporated collective enterprises" as a subspace in the larger design space of "all collective enterprises." The decision to incorporate then becomes a decision about "where" to conduct a search for value (in other words, in which subspace is the highest value likely to

The first item in the definition addresses the enterprise's need to forestall disruptions, and to prevent "holdups" based on threats to withdraw assets critical to the production process.[36] The second item greatly reduces the cost of buying and selling, and forming other economic relationships. In effect, the right to transact is an "asset" that is critical to smooth functioning of many production and design processes. Incorporation gives a collective enterprise "title" to that asset.

The third item gives the incorporated enterprise access to the capital markets. If necessary, then, it can fund current activities by issuing promises to pay based on anticipated returns from its future activities. (Box 4.3 describes some ways in which the claims to an incorporated enterprise may be divided up.)

The incorporated form of enterprise is an artifact, which like all artifacts has a history. It emerged as a legal entity about 150 years ago, and has universally diffused into all economies with advanced capital markets.[37] Its widespread use is testimony to the fact that this contract structure design rule can be advantageously applied to a wide range of task structures.

Task Structure and Contract Structure Interdependencies

Modern contracting technologies are so flexible that, usually, many contract structures can be effectively used to "package" any given task structure.[38] For example, in semiconductor fabrication, an efficient-scale wafer fab costs $1 billion, is physically indivisible, and should be scheduled to run constantly. One potential contract structure is for the fab to be owned by a corporation, which might issue shares to finance the construction of the facility.[39] Another potential contract structure is to make the fab a separate incorporated enterprise, and to finance its construction by issuing shares to its customers. The investors in turn may receive preferential or even guaran-

---

be found?). This framing in turn lends itself to the analysis of technological and contractual complementarities. (See, for example, Milgrom and Roberts 1990, 1994, 1995; Holmstrom and Milgrom 1991, 1994.)

[36] Klein, Crawford, and Alchian (1978); Baldwin (1983); Williamson (1985).

[37] However, this contract structure did not diffuse into socialist or communist economies. (See, for example, Shleifer and Boycko 1993, pp. 39–89.)

[38] Specifically the Modigliani-Miller irrelevancy propositions and their derivatives. (See Modigliani and Miller 1958; Merton 1990, pp. 413–427.)

[39] This is the Chandlerian solution (Chandler 1977). An example of this contract structure is the semiconductor chip fab that Hyundai recently built in Eugene, Oregon ("Hyundai Sets U.S. Fab in 1997 at $ 1.3B Cost," *Electronic News,* 29 May 1995, p. 1.).

Box 4.3 Measurable Surpluses and the Divisibility of Claims

In an advanced market economy, divisions of value may be accomplished in many different ways.[1] For example, the designers of an enterprise may

- charter a corporation;
- define a strategy for paying out the corporation's economic surplus over time (a dividend policy);
- create proportional claims to the payments;
- list those claims on an organized exchange.

This set of actions creates the contract structure known as a "joint stock corporation with publicly traded shares." Other divisions of value are possible as well. For example, claims to the surplus might be vested with customers or employees—this is known as a "mutual" type of contract structure.[2]

For these contract structures to be possible, however, the surplus of the enterprise must be measurable. Prior to the introduction of the income statement in double-entry bookkeeping, an economic surplus could only be measured at the very end of a cycle of activity (the end of a voyage or a partnership, for example). Claims to interim profits were meaningless then, for the profit stream was not measurable.[3] Once year-to-year profits became measurable, however, dividends could be paid out of retained profits, and divided among numerous claimants by applying a simple mathematical formula.[4]

1 Hansmann (1996); Hart and Moore (1990).

2 Alchian and Demsetz (1972); Fama and Jensen (1983b).

3 However, a fixed series of payments was measurable, hence a debt *was* a meaningful contract (Braudel 1979b); Johnson and Kaplan (1987).

4 Formulas used in this process are subject to an "adding up" constraint—the sum of all claims has to equal the whole surplus, no more and no less. If this condition is violated, the definition of claims is either incorrect or incomplete.

teed access to the fab's capacity.[40] Both of these contract structures (and others) exist for fabs, and we have no reason to suspect that one is better than the other in all cases.

40 An example of this contract structure is Singapore-based Chartered Semiconductor Manufacturing's (CSM's) Fab II, which was largely financed by customer investments. In one instance, LSI Logic made a $20 million equity investment in return for "[a] 10-year agreement [that] guarantees LSI a minimum capacity with the option of doubling or reducing wafer output depending on market conditions" ("LSI Puts $20 M into CSM," *Electronic News,* 6 March 1995, p.1). The combination of "fabless" semiconductor firms and specialized foundries (of which CSM is an example) is a relatively new contract structure for firms in the semiconductor

Not all contract structures are equally good ways to package a given task structure, however.[41] For example, it is well known among organizational economists that "asset specificity," and forward and backward linkages in production processes, create the potential for holdups wherein one party may seek advantage by offering a credible threat to harm the other.[42] Implied threats of holdup, as well as so-called transaction costs and agency costs, do influence observed contract structures in significant ways. Indeed, we maintain that a major consequence of modularization is to change the transactions and agency costs in ways that make new contract structures economical. We discuss how and why this happens in chapter 14.

Guidance Technologies

We have already said that, for some task structures, it is economically desirable to insulate the "interior" of a collective enterprise from day-to-day changes in external market forces. For example, designers working to create a new design may work more efficiently with few outside distractions. However, such insulated enterprises must still create and capture market value in the long run if they are to survive and fulfill their contractual obligations.[43]

This means that those working inside an insulated enterprise must have technologies that enable them to "see and seek" market value in coordinated ways. If those

---

industry. This new contract structure was made possible by Mead and Conway's specification of the design rules for VLSI chips (discussed in Chapter 3), and the subsequent development of computer-assisted design (CAD) tools based on their rules (Mead and Conway 1980).

41 In other words, there are likely to be complementarities between specific task structures and specific contract structures. On the theory of complementarities, see, for example Milgrom and Roberts (1995); Holmstrom and Milgrom (1994).

42 See, for example, Jensen and Meckling (1976) on agency costs, the location of decision rights, and optimal incentive contracts; Klein, Crawford and Alchian (1978) and Baldwin (1983) on opportunistic behavior with respect to rents derived from durable assets; Williamson (1985) on investment in specific assets; Grossman and Hart (1986) on investments in bargaining position; and Hart and Moore (1990) on necessary and sufficient characteristics of productive coalitions. Recent contributions to this literature include Baker, Gibbons, and Murphy (1997), who compare alternative contract structures when performance is observable by the buyer but not by third parties, and Rajan and Zingales (1998b), who look at the terms of access to specialized complementary assets.

43 Strictly speaking, non-value-creating enterprises can be sustained by government subsidies for many decades. However, state-owned and subsidized enterprises are not part of our theory.

coordination mechanisms are lacking, each person will tend to go his or her own way, and the result will be chaos.[44] We call the technologies of internal coordination "guidance technologies," because, like the guidance systems of a missile, they point individuals and groups in particular directions. We end this chapter by describing the guidance technologies that may be used to coordinate the design efforts of many people.

Consistent with our axiom that designers both see and seek value, there are two main types of guidance technologies. The first is technical knowledge about how an artifact works and why people value it. The second is managerial knowledge about how to motivate and manage individuals. Technical knowledge allows designers to "see" the space of designs better. Managerial knowledge can be used to muster designers' energies and talents, so that they can "seek" value more effectively.[45]

Technical Knowledge: The Language Surrounding an Artifact

One way to think about the process of design is as a simulation of future events.[46] Each design parameter choice in effect requires a designer to think through the consequences of the choice for the structure, the functions, and ultimately the value of the artifact. In order to make those judgments, a designer must mentally create the artifact, mentally produce it, deliver it to market, and "observe" its valuation.

All of this initially takes place within a single designer's imagination. However, if the artifact is complex, then as the design process moves forward, the simulation will involve more people. There will need to be conversations among designers about different possible structures and how each might perform. Structured experiments, prototypes, and other tests may be used to indicate how the new artifact will behave, and what its effects will be on the enterprise and its markets.[47]

At each round of this simulation, designers must communicate to one another what they are thinking and learning. However, they can only do so if they share a

---

[44] One might suppose, at first glance, that the markets themselves would guide members of an enterprise to take appropriate actions. Markets are indeed powerful instruments for measuring the value of existing artifacts, but they are not very good at estimating the value of things that have yet to be created. Hence market valuations are of limited use in design processes. To be useful, market behavior must be interpreted and translated into a form that is both meaningful to and can be acted on by designers (See Baldwin and Clark 1994; Iansiti 1997b).

[45] Barnard (1938).

[46] Simon (1969).

[47] Iansiti (1997); Bucciarelli (1996); McCullough (1996).

language that can express the things they know. Thus a shared technical language is the first guidance technology.[48]

That shared language will consist of statements linking the artifact's structural characteristics to its functional attributes. Call this list of statements *L*. At the beginning of an artifact's existence (when there are few working designs in its class), *L* will be both incomplete and imprecise. The propositions (sentences) in *L* will be conjectures, many of which will turn out to be wrong. Designers will also "know" by direct intuition many things about the artifact's structure and about users that will not be expressible in the language of *L*. However, the shared technical language will give designers a new mental tool kit, which they can use to refine their own thinking and to communicate with one another.[49]

A critical part of the shared technical language is the set of names given to the components of a complex artifact, and the verbs used to describe the components' functions in the system.[50] By naming components, designers begin the process of dividing a complex artifact into mental units, which may eventually become modules. This is an important milestone: in the designers' minds, the artifact passes from being a unified thing to being a set of interacting things. To use Plato's metaphor, the "Idea" of the artifact is carved "at the joints."

In the minds of designers, the named components of a system are rather complex mental objects. They often begin as images of physically discrete things—the point and the shaft of a harpoon, a particular chip, or block of code in a computer system. However component definitions are based on an analysis of the logic of a system, not the physical composition of its parts. Hence, just as users are indifferent as to artifacts that deliver the same functions but have different structures, designers are indifferent as to components that perform the same role, but have different structures. For example, today, a computer *modem* can be implemented as hardware (chips on a

[48] Brooks (1975).

[49] Complete designs correspond to true propositions within the formal language *L*. Gödel's theorem applies to this, as it does to all formal systems. It follows that there are true propositions (complete designs) which cannot be derived from (constructed within) the formal system, *L*. Therefore, whereas all complete designs are computable algorithms (by the Church-Turing thesis), the process of designing itself is not computable (by Gödel's theorem). (Computations may still be subsumed in, and thus assist design processes.) This is not a surprising result; however, it is a useful one to keep in mind when considering the possibilities and limitations of automated design processes. (See Hofstader 1979, 1995; Penrose 1989, 1994; Edelman 1992, especially pp. 211–252; and Deutsch 1997).

[50] Clark (1985); Henderson and Clark (1989).

board) or software (code in read-only-memory). This is possible because, in their own specialized language system, computer designers have created a role description of a modem that is independent of structure. *Anything* that satisfies this role description may be called a "modem" and is a candidate for admission in the "modem position" of an overall computer design.

Technical Knowledge: Principles

When it gets rich enough, the language surrounding an artifact can be used to construct logical arguments and state hypotheses about new designs in the artifact class. Designers can then begin to string together complex chains of deduction and reasoning like the following:

- Users value the *speed* of the system as a whole (defined as *time to complete a given job*)
- A *disk drive system* is a module of the overall system, which *stores large amounts of information at a relatively low cost*
- The information on *disks* must be *accessed by the disk drive system* and *transmitted to the processor* before it can be used
- The *access time of a disk drive* affects *the speed* of the system as a whole. (This may be a hypothesis, or a proposition derived from detailed knowledge of the structure of the system, i.e., how code gets executed.)
- Therefore, to increase the value of the system, one might *decrease the access time of the disk drive,* holding other things constant.

This is a fairly complex chain of logic. It absolutely cannot be expressed without an elaborate, meaningful language that relates components to their roles (disk drive, storage), and component performance to system performance (access time, speed).[51] (To see this, readers may replace the *italicized* words in these sentences with nonsense words or mathematical symbols. The logical structure of the reasoning will survive, but its meaning will disappear.)

The ability to construct logical chains like this is immensely useful to designers. The chains become the principles that tell designers where to look for value—in this case, *improve* the *access time* on the *disk drive* because that affects the *speed* of the

[51] Henderson and Clark (1989); Deacon (1997).

system and *users value speed.* The ability to develop and communicate these principles is critical to a coordinated, collective search for value.

Technical Knowledge: Metrics

Metrics are used to define and measure the performance of an artifact and its components. They are the complement of principles. The development of metrics begins with a list of attributes that users value in a particular artifact; this list becomes part of the shared language *L*. Hence, if users value the speed with which a computational task gets done, the notion of "speed" will soon enter the designers' vocabulary.[52]

Then, aided by the development of the shared technical language and its principles, the descriptors of what users value will tend to become more precise. While users may retain a pure perception of speed, designers will be motivated to develop a precise definition, for example, speed equals the time from initiation to completion of a job. Such definitions can then be tested against users' perceptions. (Is this indeed what users perceive as speed? Do users truly value this attribute?) The definitions are also a way of converting designers' conjectures about designs into testable hypotheses. (Does this new design deliver speed, defined this way?)

In the process of testing new structures, designers will develop *metrics* that allow different artifacts in the same class to be compared quantitatively along a variety of dimensions. For example, computer systems are universally compared on two dimensions: speed (called "performance") and cost to the user (often measured by "price"). The ratio of speed-to-user cost ("performance-to-price") is widely used as an approximate measure of value to the user: the higher the ratio (other things equal), the more successful a new computer system is likely to be.

Metrics thus become a way of translating market valuations into terms that designers can use. Metrics may be used to rate competing artifacts, to test hypotheses, and to measure outcomes against objectives. Over the last fifty years, computer metrics have been evolving in tandem with computers themselves. As a result of long history and wide experience (disseminated to young designers through textbooks and courses), computer designers today have a rich body of technical knowledge about how to measure the attributes of both components and systems. They are also highly cognizant of the pitfalls of different measurement systems.[53]

---

[52] Users have direct perceptions of their own wants, even before the language needed to describe those wants comes into existence. For this reason, users are often very good designers, especially early in the history of an artifact class. See von Hippel (1988).

[53] See, for example, Hennessy and Patterson (1990, Chapter 2).

Metrics are often touted to users as part of the marketing program for a system or a component. In this way, the mental structures that designers develop to categorize user preferences, and to understand purchase behavior, become part of the users' ways of conceptualizing their own wants.[54] A user's inchoate desire for "a machine that doesn't make me wait," turns into a desire for a "Pentium Pro II microprocessor rated at 350 MHz."

Motivation and Management: Incentives

A shared technical language, robust principles, and precise metrics are important guidance technologies that can point designers in the direction of higher value. But what value do designers seek? Or rather, whose value?

In advanced market economies, firms can be created for the express purpose of seeking market value. A shared technical language, principles, and metrics are then valuable tools that translate (albeit imperfectly) the valuations of the marketplace into information on which designers in the firms can act. Yet designers also come to their tasks with what we have called "personal" valuations of alternative designs. Hence, in the absence of other guidance mechanisms, technical knowledge alone will not be enough. Having a well-developed body of technical knowledge does not guarantee that a group of designers will proceed in the same direction, much less in a direction consistent with higher market value for the enterprise as a whole.[55]

To illustrate the problem, consider the following example based on a true story from the auto industry: In the mid–1980s, a young design engineer (we shall call him Jake) was assigned the task of designing a bracket connecting the muffler to the tailpipe on a new family car. Jake came up with an elegant design that used a new, ingenious clamping mechanism. It was simple, easy to assemble, and seemed to offer a considerable advantage in durability.

The only problem (which did not surface until prototype production several months later) was that the new bracket cost three times as much as the one it replaced. Caught up in his enthusiasm for the design, Jake had not worried much about cost (and in fact he didn't see the problem, since the "new bracket only costs a couple of dollars"—$3.50 to be exact). Senior executives saw the issue quite differently: Jake was not the only design engineer to make a decision like this, and the resulting cost of the car was so high that it threatened to price the car out of the market.

---

54 Clark (1985).

55 This is a classic agency problem (Jensen and Meckling 1976).

If each designer like Jake were evaluated in the marketplace every day, we might expect to find more congruence between his own valuation and the market's valuation of his design. However, when there are interdependencies in a design, the product and capital markets cannot effectively value an individual designer's decisions. Thus Jake and his peers worked in what we are calling an insulated enterprise. With no window on the marketplace, Jake's search for value led him to a design that was at variance with the market's valuation of the whole. Managerial guidance mechanisms were needed to bring market and personal values into congruence.

Incentives (broadly defined to include financial and nonfinancial rewards and sanctions) offer one such mechanism. For example, a bonus tied to the achievement of design goals for the vehicle, or stock options in the company are two forms of incentives. Well-designed incentives can play an important role in motivating designers to go where the "peaks" of market value are highest.

Incentives are not a panacea, however. To see why, consider Jake's design problem again. Even if his bonus had depended on meeting a set of cost objectives, Jake's choices were still problematic. Simple cost targets would not give him any way to make tradeoffs. For example, how should he think about tradeoffs between cost and durability, or cost and noise? For his part, Jake knew that he was working on the design of a family car within a total cost target, but he saw only a tiny part of the overall picture. He was missing a sense of how the parts were supposed to interact. To use a musical analogy, Jake's situation was like being a player in an orchestra without a conductor. He and each of the other designers knew their parts, but none of them had a sense of the whole.

Motivation and Management: Leadership

This is where guidance in the form of human leadership plays a crucial role. Simply put, if the artifact is complex, if the enterprise is trying new approaches, if customer needs and market valuations are uncertain, and if competitors are also moving in the marketplace, then the best formal guidance mechanisms—a shared language, robust principles, precise metrics, strong incentives—will only take the the design effort so far. There will still be considerable confusion over valuation and goals. There will still be iterations through the design and task structure as the designers search for the right combinations of parameters, without knowing exactly what they are seeking.[56]

[56] In formal economic models of firms, leadership often gets reduced to the concept of "a boss." See, for example, Coase (1937), or, more recently, Wernerfelt (1997).

What is then needed is a person or group of people who can see the design as a whole and relate it to the desires of users and competitive factors in the surrounding product, labor, and capital markets. This person or group may—in fact should—use all the other formal guidance mechanisms, but will augment them with a human ability to foresee, to imagine, to integrate, and to adapt. He, she, or they will be able to spot the difficult, unresolved issues in the task structure, and direct attention and resources there, even when (especially when!) the established design principles and problem-solving strategies fail to address the need.

The ability to marshall and direct human beings effectively when there are no fixed routes to success is the quality known as "leadership" in human affairs. Leadership does not replace, but complements the other guidance mechanisms. When the environment is dynamic, the design complex, and the enterprise attempting new solutions, stronger, more direct leadership becomes critical. Thus the challenge for the designers of a collective enterprise is to mesh human leadership with other guidance mechanisms so that the enterprise may see and seek value as an integrated and effective whole.

Artifacts and Designs in an Advanced Market Economy

We have gone to some lengths in this chapter to describe the various ways in which designs interact with the larger economic system. We began by explaining how, in an advanced market economy, *value* operates "like a force" on the imaginations of artifact designers, entrepreneurs, contract designers, and financiers. The force of value originates in the minds of designers, is magnified by product markets, and further magnified in the presence of capital markets.

In addition to valuation technologies, an advanced market economy also supplies *contracting technologies*. These include preformed contract structures that can be used to form collective enterprises. Today these contracting technologies are very powerful and flexible, and as a result, enterprise designs that were impossible or unthinkable one or two hundred years ago are common today.

Individuals working together within insulated enterprises also need coordinating mechanisms that allow them to see and seek value in coherent, effective ways. *Guidance technologies,* which encompass both technical knowledge and managerial knowledge, address this need. Roughly speaking, where valuation and contracting technologies leave off, guidance technologies take over, and provide a consistent direction for day-to-day work on a complex design.

In chapter 5 we examine the dynamic behavior of modular designs in greater detail. We first define six modular operators—these are relatively simple changes, which can be applied to different parts of a modular system. Then in chapter 6 we look to see how the concept of modularity emerged as a design possibility in the very early stages of the development of computer designs. Chapter 6 ends on the eve of the creation of System/360, the first "truly modular" computer design.

5 The Modular Operators

Modularity is a structural fact: its existence can be determined by inspecting the structure of some particular thing. If the structure has the form of a nested hierarchy, is built on units that are highly interconnected in themselves, but largely independent of other units; if the whole system functions in a coordinated way, and each unit has a well-defined role in the system, then, by our definition, the thing is modular. This is true whether we are speaking of a brain, a computer, or a city.

In this chapter we look at the dynamic possibilities that are inherent in modular structures and, especially, modular designs. Strikingly, the changes that can be imagined in a modular structure are spanned by six, relatively simple *modular operators.* These operators, applied at various points and in different combinations, can generate all possible evolutionary paths for the structure. Thus the operators are a powerful set of conceptual tools that are implicit in the logic of modular designs. We define and describe the six modular operators as follows:

1. *splitting* a design (and its tasks) into modules
2. *substituting* one module design for another
3. *augmenting*—adding a new module to the system
4. *excluding* a module from the system
5. *inverting* to create new design rules
6. *porting* a module to another system

Taken as a whole, these modular operators provide a parsimonious list of "things that designers can do" to a modular system. The first two operators—splitting and substitution—can be applied to nonmodular designs; the last four cannot. Having such a list gives us a way to classify and categorize design changes past, present, and future.

We begin our discussion of the operators with a brief overview of the way designs change, and the sense in which one design may be "descended" from another.

The "Descent" of Designs

Designs differ from other artifacts in that each one is unique. There may be many identical copies of an artifact, but there will be only one design. If someone says, "I have two designs for you to look at [of a mug, a computer, or a manufacturing process]," we expect to see two different versions of the artifact in question. Similarly, if we copy a set of blueprints or duplicate a CAD file, we are copying a design, not producing one.

Designers will usually construct new designs in the presence of knowledge about previous ones. A new design therefore generally solves a set of problems identified in the course of realizing the designs that preceded it. As Gordon Bell and Allen Newell described the process: "[T]here is at any moment a 'standard' design which is seen as emerging from the just prior 'standard' design."[1] This is the notion of descent as it applies to designs.

The mechanisms of descent in designs differ from mechanisms of descent in other evolutionary processes. Many modern theories of evolution, for example in biology, have at their core an action or event called "replication," which means "make a copy of <X> and convey it to the next generation."[2] In general, replication plays a much less important role in the evolution of designs than in other types of evolution.

The copying or replication of elements of prior designs does occur, but it is not the only way a design gives rise to descendants. It is also common for descendant designs to solve problems that are revealed in the course of realizing prior designs.[3] For example, as users work with a particular computer system they may find that the speed of data transfer is too slow. In the next generation of those machines, designers would have reason to pay attention to the memory management function. We would then expect new elements (e.g., cache memory) to appear in the next generation. The new elements would not necessarily replicate, or recombine, elements of the prior designs. Instead, the new elements would represent new solutions to the problems posed by the prior set of designs.

---

1 Bell and Newell (1971, p. 87).

2 Dawkins (1989); Edelman (1992); Plotkin (1993).

3 Marples (1961); Bell and Newell (1971, all chapters).

Designers' knowledge of prior designs includes things such as

- copies of older artifacts in the class or category;
- detailed information on design parameter choices for those artifacts;
- known problems or shortcomings of those artifacts;
- results of various functional tests (e.g., megahertz ratings or access times);
- data on user acceptance and overall market value (survey evaluations, product reviews, estimates of units sold, and prices).

Less easy to pin down, but also available to designers, are the *remembered task structures,* which were used to create the prior designs. Thus, if past experience showed that a particular set of task interconnections yielded "pretty good" designs, the designers of the design process can build those same interconnections into the organizational framework of the next round of the process. In fact it is logical to do so, because (as we saw in chapter 3) omitting interconnections can be disastrous if the omissions cause key interdependencies to be overlooked.

Changing Design Parameters

Thus, one way to create a new design is to put knowledgeable designers to work within a preexisting task structure. Figure 5.1 illustrates this approach. The process as drawn is a variant of the "generate-test" cycle of development proposed by Herbert Simon, and is also consistent with the "generate-test-regenerate (g-t-r) heuristic" of evolutionary biology.[4]

A new and incrementally better design can almost always be created in this way, but the approach is also subject to severely diminishing returns. The reason is that using the same task structure, by definition, causes the same design parameter interdependencies to be revisited over and over again. This may be useful if the nature of the interdependencies is not well understood (e.g., if there is no scientific theory to explain them). However, after some amount of experimentation, designers often will have identified most of the possible solutions to a particular interdependency (like the graphics controller interdependency discussed in chapter 3). The relevant task

[4] Scholars in a number of fields believe that a cycle like this is the core of "universal Darwinism," that is, it forms the basis of a broad family of theories of evolution that are similar in structure and descended from Darwin's original formulation. Our theory of design evolution is a member of that broad class of theories. (See Simon 1969; Holland 1976; Edelman 1992; Plotkin 1993.)

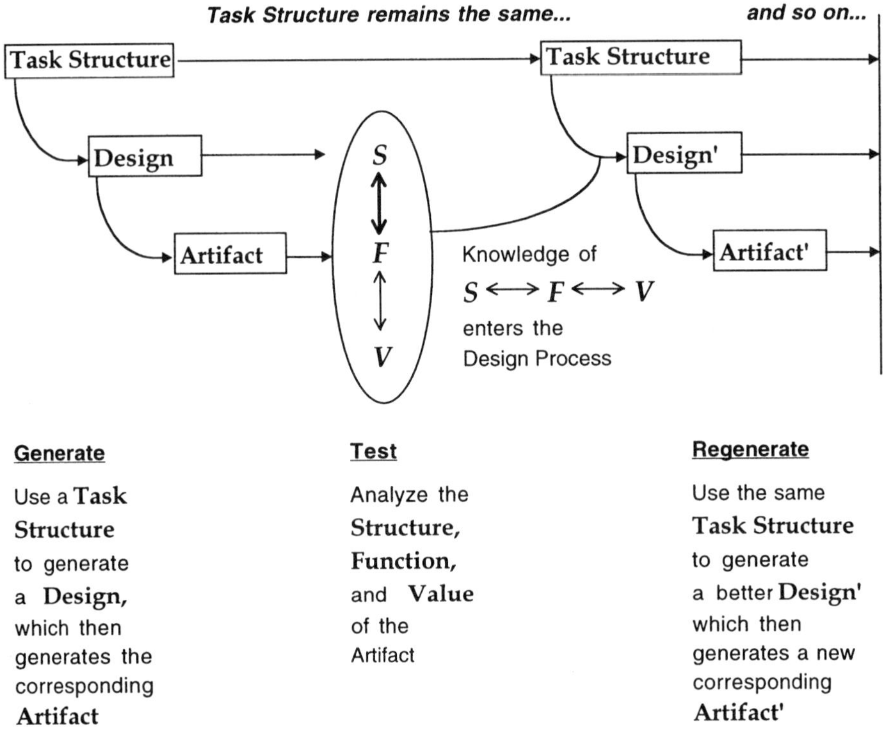

Figure 5.1 The descent of designs—method 1: new designs from an old task structure.

then becomes to choose a solution from a known set, which is quite different from finding a solution in an unknown set.

Changing the Design and Task Structure

If the dependencies embedded in a set of designs are well understood, then a designer with knowledge of those dependencies may conclude that some pathways are not worth going over again. In that case, a change in the design structure and the corresponding task structure is in order. Some possible changes include

- adding and subtracting tasks;
- changing the order of tasks (moving corresponding rows and columns of the TSM);
- adding a connection (thereby creating a coordination link);

- subtracting a connection (removing a link);
- starting an interconnected block of tasks at a particular point.[5]

Determining which among these actions are worthwhile depends on what the designers have learned from their prior experience. For example, a densely interdependent design structure may be called for during the early stages of developing a class of artifacts. As knowledge accumulates, however, links can be removed and tasks reordered. Thus in the middle stages of an artifact's development, the design structure may become sparser and more block-triangular (cf. the Intel TSM, figure 3.7).

"Design rationalization" is the name we have given to changes in a design structure that reflect growing knowledge of how to resolve interdependencies. We described this process in some detail in the graphics controller example of Chapter 3. When design rationalization is going on, design changes are taking place at both the level of design parameters and the level of the design and task structure (see figure 5.2).

Design rationalization may take place slowly or suddenly. Sudden change can occur if an architect (or a team of architects) sees the possibility of a wholly new design and task structure, and has the power to implement it. We also saw an example of this type of change in Chapter 3, when Mead and Conway modularized chip designs, and we will see other examples in chapters below.

There is a wide and continuous spectrum of rationalizing moves, ranging from minor tweaks in coordination links to dramatic reconceptualizations of how an artifact works and should be configured. The "tweaking" type of change goes on constantly, in response to random events and very local calculations of value (Should John fly in for the meeting?).[6]

At the other end of the spectrum are changes so significant that they must be conceived of and implemented by designers who can see the whole design and task structure and who know how the parts are related. These are "architectural changes." Architectural changes do not arise "naturally" from incremental analysis and

---

[5] The act of beginning to resolve a set of interdependencies at a particular place in the TSM is called "tearing" the matrix. Tearing techniques are a set of problem-solving strategies—in effect, they make certain paths through an interconnected TSM more likely than other paths. See Steward (1981a, b); Eppinger (1991); Eppinger, Whitney, Smith, and Gebola (1994); Smith and Eppinger (1997b).

[6] A metaphor that captures reality very well is to think of the off-diagonal **x**s in TSM as flickering randomly in response to small shifts in the day-to-day working environment.

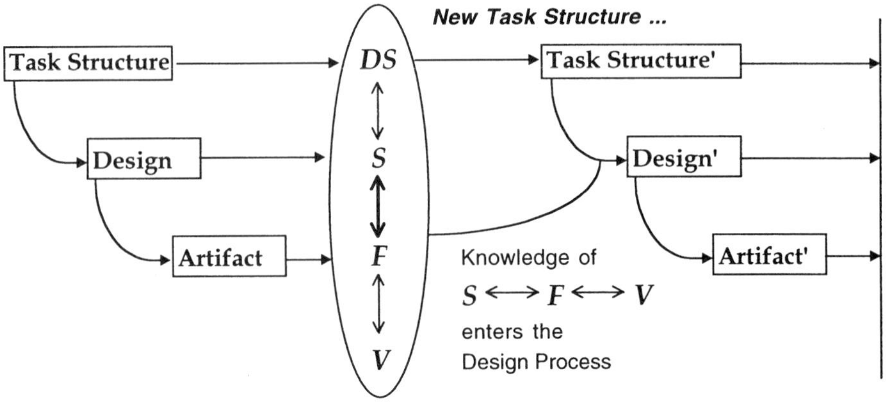

Generate

Use a **Task Structure** to generate a **Design,** which then generates the corresponding **Artifact**

Test

Analyze the **Design Structure** in relation to **Structure, Function,** and **Value** of the **Artifact**

Regenerate

Use the new **Task Structure'** to generate a better **Design'** which then generates a new corresponding **Artifact'**

Figure 5.2 The descent of designs—method 2: new designs form a new task structure.

improvement at the level of parameters and tasks. However, detailed knowledge of parameters, tasks, and interdependencies is what makes architectural change possible.

An architect, we have said, is a designer who can see different ways of approaching a design and a task structure, and can evaluate alternatives and make appropriate tradeoffs.[7] In effect, an architect says "this communication link is necessary," or, "this interdependency can be addressed by a design rule," with reasonable confidence. The knowledge needed to make these judgments is gained from experience with prior designs, combined with a detailed understanding of the artifact's intended functions and of its physical and logical structure.

In some cases, as we saw in chapter 3, architects may see the possibility of creating a modular design and task structure. If they choose to do so and are successful, the

[7] The fundamental tradeoff is the *cost* of extra tasks caused by cycling and internal debate vs. the potential *benefit* of a new solution to the interdependency.

design and task structure matrices will come to have the nested box configuration that is characteristic of modular designs (see figure 3.4). Moreover, the design hierarchy will become a true hierarchy—information will be separated into visible and hidden subsets, and tasks will be partitioned into discrete, well-defined modules. It is exactly at this point that new things can begin to happen to the dynamic trajectories of the designs.

We call the set of actions that creates a modular design the "splitting operator." Splitting is the first and most fundamental of the six modular operators, which are the primary mechanisms of change in our theory of design evolution. We will describe each of the six operators in detail in sections below. But before we launch into their descriptions, we will explain how the operators work as a group, and the role they will play in our theory.

The Concept of Operators

What are operators and why do we need them? As to what they are, we cannot improve on the definition given by John Holland at the very beginning of his book, *Adaptation in Natural and Artificial Systems:*

> *[A]daptation, whatever its context, involves a progressive modification of some structure or structures. These structures constitute the grist of the adaptive process, being largely determined by the field of study. Careful observation of successive structural modifications generally reveals a basic set of structural modifiers or* operators; *repeated action of these* operators *yields the observed modification sequences. . . . A system undergoing adaptation is largely characterized by the mixture of operators acting on the structures at each stage.*[8] *[Emphasis added.]*

Thus in complex adaptive systems, operators are *actions* that change existing structures into new structures in well-defined ways. They are like verbs in a language or functions in mathematics: by their powers of conversion (*this* turns into *that*), they define a set of trajectories, paths, or routes by which the system can change and grow more complex.

---

[8] Holland (1992, p. 3). In his later book, *Hidden Order* (Holland, 1995), John Holland moves away from the distinction between structures and operators, lumping both under the category of "building blocks" of complex adaptive systems. Indeed, for purposes of generalization, both structures and operators have a combinatorial property wherein very complex things may be constructed from a relatively small number of simple things. In this sense, structures and operators can be viewed as different aspects of the same phenomenon. However, in building a theory of a specific complex adaptive system, we find the distinction between objects (structures) and actions (operators) to be useful, and so we use it here.

As new levels emerge in a complex system, new structures and new operators may emerge as well. This is what happens when an artifact's design is modularized. Before modularization, there are two levels in each design: the whole design and the parameters (and tasks) of the design. After modularization, in addition to these, there will be a new level—modules that are coherent units, but are less than the whole.

Operators on modules are made up of very long and complex strings of actions on parameters and tasks. For example, as we saw in chapter 3, the operator we call "splitting" requires many actions and rearrangements at the level of design parameters. But splitting is also a well-defined, discrete event at the module level. And it is sometimes useful to consider just the event of splitting, without going into the details of what is happening at the level of the individual parameters and tasks.[9]

Our Criteria for Selecting Operators

Operators should be a useful way to organize knowledge pertaining to the dynamics of a complex system. With that goal in mind, we sought to define a set of modular operators that were

- simple;
- parsimonious;
- complete;
- faithful to underlying structures;
- verifiable by direct observation.

The first three criteria are familiar goals for any theory. A theory explains a complex phenomenon in terms of simple elements and rules. It uses the smallest number of elements and rules possible, but the total must be capable of explaining all cases that arise in the relevant domain.

In our case, we wanted the operators to be simple at the level on which they operated (the level of modules), even though they might be formed of complex sequences of actions at the level of design parameters and tasks. And, although we wanted the list of operators to be short, we also wanted combinations of operators to be capable of generating any structure we observed in the domain of computer designs.

The last two criteria are less familiar, and yet we think they are essential to the construction of a good model of a complex system. What we mean by "faithful to

---

[9] Note that "aggregate" operators perform the same functions of abstraction and information hiding that modules themselves do.

Box 5.1 The Structural Basis of the Genetic Operators

In 1953, Watson and Crick showed that DNA was made up of long strings of amino acids, arranged in the famous "double helix" configuration. The DNA strands would separate and recombine in the act of reproduction. Following this breakthrough, biologists had a set of structures on which to build a theory of the mechanisms (the so-called genetic operators) for recombination and reproduction. The mechanisms suggested by the structure were simple acts one could imagine performing on two strings of beads (the conceptual analog of a pair of recombinant chromosomes). For example, one could break the strings apart and join the opposite ends together at the breakpoint—this became the genetic operator known as "crossing over." Or one could cut a string at two points and invert the middle section—this was "inversion."[1] Or one could switch one bead for another one—this was "mutation."

The important thing to note is that these operators were implicit in the physical structure of chromosomes. Crossing over, inversion, and mutation would not have been the operators used in subsequent theorizing if the structure of chromosomes had been "more like" towers of blocks, arrays of tiles, or pipes containing liquid—to name some other possibilities. But the double-string structure of DNA *suggested* a set of "simple" and "easy" ways in which the structures might change. Formulating and testing those hypotheses became the agenda for research in the field of molecular biology, then in its infancy.[2]

---

[1] This "string" inversion is not the same as the modular inversion we describe below.

[2] Over the next thirty years, knowledge of biological structures and operators at a molecular level accumulated at a rate "probably without parallel in human history." (Plotkin 1993, p. 39.) Along the way, molecular biologists discarded the "beads on a string" concept as too simplistic. But the simple structural model was still a useful stepping stone to the more complex structural models used today.

structure" is that in some sense—physicial, chemical, or logical—the operators should be simple, easy changes to the structures in question, not tortuous or nonsensical changes.[10] (Box 5.1 describes how the so-called genetic operators reflect the physical and logical structure of chromosomes and genetic recombination.)

The structures we are working with are modular design and task structures; they are like blocks (or tiles) arranged in hierarchical patterns. Our modular operators are therefore actions that make sense for hierarchical arrangements of blocks. Except for splitting (the operator used to create modules), the other operators all "work" on only whole modules or groups of modules within a hierarchy.

---

[10] This argument was initially made by D'Arcy Thompson (1942), with respect to biological organisms, and more recently by Gerald Edelman (1992) and Stuart Kauffman (1993).

This brings us to the following question: How does one validate (prove the existence) of a given modular operator? It is common in science to say that one cannot prove theories but only disprove them.[11] What this means is that, given a well-constructed theory, one should be able to devise a test (statistical or otherwise) that can refute the theory if it is wrong. However, a positive result on the test is not conclusive: there may be other theories, as yet unformulated, that are also consistent with the result. As these alternative theories appear, new tests must be devised to "try" the old theory against the new. Hence scientific theories proceed from one "does not reject" to another, getting closer to truth all the time, but never reaching it.

This is true for theories in general, including ours. But for the *elemental parts* of a theory—the structures and the operators—another test may be possible: the test of direct observation. That is, given the right tools of observation, one should be able to "see" an operator work on a particular structure.[12] Thus, if we can observe designs and task structures using tools such as the design and task structure matrices and design hierarchies, and if there are operators called "splitting" or "substitution" that affect these structures, then we should be able to watch splitting or substitution occur. Thus, we believe, the best proof of the existence of an operator is to see it actually operate. For this reason, we have gone to great lengths in this book to find and describe real examples of each of the operators we are proposing.[13]

In summary, our theory of design evolution rests on six operators that can be applied to modular designs. We believe the existence of the operators can and should be verified by direct observation of designs and design processes. The evidence we offer is simple and straightforward—in engineering documents, in organizational plans, and in our direct observation, "this is how the designs have changed."

The Modular Operators Defined

Splitting

To explain the modular operators in detail, we begin by looking at a generic modular design structure. For example, figure 5.3 shows a two-level, four-module design hier-

[11] Popper (1963, 1972).

[12] The right tools may not exist at the time a theory (of operators) is proposed, in which case one has to fall back on indirect observation and inference.

[13] See chapters 10–13.

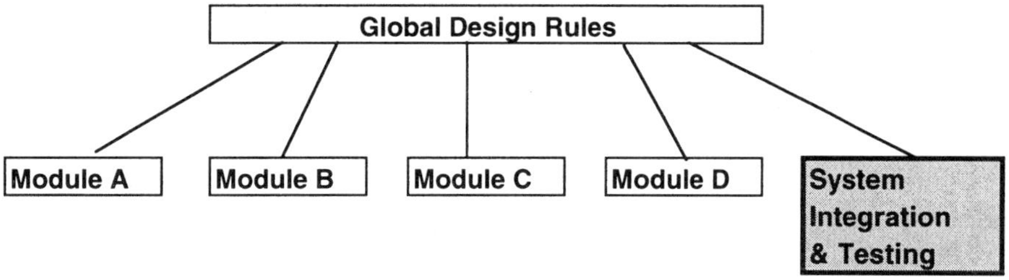

Figure 5.3 A generic two-level modular design hierarchy. The system comprises four "hidden" modules and a "system integration and testing" stage. From the standpoint of the designers of A B, C, and D (the hidden modules), system integration and testing is simply another module: they do not need to know the details of what goes on in that phase as long as they adhere to the global design rules. Obviously, the converse is not true: the integrators and testers have to know something about the hidden modules in order to do their work. How much they need to know depends on the completeness of the design rules. Over time, as knowledge accumulates, more testing will occur within the modules, and the amount of information passed to the integrators will decrease. The special, time-dependent role of integration and testing is noted by the heavy black border around and gray shading within the "module."

archy, with a separate (special) module for testing and integration. (Figure 3.4 shows the task structure matrix that corresponds to this design hierarchy.)

This design structure, we saw in chapter 3, can be created by *splitting* a previously interdependent set of design parameters and corresponding tasks. Splitting works the same way wherever it occurs. To begin with, the predecessor design structure needs to be "block-interconnected"—there have to be components or protomodules that suggest where to split. (For example, see the McCord-Eppinger laptop computer example in figure 2.5). Designers then need to go through the following four steps:

1. *Accumulate* and *organize* knowledge about specific parameter interdependencies.
2. *Perceive* that "the time is right" to create a modular design.
3. *Formulate* design rules specifying architecture, interfaces, and tests.
4. *Enforce* design rules for the duration of the design process, so that the structure does not collapse back into interconnectedness.

Splitting takes a single-level design with interdependent parameters and converts it into a hierarchical design with a core set of independent modules. After the split, however, each module will constitute a separate design in its own right. And so, what has been split might be split again.

If a module is split, only its internal design structure will change. Thus "subsplitting" or "modularizing a module" creates a three-level design hierarchy, as shown in

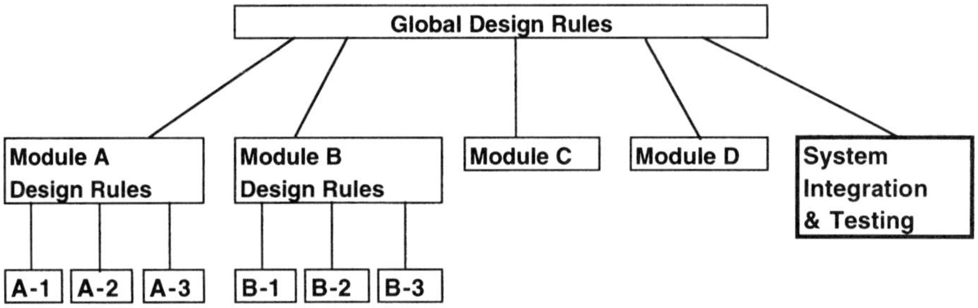

Figure 5.4 The subsplitting of modules: a three-level design hierarchy. Modules A and B may be split into subsidiary modules via the same process that created the initial modular design. The design/task structure is now a three-level hierarchy, although, by assumption, the "subsplitting" of modules is hidden from the architects (because the global design rules did not change). Subsplitting may or may not be apparent to the systems integrators and testers.

figure 5.4. (Readers may visualize or sketch the TSM corresponding to this design hierarchy.)

Substitution

Our second operator is *substitution.* The possibility of substitution forms the basis for economic competition, hence this operator, more than any other, affects and is affected by the surrounding economic system. Indeed, competition is economically justified only if two different designs can serve the same ends, but not equally well. If that is a possibility, and if it is not known (in advance) which design will be better, then it makes sense to create both designs and test them against each other. The idea that the better design will win this competition captures the essence of substitution.

Splitting dramatically changes the opportunities for substitution in a design. In an interdependent design structure, the only possible substitution is for the design-as-a-whole. In effect, because of the interdependencies, each whole design is a take-it-or-leave-it proposition, and any change in a parameter creates a whole new design. In contrast, after splitting, a designer can stay within the framework of the design rules for the system, but nevertheless substitute one module design for another. Thus splitting and substitution are natural complements. (We formalize this assertion in chapter 10.)

However, a new set of questions having to do with the *naming, categorizing,* and *testing* of modules arises when substitution is possible at the module level. This is because, in order to have efficient competition between modules, it is not enough for

Box 5.2 System Functions and Module Roles

Module functions are not simply a parsing out of system functions. For example, the system functions of a personal computer are things that matter to the user: the programs it can run, its speed, reliability, capacity to manipulate and store data, size, and price. The principal modules of the same computer are the microprocessor, other chips, one or more disk drives, a monitor, keyboard, mouse, and many software packages.

These modules all contribute to the functionality of the system, but in different ways. A module's *role* therefore describes the functions it performs within a system. For example, the role of a disk drive is to provide storage capacity and access to data. The role of a microprocessor is to compute. The role of a program is to direct computation, input, and output to achieve specific ends. These different, module-specific roles all contribute to, but are not the same as, the system's overarching functions.

Over time, knowledge about the relationship between system functions and module roles will accumulate around each module. As we discussed in chapter 4, this process actually begins the moment that designers see that an artifact can be decomposed into a set of components, and give the components special names. When a system is modularized, however, the role of a hidden module can be translated into a set of well-defined objectives for the module itself. These objectives in turn may be quantified, and transformed into a set of formal metrics or tests of module performance.

architects merely to segregate design information and partition tasks. To compete with one another efficiently, hidden-module designers must know what their module does and what it might do better.

In other words, for each hidden module, designers must have a theory of how the module functions within the larger system, and ways of measuring how well it performs. (See box 5.2.)

In summary, competition among alternative module designs can arise only after *splitting* makes possible module-level *substitution.* Beyond that, efficient competition at the module level requires that designers develop knowledge about what a module does (its role in a system), as well as low-cost ways to test a module's performance in its role. In part III we show how the value created by these two operators, in the presence of module-level tests, may "pull" a design and an industry into entirely new configurations.

Augmenting and Excluding

Augmenting means adding a module; *excluding* means leaving one out. Like splitting and substitution, these two operators are complementary. However, in contrast to splitting and substitution, these operators apply only to systems that have been modularized, and thus already have design rules and a modular structure. We will discuss exclusion first, and augmentation second.

The *exclusion* of modules can take place in two different ways. First, if a modular system is broad, a user may not need all the modules that are available. The user may then select a subset of modules, excluding those that he or she doesn't want.[14]

Systems in which a user can initially choose a set of modules to match his or her needs are called "configurable." If the user can also subsequently add or subtract modules as his or her needs change, the system is "reconfigurable," or modular in use (see box 3.2).[15] Reconfigurations may take the form of *substitutions* (e.g., upgrades of existing modules), *augmentations* (the user adds a module to give the system some new type of functionality), or *exclusions* (the user gets rid of a module no longer needed).

Users are often willing to pay for the opportunity to augment or exclude modules from their systems.[16] Modular designs, because of the way they are structured, provide users with these options. Thus conveying exclude-and-augment options to users may in itself justify an investment in modularity, quite apart from any considerations of design improvement. (Indeed, as we will see in Chapter 7, it was IBM's customers' desire to reconfigure their systems without rewriting code that, in the late 1950s, led IBM's top managers actively to seek a new approach to product design. This search eventually led to the creation of System/360, the first modular computer system.)

Architects of systems, as well as users, can use the exclusion operator, too. They do so by first specifying the design rules for a flexible system that "has room" for a large set of modules. But they do not design all the modules at once. Instead, they

---

[14] Systems like this are often called "families." IBM's System/360, whose creation is described in Chapter 7, was the first large computer family, as well as the first "truly modular" computer design.

[15] Configurability is a type of modularity in design—the user becomes the designer of his or her own system. A configurable system does not have to be modular in use—for example, a set of physical modules can be welded together (think of kitchen cabinets), or pieces of software may be so deeply embedded in system code that they can never be removed. Conversely, modularity in use does not necessarily imply that the system is modular in design. A sectional sofa, for example, is modular in use, but not in design.

[16] Technically, the "rights" to configure and reconfigure a system are options. The options will be valuable to users with heterogeneous needs, and to those whose future needs are uncertain. Garud and Kumaraswamy (1993) discuss "modular upgradability" as a competitive weapon used by Sun Microsystems in the late 1980s.

select a core set of modules that constitutes a functioning system for some set of users. They then introduce the minimal system; if it succeeds in the marketplace, they go on to design and introduce other modules as augmentations of the initial design.

The virtue of this "exclude-then-augment" design strategy is that a relatively large and complex system can be created by a small design team with limited financial and human resources. In short, exclusion makes bootstrapping possible. Early sales serve to establish the system in the marketplace, and pay for the design of modules that come later. For this strategy to work, however, the system as a whole must have a robust, modular architecture. From the beginning, the architects must have in mind that they are designing a system for later augmentation, and they must set the design rules accordingly.[17]

It should be clear from the above discussion that augmentation is naturally paired with exclusion. Users who initially purchase minimal systems may want to augment them later. The same holds true for designers who are following an "exclude-and-augment" development strategy. However, augmentation has another very important dimension. If the design rules provide for strict information hiding, then a modular system can "play host" to totally new modules that perform functions never dreamed of before.

As one might expect, totally new augmentations are quite rare. Nevertheless, in the fifty-plus years that modern computers have been in existence, augmentations have had a large effect on our concept of this artifact. Table 5.1 is a list of "great augmentations" of the computer: We invite readers to add their own favorites to the list, *and then imagine what our collective experience of this artifact would be if none of these augmentations had taken place.*

All the augmentations listed in table 5.1 were initially implemented on a general-purpose computer that was modular to some degree. The designers of the host systems did not foresee these augmentations. And yet, it is in the nature of modular designs to tolerate the new and unexpected as long as the novelty is contained within the confines of a hidden module. Thus modular augmentations have been a persistent theme in the history of computers. They are an aspect of these artifacts that continues to surprise and delight us.

---

[17] Of course there are complexities involved in the implementation of this strategy. We discuss them in more detail in chapter 12, when we value the options embedded in the exclusion and augmentation operators and look at the experience of Digital Equipment Corporation.

Table 5.1 Great augmentations

Mouse	Word-processing programs
Text-editing programs	Graphics programs
Bit-mapped graphical user interface (GUI)	Translators and emulators
Spreadsheet programs	Boot ROM
Hierarchical file structures	Games
E-mail	Space games
World Wide Web	Story games
Web browsers	Simulation games
Automated teller machines	Hypertext
Newsgroups, bulletin boards, and chatrooms	ARPANET/Internet
Artistic screensavers	Checkbook programs
	Alert sounds

Our thanks to Clare and Nicholas Hawthorne for helping us to compile this list.

Inversion

The operator *inversion* describes the action of taking previously hidden information and "moving it up" the design hierarchy so that it is visible to a group of modules.

As an example, consider the case of a subroutine that manages the printing of data. Printing occurs in many different programs, and each time it occurs the printing process must be managed. Thus, in the early stages of design development, there will be many versions of the "print manager" subroutine buried in many programs. Such redundancy is inefficient, however. Clearly, at some point, it makes sense to think about moving the best "print manager" subroutine to a higher level in the design hierarchy, thereby making it visible to and usable by a range of programs. Those actions, which are depicted in figure 5.5, are the essence of *inversion.*

Inversion also occurs in hardware, as for example when circuit designs are standardized. Just as standard system software routines like "print manager" are visible to programmers, standardized circuits are visible to and usable by hardware designers. Mead and Conway's standardization of VLSI circuits, discussed in chapter 3, was an inversion.

Over time, it is natural for a good solution to a common problem to be inverted and thus migrate up the design hierarchy. The solution may begin as part of the design of a specialized component within a hidden module. But if the component solves a general problem, its design will be separated from its initial context, and

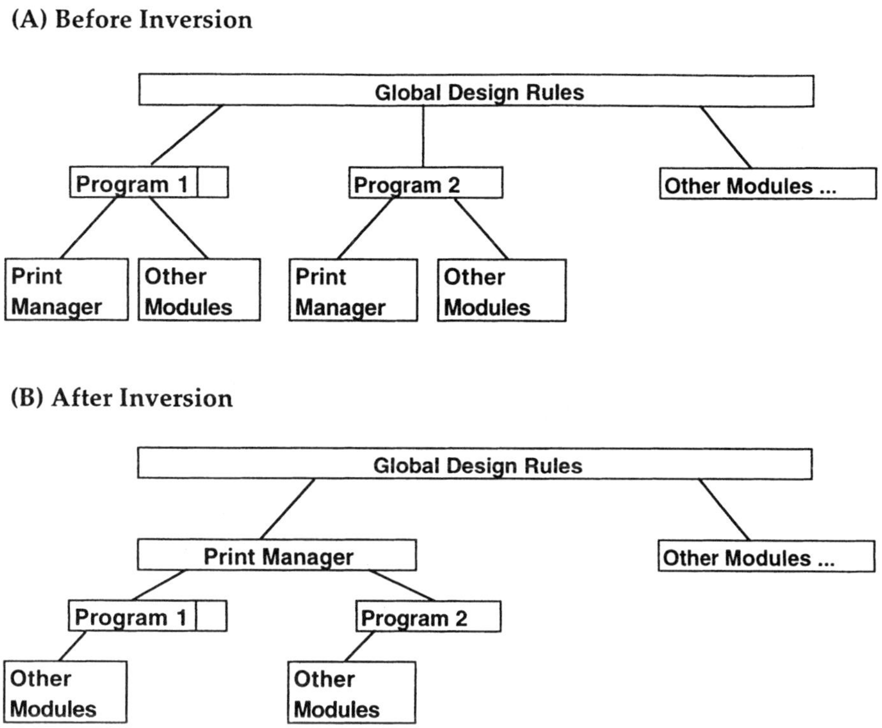

Figure 5.5 Inversion. Before inversion, every program needs to have a "print manager" subroutine. These subroutines are redundant, since they all perform essentially the same role. The "inversion operator" takes the "print manager" role and pulls it "up" to a new level in the design hierarchy. Programs written for the system can call a single "print manager" routine instead of incorporating their own routines. As the diagram shows, elements of the system's "print manager" are visible to all programs. Operating systems got their start in this fashion.

"published" for wider use. In this way, through inversion, what was hidden becomes visible.

The visibility of an inverted module may not be total, however. Hardware designers usually have to know only a subset of what goes into designing and producing a standard circuit. They can manage with an abstraction as long as it contains the information that is relevant to their task. The same is true for programmers using routines like "print manager"—they do not want to know how the routine was coded, but only that it works and is efficient.

Standard circuit designs and software utility programs are thus an interesting type of module—they are partly visible but partly hidden as well. Because they combine aspects of visible architecture and hidden modules, we call them *architectural modules.*

In an advanced market economy, enterprises that control key architectural modules can exercise great power and thereby capture large amounts of economic value within a specific design hierarchy. In chapter 13, we look at the value created by inversions as well as the cost of implementing them. Later, in volume 2, we shall look at the competitive jockeying that goes on as some firms seek to control architectural modules, while others try to prevent them from doing so.

Porting

Our last operator is *porting.* Porting occurs when a hidden module "breaks loose" and is able to function in more than one system, under different sets of design rules. Like inversion, porting causes a hidden module to move up the design hierarchy. But porting differs from inversion in that it is invisible. The architects of the system and designers of other modules do not have to know that a port has taken place.

As with inversion, the first step in porting is to modularize a module. However, to make a module portable, designers must first partition its design into (1) those parts affected by the surrounding system and (2) those not affected. They must then create a "shell" around the parameters not affected.[18]

Once the shell exists, the interior parts of the module are doubly hidden: they will not affect the surrounding system, and the system will not affect them. This "two-way invisibility" is potentially advantageous, for not only can these parts of the module be changed without changing the larger system but they can also migrate from one system to another without having to change their own inner structure.

The "exterior" parts of the ported module's design must then translate information from the external system into data usable by the interior sections, and vice versa. The design rules of the surrounding system and of the ported module are both equally visible to these *translator modules.*

As their name suggests, then, translator modules are a special type of hidden module, whose role is to convert one system's information into a form usable in another system. Examples of translator modules include (1) compilers that translate high-level languages into machine instructions; (2) printer drivers that translate machine commands into printer commands; and (3) read-only-memory instruction sets that translate machine instructions into microcode.

---

[18] Among other things, the shell provides a uniform way for the "interior" portions of the module to interact with the "exterior" parts.

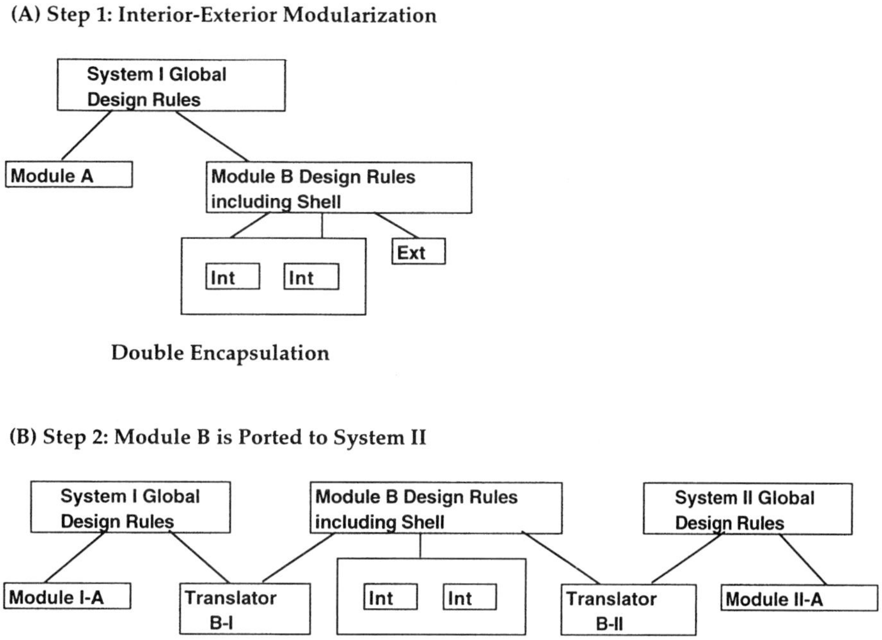

Figure 5.6 Porting. (A) Before porting, module B is a hidden module "looking up" to the design rules of system I. Before porting can occur, module B must be modularized to have at least two submodules: one "interior" and one "exterior." This interior-exterior dichotomy is an essential feature of a portable modular architecture. The "exterior" submodule must be capable of translating signals from the external environment (system I, in this case) into signals that can be processed by the interior components of module B. (B) To port module B to system II, a new translator module must be designed—one that is capable of translating system II's signals into information that can be processed by the interior components. If two translator modules are functional, module B has been successfully ported. At that point, as the design hierarchy indicates, systems I and II, and erstwhile "module" B, are all apexes of visible information. Moreover, they are linked by the hidden translator modules.

If all of these actions—subsplitting, defining a shell, and designing translator modules—can be accomplished, the ported module will be able to function in two or more otherwise incompatible systems. The module then moves "out from under" all systems' design rules and becomes a source of visible information (for the translator modules) in its own right. (See figure 5.6.)

However, in order for a hidden module to function in more than one system, two levels of modularization are required, and strict two-way information hiding must be maintained. This inevitably increases the costs of the overall design. In addition, when translation goes on in real time, it can impose a severe performance penalty on the system as a whole. For both these reasons—the extra design cost and because of the performance penalty—it may not be cost-effective to make a particular hidden module portable. Often it is easier to redesign a hidden module from scratch rather than to port it.

The Six Modular Operators' Effect on the Whole System

Once a system has been modularized, the six operators just described are available to designers who want to improve the system. Aided by the language, principles, and metrics they develop in parallel with the design, and given appropriate incentives and leadership, the designers will be able to see new opportunities and thus seek value in new ways.

An important feature of all the modular operators is that they can be implemented locally. As long as the initial modularization creates true information hiding (strict partitions and clean interfaces), the original architects do not have to foresee—or consent to—subsequent changes in the design. This tolerance of "local rule" is a property unique to modular designs.

Figure 5.7 shows how the six operators, applied locally, can affect the structure of the system as a whole. In constructing the figure, we began with a generic two-level modular design, similar to that shown in figure 5.2, but with six modules instead of four. We then applied each of the operators to a different module. The resulting system looks quite different from the one we started with. In the new structure, there are three hierarchical levels, two modular subsystems, many hidden modules, and some new "special" modules.

The Complexity of the Operators

In fact figure 5.7 only begins to indicate the possibilities inherent in the system as a whole. We applied one operator to each preexisting module, but there were five possibilities for each, plus the option to do nothing.[19] This means there were $[(6^6 \times 2) - 1] = 93{,}311$ different design structures that were one "operator-step" away from the system that we started with.[20] Figure 5.7 shows only one of these possible steps.

---

[19] Augmentation adds new modules to the system as a whole.

[20] The formula is: $[(j^6 \times 2) - 1]$ where j is the number of modules. This number understates the actual number of one-step-away designs if there are multiple substitution and augmentation

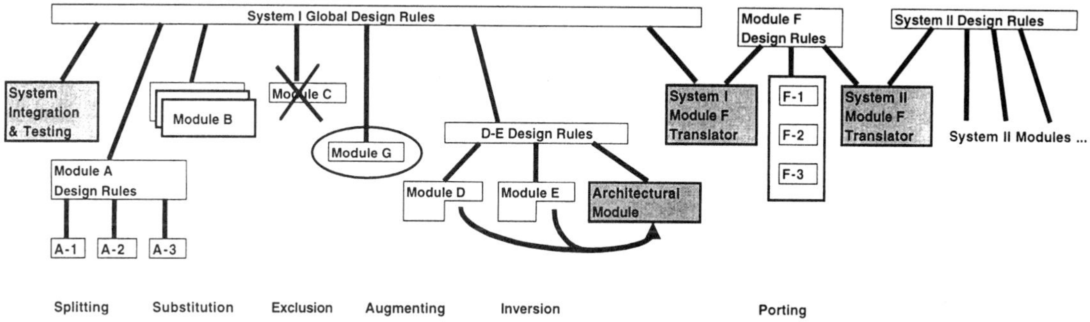

Figure 5.7 The effect of the six operators on a modular system. We started with a generic two-level modular design structure, as shown in figure 5.3, but with six modules (A, B, C, D, E, F) instead of four. (To display the porting operator, we moved the "system integration and testing module" to the left-hand side of the figure.) We then applied each operator to a different set of modules. Module A was *split* into three submodules. Three different *substitutes* were developed for module B. Module C was *excluded.* A new module G was created to *augment* the system. Common elements of modules D and E were *inverted.* Subsystem design rules and an architectural module were developed to allow the inversion. Module F was *ported.* First it was split; then its "interior" modules were grouped within a shell; then translator modules were developed. The ending system is a three-level system, with two *modular subsystems* performing the functions of modules A, D, and E in the old system. In addition to the standard hidden modules, there are three kinds of special modules, which are indicated by heavy black borders and shaded interiors: (1) system integration and testing module, (2) architectural mod ule, (3) translator module(s).

The operators can also be applied in combinations and in sequences. This means that at each of the ninety thousand endpoints after the first move, there are approximately ninety thousand potential "next steps." Even though some of the two-step endpoints overlap (i.e., one can get to the same structure using different sequences of operators), the "space" of designs opened up by the modular operators is very large indeed. And within that space, the number of routes to any particular design is quite large also.[21]

---

possibilities. This construct is similar to the sequence-space concept used in biology and complexity theory. See, for example, Waddington (1977); Kauffman (1993); Holland (1995); Dawkins (1996).

21 Formulas can be derived only for highly regular (i.e., symmetric) structures. However, the number of points in the design space is on the order of the number of module-operator pairs (36 in this case) to the power of the number of steps. The number of routes to a given point in the design space is on the order of the number of permutations of sets of operators used

Thus, by applying all six operators at different points and in different combinations, the structure of the system as a whole can change drastically. Moreover, all of this can occur without any change in the original design rules. In this very real sense, the final form of a modular design lies outside the original architects' control.[22]

What creates order in the space of potential designs, and ranks the myriad routes within that space, is value. We discussed the concept of value at some length in chapter 4. To reiterate what was said there, designers do not randomly explore the space of potential designs. If that were all that was going on, human effort most likely would lead to only "restless, minor, directionless changes," and we would still be using Stone Age artifacts. Instead, human beings are capable of seeing and seeking design improvements that "move" an artifact in the direction of greater functionality and therefore higher value. Apparently, once the goal—value—and the constraints—structures, context, and operators—have been appropriately brought into focus, humans are capable of selecting good solutions from a vast number of alternatives with very little apparent effort.

The Completeness of the Operators

Taken as a group, the six operators are *complete.* By this we mean that any conceivable modular design or task structure can be generated from a set of earlier structures via some sequence of operator steps. We will not prove this statement formally, but we offer the following intuitive argument.

An arbitrary modular structure can be represented as a set of interlocking hierarchies, with an arbitrary number of *apexes,*[23] any number of levels, and any number of connections. The only constraint on the structure is that it must be a true hierarchy: a module cannot be both above and below another module.

---

to get to the point. These calculations assume that operators are commutative, which is not always true.

[22] However, while we can be impressed by the potential complexity of the system, we should not be overly impressed. The job of the designers in a six-module, six-operator system may be likened to that of a writer given six nouns (the modules), six verbs (the operators, excluding augmentation but including the "null" operator), and two different endings (to augment or not). The writer's job is to combine each noun with one of the verbs, string the resulting sentences together, and add one of the endings. There are indeed over ninety thousand possible paragraphs that can be constructed with these resources, but most first graders can perform the job without difficulty.

[23] An "apex" is the highest level of visible information in a particular system. It is distinct from a "peak," which is a point of high value in the "space" of designs.

Now consider another, equally arbitrary, but different structure. The question is, can one be generated from the other by applying a finite sequence of operators? Splitting and substitution alone cannot take us from any modular structure to any other structure. For example, these two operators alone cannot generate new functions, link systems, or create new apexes in a design hierarchy. Hence splitting and substitution by themselves are not a complete set of operators.

However, augmenting allows new modules to appear, while exclusion allows them to disappear. Inversion increases the number of levels in a design hierarchy. Porting permits different systems to be connected via hidden translator modules. And both porting and inversion permit new apexes to emerge. Thus, with all six operators, it is always possible to construct a hypothetical sequence of steps that would lead from one system to the other (and vice versa).

The completeness of the operators is important because up to this point we have been striving to construct a theoretical framework that was rich enough to encompass the phenomenon we are studying. The objects of our study are computer designs over the last half century, their changes, and the related changes in the structure of the computer industry. Taken as a whole this is a very large and complex phenomenon. To describe it completely, we would need to reconstruct the designs of every computer system that existed in this time period, trace their lines of descent, and correlate them with the structure of firms that made the actual computers. We simply do not have that kind of data and if we did, we would not be able to process it.

However, we do have some interesting partial snapshots of the system. For example, it is relatively easy to construct a picture of the "computer scene" in 1944. At that time, there were three ongoing projects of interest: (1) Howard Aiken's Mark series at Harvard; (2) the code-breaking efforts at Bletchley Hall in the United Kingdom; and (3) Ekert and Mauchly's ENIAC project at the University of Pennsylvania.[24]

We also have a snapshot of the industry as it stands today:

- There are more than one thousand firms with publicly traded stock in existence, plus on the order of 200,000-plus private firms and sole proprietorships.[25]
- There are several hundred thousand modules (one firm can make many modules).
- There are at least three and perhaps as many as five levels in the design hierarchy of most computer systems (but overall the hierarchy is relatively shallow).

---

[24] Bell and Newell (1971, chapter 3).

[25] Compustat data for sixteen four-digit SIC codes in the greater computer industry.

- There are many sources of visible information, hence many apexes in the design hierarchy.
- There are many incompatible systems, but they are not wholly isolated from one another (e.g., e-mail can get through).

Our theory rests in part on the proposition that there is a "line of descent" in computer designs and task structures that connects the artifacts of 1944 to those of today. One way to test this proposition is to ask, Is it possible—in theory—for today's designs to have descended from the originals via the mechanisms we have identified (the operators)? If the answer is no, then our theory is refuted before we even begin.

However, the fact that the class of operators is complete means that it is possible for today's designs—as complex as they are—to have descended from the originals in the manner we propose. We do not have to map all the paths from the original starting point to today's designs in order to conclude that such descent was possible. We can deduce the possibility from the completeness of the operators as a class.[26]

Once we know that descent via operators is possible, then the relevant question becomes, *did* designs and the industry in fact develop in this way? This question cannot be answered by rational deduction alone. To address it, we must deepen our empirical knowledge of how the system actually behaves. In other words, it is time to switch from defining the structures, context, and operators of our complex adaptive system, and proceed to describing its behavior.

Using the conceptual tools developed in part I, we can begin to understand the historical development of computer designs, and the industry that grew up around them. Therefore, in the next three chapters, we will look at the early history of computer designs. We will focus first on the industry's "premodular" era (1944–1960), and then look at the period in which designers and managers at IBM conceived of, designed, and introduced System/360 (1960–1970).

---

[26] The fact that the operators are complete in the sense of spanning the space of possible designs does not mean that other operators may not exist. The nature and number of operators pertaining to a complex system is inherently an empirical question. One must look, in detail, at how the systems in question actually behave, to know what operators it is useful to define.

Part II The Creation of Modularity in Computer Designs

6 The Origins of Modularity in Early Computer Designs

Many complex adaptive systems, both natural and artificial, have characteristically modular structures. However, in the domain of computer designs, modularity does not arise by chance, but is the intentional outcome of conscious design effort. Therefore, the concept of modularity must exist as a possibility in the minds of designers before it can appear in the designs of the artifacts themselves. In addition, as we have said, the possibility of modularity does not spring up overnight, but requires a steady accretion of knowledge over many design cycles.

Thus the modularization of computer designs—a process with profound economic consequences—has also been a historical and intellectual process. It has taken place in stages, with each stage building on knowledge accumulated in the stage before. For a long time, moreover, it was an economically invisible process, whose milestones were obscure, technical events.

In this chapter we begin our story of the evolution of computer designs, by focusing on the "premodular era," which lasted from 1944 to 1960. During this time period, actual computers had complex, interdependent designs, and the task structures of design, production, and use were all highly interconnected. Modularity and the concomitant modular operators did not exist in practice.

However, the modular operators did exist in the mental tool kits of key designers and architects, including John von Neumann of the Institute of Advanced Studies, Maurice Wilkes of Cambridge University, and Ralph Palmer of IBM. In the minds of these individuals, the operators took the form of value-enhancing, imaginary changes in the artifacts and their production systems. In their imaginations, the designers could visualize splitting systems into components, or substituting one component for another, or inverting a good solution, or porting a component from one system to another.

Unfortunately, however, such improvements could not be implemented on the systems then in existence: each round of improvement meant designing a whole new system from scratch. Frustration with this state of affairs in turn drove some—like Maurice Wilkes—to search for design rules that would support the creation of more flexible, modular systems. Those searches culminated in the design of System/360, which we describe in detail in the next chapter.

The Original Design and Its Descendants

Our story begins in 1944. At this time, computers existed mainly in the heads of a few visionary designers, production was by trial and error, and uses were all experimental. The tasks of designing, producing, and using a computer were overlapping and interconnected with one another. Essentially everything—every design parameter, every production constraint, every idea for a new use—could and did influence everything else. Small wonder that making a computer then was a heroic undertaking, even though by today's standards the machines were simple and primitive.

We have said that before a design can be split apart in fact, it must be broken into discrete, named components in the minds of designers. Computers are a fairly recent invention. As a result we have a record of when the mental decomposition of the artifact took place, although who deserves credit for the achievement is in dispute.

In 1944, a team directed by J. Presper Eckert, Jr. and John W. Mauchly built a general-purpose computer, named ENIAC,[1] at the University of Pennsylvania under the auspices of the War Department. By the time ENIAC was up and running, Eckert and Mauchly knew how to build a much better (faster, smaller, cheaper, simpler) machine. But ENIAC was too interconnected to be redesigned piecemeal, so Eckert and Mauchly planned to start over from scratch. The new machine they envisioned was called "EDVAC."[2]

ENIAC programming consisted of manually plugging in cables and setting switches—it was similar to a telephone switchboard of that time, but much more complicated.[3] John von Neumann, the mathematician, did not participate in the design of ENIAC, but he visited the project several times, and became especially intrigued by the problems of programming and control of the machine.

---

[1] ENIAC stood for *E*lectronic *N*umerical *I*ntegrator *a*nd *C*omputer.

[2] EDVAC stood for *E*lectronic *D*iscrete *V*ariable *A*utomatic *C*omputer.

[3] Burks (1987).

Von Neumann—First Draft of a Report on the EDVAC

In March and April of 1945, von Neumann spent a few days with Eckert, Mauchly, and other members of their team, discussing the specific design of EDVAC. During the next eight weeks, von Neumann worked out a set of recommendations for the logical structure and control of this machine. He wrote up his thoughts in a "first draft" report, which he sent (in manuscript) to the team at the Moore School. Herman Goldstine, a liaison officer to the ENIAC project, arranged to type and duplicate the memo and circulated it quite broadly in military and academic circles.

Eckert and Mauchly were furious on two counts. First, although only von Neumann was listed as the author, the memo discussed design solutions they had worked on well before von Neumann appeared on the scene. Second, the circulation of the memo amounted to publication of the ideas, which greatly compromised their patentability.[4]

It is quite possible that von Neumann "merely" observed what Eckert and Mauchly were doing, and wrote down many things that the engineers already knew or found obvious. But his writing down of the principles brought crispness and clarity to the design, and so the basic structure of the artifact came into focus for the first time. Maurice Wilkes, who built the first operational, EDVAC-type computer, described the significance of the report in the following way:

In it, clearly laid out, were the principles on which the development of the modern digital computer was to be based: the stored program with the same store for numbers and instructions, the serial execution of instructions, and the use of binary switching circuits for computation and control. I recognized this at once as the real thing, and from that time on never had any doubt as to the way computer development would go.[5]

---

[4] This account is taken from that of Arthur Burks, who was a member of the group working with Eckert and Mauchly at the time. Burks went on to say:

In my opinion, Herman Goldstine was wrong to issue [the] report as he did. He should at least have given credit to Eckert and Mauchly as originators of much of the electronic design out of which the report arose. . . . EDVAC arose out of ENIAC and the mercury delay-line register [a form of memory invented by Eckert and Mauchly]. . . . It was later ruled that the EDVAC was not patentable because an application had not been filed within one year of von Neumann's manuscript (Burks 1987).

[5] Wilkes (1985, pp. 108–109). But he also said:

[T]he fact that it bore von Neumann's name only led to a grave injustice being done to Eckert and Mauchly, since many people assumed that the ideas in the draft report were von Neumann's own. It is now clear that Eckert and Mauchly had advanced a substantial way in their thinking before von Neumann was given permission to visit the ENIAC project.

The von Neumann "first draft" contains the seminal definition of a computer system. Significantly, it defined a computer system as a "composite" artifact and labeled its critical parts—primary memory, control, arithmetic "organ," input and output, and secondary memory. The memo also stated that data and instructions would be coded in the same way—as binary numbers in a single memory hierarchy.

In this categorization and grouping of problems, we see for the first time the glimmerings of a design structure and a related task structure for designing and building a computer.

The Institute for Advanced Studies (IAS) Machine

Von Neumann's first draft was only that—historical, controversial, but not complete or clearly thought out in all areas. Von Neumann soon had better ideas. In 1946, he conceived of building a parallel machine at the Institute of Advanced Studies in Princeton, New Jersey. Together with Arthur Burks and Herman Goldstine, he wrote out the specifications for the new machine, and submitted the proposal to the U.S. Army Ordnance Department.[6] This "Preliminary Discussion" illuminated the basic structure of the computer and described the problems that would occupy computer designers and architects for the next five decades. Writing in 1990, John Hennessy and David Patterson said of it: "Reading it today, one would never guess that this landmark paper was written more than 40 years ago, as most of the architectural concepts seen in modern computers are discussed there."[7]

Burks, Goldstine, and von Neumann (BGV) began with two overarching design goals: their machine would be both *general-purpose* and *fully automated*. "General-purpose" meant that the machine would not have special circuitry designed to solve particular problems, but instead would compute solutions to difficult problems via a sequence of simple calculations. This goal was not new: Eckert and Mauchly's ENIAC was a general-purpose machine and EDVAC was envisioned as such.[8]

---

[6] Burks, Goldstine, and von Neumann (1946), reprinted in Bell and Newell (1971, pp. 92–119).

[7] Hennessy and Patterson (1990, p. 24).

[8] The EDVAC "preliminary memorandum" preceded the BGV "Preliminary Discussion." When Burks, Goldstine, and von Neumann made their proposal, the EDVAC was under construction at the Moore School, but was not yet running. Hence the true predecessor of the IAS machine was ENIAC, not EDVAC. The BGV proposal reads like a finished version of the "first draft." The main difference in the designs of the EDVAC and IAS machines was that while EDVAC was a serial "one bit at a time" processor, the IAS machine was a forty-bit

However, it could take anywhere from half an hour to a day to set ENIAC up for a series of calculations. Having watched ENIAC in action, BGV's second goal was to have a machine that was "independent of the human operator after computation starts." To achieve this goal, each instruction needed to follow the next in a completely predetermined fashion. In other words, the computational process itself would have to be completely sequential in the sense we defined in chapter 2 (see under Strictly Hierarchical or Sequential Structure and figure 2.8).

Thus BGV proposed (1) a fully modular partition of tasks between humans and machines, and as a corollary, (2) a completely specified sequence for any set of machine tasks. They were not the first or the only ones to have this vision. But they went further than others in working out how to design a machine that would work in this way.

First they rationalized the design by reducing all possible sets of valid instructions (all possible problems the machine might be asked to solve) into a single, generic sequence as follows:

- Put instructions and data in.
- Store instructions and data.
- Fetch and decode instruction.
- Store temporary results.
- Execute arithmetic operation.
- Communicate results.

Human operators were needed at the beginning and end of this sequence, but BGV did not want them to be involved in any of the intermediate steps.

From this generic sequence of tasks, BGV reasoned backward and described a special-purpose machine to perform each of the necessary functions. The trick was to understand *all* the functions that needed to be performed to automate a sequence of operations. In this respect, it was undoubtedly very helpful to have seen firsthand the bottlenecks and inefficiencies of the already up-and-running ENIAC machine. (This is classic factory analysis, applied to the inside of a machine. It is strikingly similar to Samuel Colt's rationalization of small arms manufacture at his Hartford

parallel processor. Parallel processing vastly simplified control of the machine, and von Neumann could count on generous funding from the War Department to support construction of a large machine.

armory in 1855. Colt, too, carefully analyzed every task that went into the production of a handgun, and built a special-purpose machine for each function.[9])

The task structure implicit in BGV's analysis appears in figure 6.1. As we discussed, each generic step in the automated computational process called for a specialized machine. This relationship between operational needs and the individual components is indicated by the back-bending arrows in the figure.

There are a number of points to make about this diagram. First, the BGV task structure was a drastic simplification compared with the interconnected, and highly inefficient task structure that surrounded ENIAC. With ENIAC, as with many experimental prototypes, the tasks of design, production, and use were intertwined. When posed with a new problem, the designers would figure out which plugs to plug and what switches to switch. The team was not averse to adding a few new vacuum tubes or copper wiring if those were called for.

Thus the ENIAC design-production-use structure would have had points of interaction (**x**s) in every quadrant of figure 6.1. In contrast, BGV envisioned a *complete separation* between the tasks of coding (programming) and the tasks of computation, and an equally strict separation between the design and production of machines and their programming and operation.

The design that von Neumann and his colleagues perceived to be right for their new machine was modular at its core. In addition to software (which was treated extensively in later memorandums), they defined six basic elements in the design—memory, control, arithmetic unit, input, output, and storage.[10] Each element had a well-defined function and needed to be designed according to its own internal logic. Thus:

- [T]he memory organ can be used to store both numbers and orders.
- [T]here must exist an organ that can automatically execute the orders stored in the memory. We shall call this organ the *Control*.
- [T]here must be an arithmetic organ . . . capable of adding, subtracting, multiplying, and dividing.
- There must exist devices, the input and output organ, whereby the human operator and the machine can communicate with each other.

---

9 Hounshell (1984).

10 Many people count these as five elements, by lumping together input and output or by counting storage as part of memory.

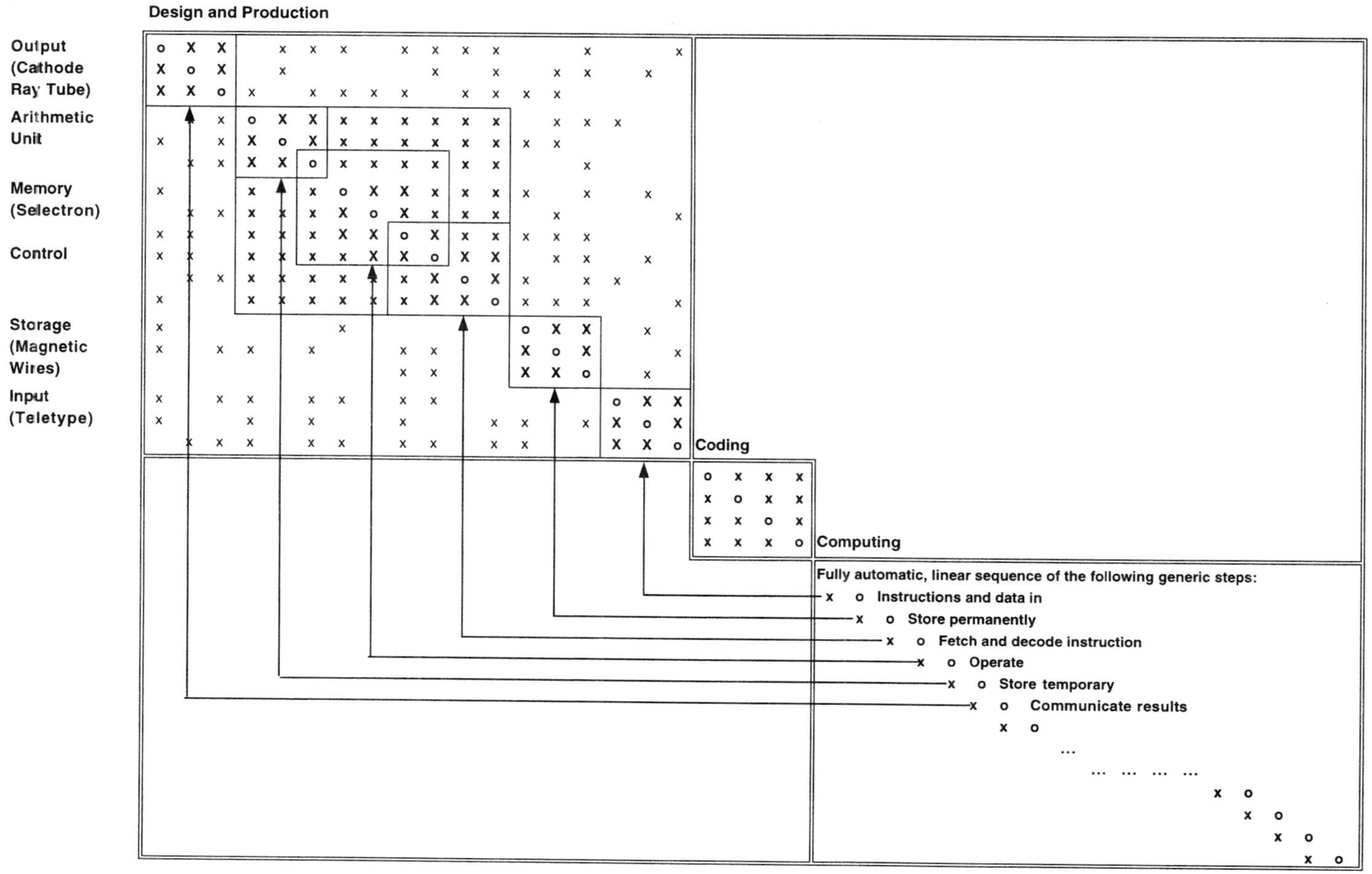

Figure 6.1 Task structure implicit in the Burks, Goldstine, and von Neumann design of an "electronic computing instrument" circa 1946.

- [W]e propose that we have a subsidary memory of much larger capacity, . . . on some medium such as magnetic wire or tape.[11]

To a large extent these labels survive to this day. (The arithmetic "organ" is now called the arithmetic logic unit, or ALU; it is also sometimes called the datapath. Secondary memory is now usually called "storage." The terms *storage* and *memory* were used interchangeably at the time the report was written.)

This report marked a significant milestone in the modularization of computer designs for several reasons. First, it defined a computer as a complex system of interacting parts, each with a specialized function. Second, it categorized the engineering problems of computer design according to which part of the system they affected. For example, the section on the arithmetic unit discusses number representation and how the machine would add, subtract, multiply, and divide. The section on control discusses the encoding of operations as numbers, timing, error detection, and the movement of bits between memory and computational registers. The sections on memory discuss capacity, cost, parallel vs. serial access, speed, and the relationship between different levels of the memory hierarchy.

There is also a strong sense of concurrency in the way problems were laid out in the report. The authors did not plan to attack the design in an interconnected fashion. Instead, the report was organized so that different designers might set out to construct each of the functional units, solve the detailed problems for each one, and bring them all together to make a functioning whole.

The authors also envisioned that designers might discover that some of their original proposals would not work out as planned. Hence they discussed second-choice options in some depth, and gave their reasons for initially choosing one path over the other. They clearly assumed that if a first-choice option did not work, some other approach would be used. Thus failure to solve problems in one subunit would not hinder problem solving in other units, nor would it jeopardize the ultimate success of the venture. (In this respect, the authors were probably too optimistic. But, remember that the Institute for Advanced Studies was seeking government funding for this machine, and so the authors needed to project both competence and confidence.)

Where there appeared to be a "prepackaged" solution to one of the components, the authors were quick to adopt it. For example, they wanted to use a cathode ray tube to display the computational output in real time. (This was a good idea, but ahead of its time: such displays did not become reality until the late 1960s.[12]) For

---

[11] Burks et al. (1946), reprinted in Bell and Newell (1971, pp. 92–93).

[12] Bell and Newell (1971, p. 89) say that CRT real-time displays were then "almost a reality."

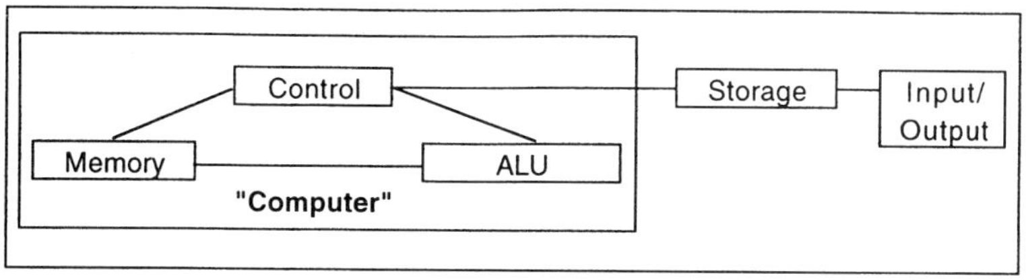

Figure 6.2 Design hierarchy of the Institute for Advanced Studies (IAS) machine. ALU, arithmetic logic unit ("arithmetic organ").

input and printed output, they planned to use a Teletype machine connected via paper tape to a magnetic storage device. The paper tape was an inconvenience dictated by the design of existing teletypewriters. The authors indicated that they planned "to design equipment that will eliminate the necessity for using paper tape." This was an early example of interface design rationalization—an ongoing activity in changing modular designs.

The implications of the BGV memo for the design of computers were far-reaching and profound. But their decomposition fell short of a true modularization. BGV were striving for modularity in use, not modularity in design or production. In no sense was the detailed design of one component going to be hidden from the others: all pieces of the system would be designed and produced "in full view" of the others.[13]

However, their analysis of computer operations did suggest certain fundamental divisions in the structure of a machine. Certain problems were more intertwined than others. For example, the relationship between memory and control was especially tight; the authors cycled around these issues many times in the report. In other areas (like input and output), the linkages were not so close, and it appeared that existing technology could supply the necessary functions.

Thus, although modularity in design and modularity in production were *not* their goals, we can see a set of components and a corresponding task structure for design and production begin to emerge from BGV's analysis. That emerging structure is depicted by the outlined blocks inside the design-and-production TSM in figure 6.1.

Figure 6.2 then depicts the design hierarchy of the BGV machine. In fact, there was no "hierarchy": all components were interdependent, and there was no significant amount of information hiding. The solid lines between components and the boxes around the "computer" and the system as a whole indicate that these components

[13] The one exception was the teletypewriter, and even its internal workings would be exposed to get rid of the offensive and unnecessary paper-tape interface.

were all visible to one another, and the components within the "computer" itself were more tightly linked than those outside. But BGV's conceptual breaking apart of the computer into functional components was a necessary first step toward the full-scale modularization of the design of the machine itself.

Yet to realize a modular design, those who followed the path laid out by BGV had to solve several difficult problems. Two in particular stand out, both because of their impact, and because of what they teach us about modularity and the modular operators. The first is *microprogramming;* the second is *standardized circuits.*

Microprogramming

Between 1945 and 1950, following the distribution of von Neumann's "first draft" and the BGV "Preliminary Discussion," nine experimental computer projects got underway at universities and research centers in the United States and England.[14] In all of these experimental efforts, the most problematic part of the design was the control unit.[15]

In the early days of computer design, the control unit was particularly complex. The arithmetic units were spread over several circuit boards, and each one needed its own control wire to turn it on or off, according to whether it was called for at that particular time in the sequence of instructions. Controls were hardwired in complicated logic circuits. Despite designers' best efforts to keep the wiring simple, these often added up to a mess of so-called random logic.

Maurice Wilkes, the principal designer of the EDSAC computer,[16] which was built at Cambridge University in the late 1940s, described the problem as follows:

---

[14] The only commercial computer project in this period was begun by Eckert and Mauchly, who left the University of Pennsylvania, and started their own company. Their efforts eventually led to UNIVAC I, the first commercial computer, which was introduced in 1951 (Bell and Newell 1971, chapter 2, figure 2a, p. 43).

[15] To see the difficulty, imagine a long hallway with several rooms on either side. Executing basic computer instructions (e.g., "add 4 and 6") is like turning on and off the lights in the rooms along the hallway in a specific sequence. In modern parlance the hallway and its rooms are the "datapath" (i.e., execution units like the arithmetic logic unit, the registers for holding addresses, counters for tracking progress, and the links among them). The component of the computer that manages the light switches is the "control."

[16] EDSAC was the first full-scale stored-program computer to become operational. It was built at Cambridge under Wilkes's supervision beginning in 1946, and became operational on May 6, 1949. (Wilkes 1985; Bell and Newell 1971; Hennessy and Patterson 1990.)

For each different order in the code some special piece of equipment must be provided and the more complicated the function of the order, the more complex this equipment. . . .[17]

We had tried to make the design of EDSAC as systematic as possible, but it contained a good deal of what is now called random logic. I felt that there must be a way of replacing this by something more systematic. . . .[18]

Wilkes's first insight (which was not his alone) was to see that the state of control, in other words, which switches were on and which were off at any particular time, could be represented as a logical, binary expression—a string of 0s and 1s. That in turn meant that all states of control could be represented as an array—a matrix of 0s and 1s.

What mattered for effective operation of the machine, however, was the flow of control as instructions were processed. Wilkes had observed that in all the computer projects underway in the early 1950s, the flow of control—the sequence of changes in the state of the machine associated with a given instruction—was fixed. There was no way to condition the flow of control on outcomes within the sequence.

But Wilkes also recognized that the flow of control was a program in miniature being executed deep inside the machine. That is, the changes in the electrical control settings required to execute a machine instruction were themselves just instructions (Wilkes called them microinstructions). Sequences of microinstructions could be made perfectly flexible by writing them in software. Thus, if the control settings were microinstructions, then the machine instructions themselves were programs—microprograms. This was Wilkes's second key insight, and it was his alone.

In 1951, when there were only about four operational computers in all the world, Wilkes described how it would be possible to convert machine instructions into microprograms.[19] The microprograms could then be stored in some form of (very fast) memory. In executing a set of machine instructions, the computer would first translate each instruction into the corresponding microprogram, and then execute the microprogram. (Fast memories of the type needed to implement this idea did not exist when Wilkes gave his first talk in 1951. By 1953, however, when Wilkes's team began to build EDSAC II, fast core memories had been invented. The Cambridge team designed EDSAC II to be a microprogrammed machine using a core memory.)

Microprogramming was in effect a modularization of the highly interconnected control logic of a computer. It divorced control logic from the hardware and thus

17 Wilkes and Stringer (1953), reprinted in Bell and Newell (1971, pp. 335–340).

18 Wilkes (1985, p. 178).

19 Wilkes (1951).

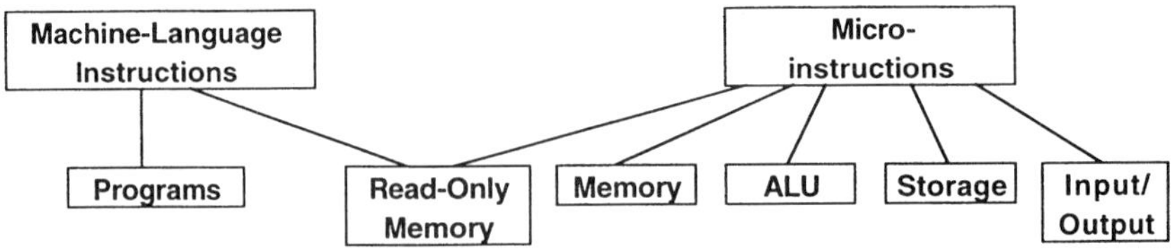

Figure 6.3 Design hierarchy for a microprogrammed machine. ALU, arithmetic logic unit ("arithmetic organ").

split the machine into three separate modules: the instruction set, the microprogram, and the datapath. Only the datapath needed to be implemented in hardware.[20]

Figure 6.3 shows the design hierarchy of the machine Wilkes was proposing. It has the classic pattern of a ported design—there are two visible *apexes* connected by a *translator module*. In this case, the microinstructions of the hardware and the "machine language" instructions of the software constitute the two apexes of visible information. The microprogram or microcode, stored in read-only memory, serves as the hidden translator module.

Despite its elegance and flexibility, the concept of microprogramming did not take the industry by storm. The greatest criticism leveled against the idea was that it was inefficient. Wilkes's approach was more complicated and circuitous than the usual ways of building machines. There was also a performance disadvantage implied by the double decoding of instructions. Finally, fast memory, which was critical to implementation of the concept, was expensive. Thus many designers believed that they could build faster machines at lower cost by hardwiring the "right" set of instructions in the first place.

These criticisms were fair but missed the essential point. When Wilkes asserted that microprogramming was the "best way" to design a computer, he did not mean it was the "highest speed" or "lowest cost" approach. What Wilkes sought above all were flexibility and ease of improvement. More than anyone else of his generation, Wilkes expected all the components of the computer to get better over time. He was especially aware of the shortcomings in the state of the art of programming, because he wrote his own code (in machine language). He fully anticipated that there would be a lot of experimentation centered around instruction sets, and he wanted to be able to switch to better sets of instructions as they came along.

However, in this respect, Wilkes was an exception. Flexibility and adaptability were not dominant themes in computer design circles in the 1950s. At that time, if

20 The datapath generally contained a number of submodules.

the goal was to build a small and cheap machine, designers would economize on memory, and hardwire a small instruction set directly. Conversely, if the goal was to build a large, expensive machine, designers would hardwire a large instruction set. Only in special circumstances—for example, if the designers wanted not to finalize instructions until late in the design process—would they find the microprogramming approach appealing.

Thus microprogramming was an oddity in computer design until IBM's designers seized on it for System/360. What attracted the IBM team to Wilkes's idea was the low cost of *porting* that was achievable through microprogrammed control. The architects of System/360 were intent on building a large family of computers, each of which had to run the programs written for any other member of the family. But that meant that all machines would need to use the same instruction set, even though some of the processors were big and some small. As we will see in the next chapter, microprogramming gave the IBM engineers a low-cost solution to this problem, and one that was based on the principles of modularity.

Standardized Circuits—Inversion Leading to Modularity in Production

BGV's "Preliminary Discussion" aimed to rationalize the design of a large, complex computer. It specified a large number of design rules (binary encoding of instructions, forty-bit parallel processing, etc.) But the Institute of Advanced Studies computer was intended to be a handcrafted machine. There was no particular need for, nor was much attention paid to, the problems of making a dozen, a hundred, or a thousand copies of the machine. Hence, while the major components of the system were defined and given roles within the system, inside of each major "box" was a mess of tangled circuitry.

The lower levels of the design also needed to be rationalized if computers were going to be produced in volume, but the impetus for *this* type of design rationalization would not come from the academic researchers. In fact, most of them doubted that it could be done. Thus, at an engineering symposium in 1948, Jay Forrester (who later invented core memory) said, "I believe that, barring an all-out emergency effort . . . large-scale computers will for several years need the sympathetic care of a laboratory crew."[21]

[21] Bashe, Johnson, Palmer, and Pugh (1986, p. 145).

But even as Forrester was speaking, events were underway at IBM that ultimately would create another level of modularity within the design. These design changes were aimed at making computers easier to manufacture and more reliable in the field. They also amounted to a classic inversion, in the sense defined in the previous chapter.

IBM's prewar tabulating machines were organized in a modular architecture, connected via the "universal interface" of an eighty-column punch card. Hence it was natural for engineers conversant with IBM's systems to apply the principles of modularity to electronic circuitry. Their aim was to divide up the tasks of designing and manufacturing computing devices in order to achieve high-volume, low-cost manufacturing.

To achieve mass production economies, it is necessary to have standardized, interchangeable parts. The basic parts of a computer are its circuits. Thus, it was obvious to IBM's engineers that computers produced in high volumes would need to have standardized circuits.

The first standardized electronic circuit to be used at IBM was a "pluggable" unit found in the IBM 604 Electronic Multiplier, which was introduced in 1948.[22] The unit consisted of a subassembly of one vacuum tube plus associated resistors and capacitors that were packaged and tested as a unit, and then inserted in the back of the machine. The pluggable circuit made manufacturing easier by providing stable subassemblies (see the fable of Tempus and Hora in box 3.1). It also facilitated field service and repair:

It became a common servicing technique to "swap" pluggable units. . . . If the trouble moved to a new location along with the pluggable unit, then that unit was likely to be defective. If the trouble remained . . . then something in the back-panel wiring or elsewhere in the machine was suspected of being faulty.[23]

The concept of a pluggable unit was later applied in IBM's first large-scale, general-purpose computer, the IBM 701. But despite all efforts to the contrary, the growing complexity of the IBM 700 series outpaced attempts to maintain design

---

[22] As its name indicates, the 604 was a multiplier. That is, it was not a general-purpose computer like ENIAC or the IAS machine, but a special-purpose calculator. However, the 604 was one of the first devices made by IBM that was based on *electronic* circuits, as opposed to electromechanical circuits. The advantage of *electronic* circuits was their very high speed; hence it was natural to use this new technology for a computationally intensive task like multiplication.

[23] Bashe et al. (1986, p. 63). At this time, and long afterward, IBM owned all of its units and leased them out to users. This was a very advantageous arrangement for IBM, but it did mean that they were clearly responsible for fixing any machine that did not run.

discipline at the level of the circuits. Thus by 1957, when transistorized circuits arrived on the scene, over two thousand different configurations of a basic eight-tube building block were used across the product line.[24]

The Standard Modular System (SMS)

Anticipating that transistor-based circuits would proliferate in the same way if nothing was done, Ralph Palmer, head of IBM's Endicott Laboratory, asked Edward Garvey to devise an economical system for manufacturing transistor circuits.[25] Garvey's group created a circuit packaging technology that became the foundation of IBM's Standard Modular System (SMS).[26]

The SMS for circuit design and manufacturing worked as follows:

- Circuits were constructed on standard cards measuring 2½ in. × 4½ in.
- Each card had sixteen contact tabs arrayed on one 2½-in. side. The cards were pluggable into sockets in a standard box.
- Circuit elements (transistors, resistors, and capacitors) were inserted automatically on one side of the card. (Specialized equipment for insertion had already been developed.)
- Wiring was inserted and soldered on the *other* side of the card. Specialized equipment allowed these steps to be automated too.
- Cross-card connections were provided by pins on the back of each socket, which could be wired together using a numerically controlled automatic wiring machine.[27]

---

[24] Bashe et al. (1986, p. 406).

[25] Garvey was the coinventor of the eight-tube unit that was the basic building block of the 700 series. It was Palmer who ten years before had imposed standardized circuits on the much simpler Model 604 (Bashe et al. 1986, pp. 59–68, 406–415).

[26] The "Standard Modular System" (SMS) technology was developed at IBM in 1957–1958. At almost the same time, newly-founded Digital Equipment Corporation introduced a line of "digital laboratory modules"—standard circuit packs that engineers could use to build logic systems for their instruments. These are the first uses of the terms "module" and "modular" that we have been able to track down in the technical literature. Before 1957, circuit modules were called "plug-in units" in technical documents. (See, for example, Olsen 1957, reprinted in Bell, Mudge, and McNamara 1978.) The terms "module" and "modular" did not move into general usage until well into the 1960s.

[27] For description and photographs, see Bashe et al. (1986, pp. 412–413), and Johnson, Palmer, and Pugh (1991, pp. 48–55).

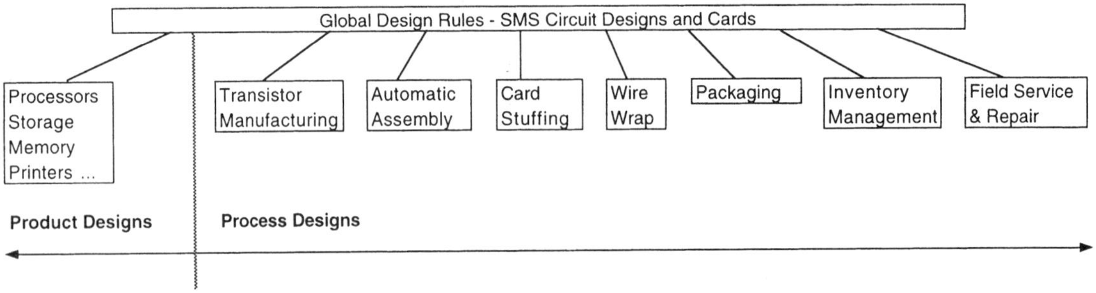

Figure 6.4 Design hierarchy of the IBM product lines after the adoption of SMS circuit designs.

Designers of new computers then had to use standardized circuits that were in the repertory of the automatic equipment. The SMS library of standard designs governed almost all of the basic physical features of a computer. Beginning in 1958, within IBM, circuits, cards, boards, power supplies, memories, frames, and covers were all standardized.

Hence, from the time of its introduction in 1958, SMS imposed a restrictive, sometimes onerous, set of design rules on IBM computer designers. One cost of this policy was that, below the level of the SMS card, design improvements essentially stopped. In addition, the size of the cards and the number of tabs and pins limited the density of the circuits, which in turn reduced the speed of some high-performance machines.

The benefits of SMS were that, for the first time, mass production techniques could be applied to the manufacturing and testing of complicated electronic circuits. The use of automatic equipment, as well as automated design and testing techniques, dramatically reduced the cost of the most basic building block of a computer. And as these costs went down, larger and more complex machines became feasible; these in turn justified investments in further automation and standardization of manufacturing processes.

The SMS was not merely a conceptual decomposition of the artifact. It was a true and effective modularization of a design and the corresponding task structure. Moreover, it involved the application of both the *splitting* and *inversion* operators, described in the previous chapter.

Figure 6.4 shows the design hierarchy of the IBM product line after the rollout of SMS. The figure underscores the fact that the modules of a complex design do not have to be tangible objects. The modules created by SMS were *process modules.* In other words, they were units that performed the tasks of production, distribution, inventory management, and field service for the company as a whole.

As indicated in figure 6.4, SMS first turned hidden information (about the physical and logical layout of circuits) into *design rules* that were visible across the whole of IBM's product line. At the same time, SMS partitioned the remaining hidden information into two subsets. Product designers were shielded from issues involving circuit layout, the design of manufacturing processes, and the testing of layouts, cards, and subassemblies. Process designers for their part could divide up their own tasks by function, and optimize their operations on the basis of a standard set of building blocks. Because SMS applied to every product made by IBM, the process engineers were able to realize significant economies of scale in component fabrication, work-in-process inventory, spare parts management, and the training of service personnel.

In this fashion, SMS established a set of manufacturing and design principles based on modularity, and these became an integral part of IBM's strategy. The principles proved to be robust even though the particular SMS circuits were soon made obsolete by the arrival of a new technology—the planar process for fabricating integrated circuits.

The Planar Process and IBM's Solid Logic Technology (SLT)

In late 1958, Jack Kilby of Texas Instruments managed to fabricate all the elements of a circuit on a single chip, and to connect them using gold wires. In May, Jean Hoerni of Fairchild Semiconductor Company sought a patent for the process of using a thin coat of oxide as an insulating layer for a semiconductor, and connecting wires to the active areas through holes etched in the oxide. Then in July, Robert Noyce, also of Fairchild, filed his landmark patent that laid out the steps of the "planar process" for fabricating a whole circuit by applying various patterns in layers on a single chip. Noyce's process promised to replace much of the circuit manufacturing technology that had been so carefully worked out at IBM over the previous decade.

IBM engineers considered using "planar silicon devices" (later called "chips") in the generation of circuits following SMS, but decided not to do so. Instead they opted to extend the basic modular structure of SMS down one more level in the design. Thus Solid Logic Technology (SLT), as set forth by Erich Bloch, the chief designer, in August 1961, was based on *four* levels of physically nested devices (SMS had three physical levels):

- At the bottom level were very small diodes and transistors: these were mounted on half-inch-square ceramic modules and wired together to make a circuit.
- Six of these modules could then be mounted on a card smaller than the smallest SMS card (which only held one circuit).

- The cards, in turn, would be mounted on printed circuit boards. To facilitate large-volume production, the boards had standard ground and voltage planes, a standard array of holes, and parallel signal lines of standard width.
- Customized patterns on the outside surface of the boards were created as the last step of the manufacturing process.[28]

As with SMS, the design and task structure implicit in SLT was highly modular. Each layer in the physical hierarchy posed its own set of problems, which could be worked on independently of the others. Once the SLT architecture was specified, transistor chip design could be assigned to one team, ceramic modules to another, cards and boards to another, and automatic wiring to yet another.

This division of effort paid great dividends in terms of speed and timeliness. The architecture of SLT was laid out in Bloch's memo of August 1961; by the beginning of 1964, a whole new set of manufacturing facilities had been constructed and was ramping up to full production. And that was not a moment too soon, for in April of 1964 IBM announced its new product line, System/360, with processors built out of SLT circuits.

As orders for System/360 poured in, the demand for circuits doubled and then doubled again. The nested circuit designs and modular manufacturing systems set up at IBM in the late 1950s and early 1960s proved to be robust building blocks for the System/360 family. In the succinct words of Gordon Bell, an IBM critic and competitor, "The production technology of the 360 series is outstanding, perhaps surpassed only by the 360 marketing plan."[29]

During the latter half of the 1960s, when System/360 was preeminent in the marketplace, IBM's modular production systems allowed the company to achieve unit costs that were much lower than its competitors ever imagined.[30] These production systems also allowed the company to make a seamless transition to integrated circuits, when it switched over to that technology in the early 1970s.

---

[28] Pugh et al. (1991, pp. 74–95). Applying customized connecting lines as the last step in an otherwise standard process is a technique used today in the fabrication of application-specific integrated circuits (ASICs) and gate arrays.

[29] Gordon Bell, chief architect of most of DEC's computers, writing with Allen Newell (Bell and Newell 1971, p. 564).

[30] Pugh et. al. (1991, pp. 109, 441–442). Unfortunately, we do not have access to actual cost data from the late 1960s and early 1970s. Official IBM histories assert that IBM enjoyed a large cost advantage throughout the 1960s and 1970s. IBM's pricing flexibility and profits during this time period give credence to this assertion.

Summary: The Origins of Modularity in Early Computer Designs

From our review of the early literature on computer designs, it is clear that the potential for modularity was present virtually from the beginning. Credit for the first conceptual "carving up" of the system goes to John von Neumann, although the ideas may not have been his alone. In the language of computer design, an abstract computer divided into its generic components is still known as a "von Neumann machine." From 1945 onward, designers used von Neumann's conceptual model to organize their work, and to communicate with one another about their designs.

This model in turn allowed designers to imagine independent trajectories of improvement in each of the different components of a particular system. For example, in the late 1940s, short-term memory technology was perceived as a bottleneck, and a number of designers set out to improve it. One of these searches paid off: core memory was invented by Jay Forrester at MIT in 1951, and subsequently became the memory technology of choice for many systems.[31]

In mental calculations, therefore, computer designers mixed and matched solutions and swapped components, even though the computers actually built at this time had highly interdependent designs, and were physically wired and soldered together. In effect, the designers tapped into the "power of modularity" in their imaginations long before modularity was feasible for real machines. The results of their mental experiments were numerous improvements in the components of computers, which led to the development of systems that every year became more powerful and complex.

However, as we said in chapter 3, to achieve true modularity in a design and corresponding task structure, the mental decomposition of the artifact is not enough. Designers must also have experience with many actual designs in order to understand precisely the nature of the underlying parameter interdependencies. Only when that knowledge is in place is it feasible to think about converting mental components into actual modules.[32]

The fact that knowledge must be built up through experience with real artifacts means that there may be long periods of time during which the fundamental design

---

[31] It was expensive, however. Drum memory was a slower but cheaper type of memory that was used in low-end machines (Bashe et al. 1986, pp. 73–101).

[32] Attempts to modularize without sufficient knowledge result in the discovery of interdependencies, which may render the system inoperable. The real design and its task structure will remain stubbornly interdependent and interconnected until the designers know *all* the points of interaction and can address them via sensible design rules.

and task structures of an artifact class do not appear to be changing very much.[33] However, it is possible to come to the end of the catalog of interdependencies and reach the point where each important **x** is known and can be addressed via a design rule. At that point, a new, modular architecture may emerge quite suddenly. This is precisely what happened to computers in the early 1960s.

The early conceptual breakthroughs in computer designs—the von Neumann decomposition, the development of microprogrammed control, and the standardization of circuit designs—were each in their own way different expressions of the principle of modularity. They foreshadowed a new, more flexible, and more efficient approach to the design of these artifacts. In 1961, these concepts, which had previously been applied independently, were brought together in a single, unified context. The result was IBM's System/360, which proved to be a turning point in the history of computer designs and in the history of the computer industry.

---

[33] At the parameter level, the designs may change quite a lot. This was certainly true in the 1950s. Essentially all computers of that era had interdependent designs and interconnected task structures, but the designs themselves spanned a very wide range of concepts. Many of these concepts (e.g., alphanumeric machines) proved to be dead ends; a few survived and are the basis of computer designs today. Bell and Newell (1971) provide an excellent survey and synthesis of the design concepts in computers of the 1950s and 1960s.

7 Creating System/360, the First Modular Computer Family

By 1960, the computer had become an important tool for processing data, for analysis, and for the storage and retrieval of information of many kinds. In enterprises ranging from government agencies to insurance companies, scientific laboratories, and commercial banks, computers were fast becoming an integral part of the work flow. At the same time, new concepts in circuit design, solid-state electronics, software, memory, and storage promised high rates of performance improvement at ever-decreasing costs.

But beneath this apparently happy confluence of supply and demand lurked a more difficult reality. The support of older applications and systems was becoming a problem of nightmarish proportions. For users, taking advantage of new technology meant writing off investments in old systems and software, and moving all their data to new formats and locations. These conversions were delicate operations, and dangerous if the data were crucial to the functioning of the enterprise (if the data included a company's general ledger or the accounts receivable files, for example).

The root of the problem lay in the growing complexity of the systems and the interdependent structure of their designs. Each new computer system had to be designed from the ground up, and only new systems could take advantage of new technologies. Thus system designs proliferated, and with them the need for applications support, software development, and service.

Across the industry in the early 1960s, designers sought answers to these dilemmas. At IBM, that search led to System/360, the first modular family of computer systems. The System/360 design achieved compatibility of software applications across a broad line of products, and thus helped to unleash tremendous demand for computing power. This in and of itself would have had important economic consequences. But the eventual impact of System/360 was even more profound.

A new design may pose a challenge to the surrounding economic system: to realize its potential value, managers, contract designers, and financiers may need to think about economic structures—firms, markets, financing conduits, and even property rights—in new ways. In other words, a change at the artifact design-and-task level of the economic system may set in motion reactions at the firm-and-market and financial levels of the system.

The modular design of computers had such far-reaching consequences. Modularity was a structural change that originated at the artifact design-and-task level, and proceeded to work its way "up" through the other levels. In this process the creation of System/360 was a landmark event. Because it was the first commercial computer to be designed as a modular system and because it was a huge economic success, System/360 created a shock wave in the economy. Firms, markets, and financial institutions all changed as they encountered the wavefront created by this new artifact and the process of designing it. These changes, in aggregate, opened up new evolutionary pathways for computer designs, and in so doing transformed the structure of the computer industry.

Our task in this chapter and the next is to describe how these new evolutionary paths came into being. We focus first on how System/360 was created. This story is an extremely well-documented instance of an actual modularization; as such, it provides us with important insights into the kinds of knowledge and actions that underlie the *splitting* of a large, complex design. That story is the focus of this chapter.

The Goal of a Compatible Product Line

IBM's top management wanted a compatible product line well before the company's engineers and designers thought this was a feasible goal. In 1959 and 1960, important customers were telling IBM's sales force and top managers in no uncertain terms just how unhappy they were with the growing costs of incompatibility and the expanding complexity of the product line.[1]

These customers were dissatisfied because before any of their powerful, expensive computers could do useful work, a series of precise instructions had to be coded in a form the machine could understand, and loaded into the machine's memory. The loading was easy, but the coding was hard. Yet coding—early-stage programming—was the necessary link between a general-purpose computing machine and any output: without it, the machines were only boxes filled with very intricate and expensive wires. Moreover, in those days, programmers had to know many specific details

[1] Pugh et al. (1991, pp. 113–121); Watson (1990, pp. 346–349); Wise (1966a).

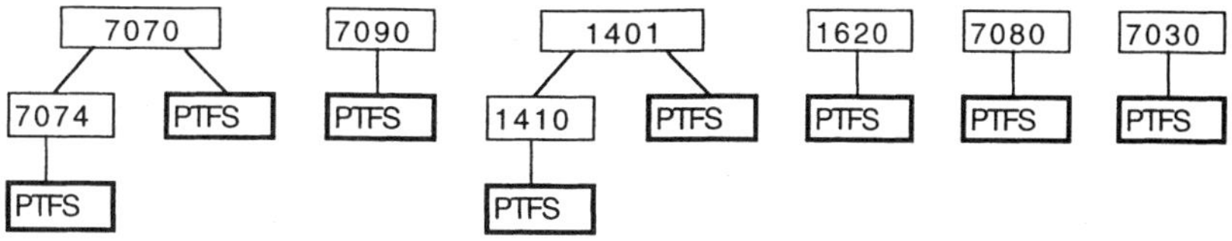

Figure 7.1 The cost of incompatibility: IBM's transistor-based processors in 1960. PTFS, programming, training, and field service. These activities were designed anew for each processor.

about the designs of the machines their programs would run on. Machine designs, in turn, were complex and varied in arbitrary and nonobvious ways. Thus the knowledge needed to program any particular machine was highly specific, idiosyncratic, and not transferable to other types of computers.

Figure 7.1 shows condensed design hierarchies for the computers in IBM's product line in 1960. At this time, IBM sold systems based on eight different processors. Six of them used different data formats and instruction sets, hence were incompatible. As a result, the company suffered from two problems that are common in expanding, incompatible product lines: (1) redundancy of components; and (2) legacy design constraints. We will discuss these problems in the next two sections.

Redundancy in Components

In the 1950s, virtually all computers had instruction sets that were literally hardwired into their control logic and arithmetic units. This high degree of interdependence also characterized input/output design, the processor-memory linkages, and all hardware and software interactions. Because the design parameters were so interdependent, each new system had to be designed from scratch; hence every market niche had a different whole system matched to it.[2]

However, as programmers gained experience across a wide range of systems, it became clear that many tasks—editing, testing, debugging—arose again and again. The programmers and their managers realized that a great deal of effort was expended in performing tedious, minor variations of similar tasks. Yet these repetitive tasks were inextricably tied to the data formats, instructions, and conventions associated with different processors, and thus could not be standardized.

[2] Proliferation in the marketplace was worse than this figure indicates. Bell and Newell (1971, chapter 2, figure 2a, p. 43) counted twenty commercial whole systems in production or under design in 1960, in addition to ten whole systems in research institutes and universities.

Discoveries of common elements and similar tasks buried deep in a design are part of the normal buildup of knowledge about a complex system.[3] They call for the application of the *inversion operator* described in chapter 5. Recall that IBM inverted the design hierarchy of its manufacturing systems when it implemented the Standard Modular System (SMS) in 1958. The analogous move in product design would be to create standard data formats, addressing conventions and a single instruction set, and to add those to the circuit-design rules that the product designers already had to obey.

Such a move held risks, however. New data formats, addressing conventions, and instructions would be visible not only to IBM's own programmers but to its customers' programmers as well. When IBM adopted SMS, Ralph Palmer, head of the Endicott Laboratory, could order the company's designers to use standard circuits—they grumbled, but they complied. IBM could not count on having such power over its customers: they might grumble and then switch to another vendor's machines.

Legacy Design Rules

It is possible to achieve compatibility without changing the customers' view of the product by using legacy design rules. Unlike modular inversions, which generally require redesign of a whole system, legacy design rules leave existing products as they are—although later products must be compatible with their (successful) predecessors. The advantage of legacies is that a company can preserve its existing products and build on a loyal base of customers. The disadvantage is that both good and bad features of earlier designs are preserved, and thus legacy design rules generally impede design rationalization. Designs so constrained tend to become interdependent and complex, with much "random logic" inherited from the past.

As figure 7.1 shows, in 1960 IBM had two processors—the 7074 and the 1410—that were designed to be compatible with earlier products. Even this slight degree of compatibility was not easily won. For example, the 1410 was initially not going to be compatible with the 1401: its designers wanted instead to develop a new, more versatile instruction set. According to Robert Evans, manager of this project, some designers even threatened to quit if the two machines had to be compatible. However, after a detailed examination of the technical issues, Evans became convinced that

[3] This is the property of "self-similarity" in the structure of the complex system. In natural systems, it arises from the interaction of physical laws. In artificial systems, self-similarity is a possibility, which needs to be recognized and then created. Systems with this property economize on scarce design resources.

compatibility was an attainable goal and decreed that the 1410 should run "almost all" programs written for the 1401. As a result, the 1410's instruction set was recast as a superset of 1401's, and a number of its more ambitious design goals were abandoned.[4]

When the 1410 came to market, customer demand was unexpectedly strong. This enthusiastic reception, coupled with strong criticism from large customers of the existing product lineup, convinced IBM's top managers that more compatibility was the order of the day. Therefore, in 1960 and early 1961, IBM's Management Committee systematically cut off the funding of all R&D projects aimed at developing processors that were not compatible with something already in the line. In this heavy-handed fashion, they signaled "a desire for new thinking" about the problems of system proliferation and incompatibility.

IBM's top managers could not command the engineers to do the technically impossible, however. On their own, IBM's top management could not answer the following critical questions: How much compatibility was really feasible? Should IBM perpetuate the instruction set legacies of its 1400 and 7000 machines? Or should the company start afresh and develop a new line of compatible processors from scratch? These questions could only be answered by engineers working at what was then the edge of the technical envelope.

Creating System/360

The SPREAD Report

In the fall of 1961, Vincent Learson, vice president and group executive in charge of IBM's North American computer divisions, established a task group of thirteen people drawn from the company's research, development, and marketing units. The group, code-named SPREAD (for Systems Programming, Research, Engineering, and Development), was charged with developing "an over-all IBM plan for data processor products." The group withdrew to the Sheraton New Englander motel in Cos Cob, Connecticut, in the first week of November 1961.[5] On December 28, they

[4] Evans, in Aron, Brooks, Evans, Fairclough, Finerman, Galler, Heising, Johnson, and Stern (1983, p. 30); see also Bashe et al. (1986, pp. 475–479). It is not recorded whether anyone actually quit.

[5] Aron et al. (1983, p. 35).

submitted their report, which recommended that IBM develop a new family of compatible processors to replace its entire product line. Thus was System/360 born.

As we shall see, the *design rules* developed for System/360 placed computers and the computer industry on a new evolutionary path. Moreover, the group understood the significance of what it was proposing to do, as Robert Evans, a key member of the group, much later recalled:

[F]rom the moment the SPREAD study convened we knew it was an extraordinarily important assignment. . . . There was early skepticism among all of us . . . as to whether the goals were attainable. But the deeper we got into the design of System/360 across the company, the more we were convinced it was fundamentally the right direction and that it would succeed. We had our hearts in our mouth all the time. . . . But I would say, from that early sense of importance and skepticism, our confidence grew and grew. . . .[6]

The group recommended that IBM turn its back on all its existing processor products and begin development of five new processors:

IBM's customers' needs for general-purpose processors can be most profitably met by a single compatible family. . . . Each processor is to be capable of operating correctly all valid machine-language programs of all other processors with the same or smaller I/O and memory configuration.[7]

"Machine-language" compatibility implied that programs running on different machines would be similar in terms of their *binary* representation. This meant that a program written for one configuration could be automatically assembled, compiled, and run on any equal or larger configuration; also, a program could be compiled on one machine and executed on another.[8]

The report also called for standard interfaces between the processor and other parts of the system:

- To any I/O device type, all channels shall appear identical except in data rate.
- A single detailed method for memory-CPU coupling shall be developed and applied to all processors.

---

[6] Evans, in Aron et al. (1983, p. 37).

[7] Haanstra, Evans, Aron, Brooks, Fairclough, Heising, Hellerman, Johnson, Kelly, Newton, Oldfield, Rosen, and Svigals (1983, p. 7).

[8] The report noted that IBM's competitors were attempting to achieve compatibility through the use of common programming languages. Binary compatibility was a more rigorous standard. However, because it made conversions automatic, it was commensurately more valuable to customers.

- A single I/O control system (IOCS) structure will be provided for the entire processor family.[9]

Finally, the new machines would be compatible with one another, but would not be compatible with any existing machines.[10]

These were ambitious, even extravagant, goals. No project like this had ever been attempted in the domain of computer designs. To make the case for feasibility, the report went to some lengths to explain how the goals were to be achieved. In that respect, the SPREAD report can be read not simply as a proposal to develop a revolutionary new product but as an essay on the twin topics of modular computer designs and modular task structures. (Box 7.1 describes what the SPREAD group elected *not* to do.)

A New Task Structure

To understand the significance of the SPREAD report, we must bear in mind that up to that point computers were designed by teams of people working in close proximity. The basic task structure of the design process was interconnected and highly iterative. Very little in the design of any piece of equipment was standardized, and designers were accustomed to making all the necessary parameter decisions without interference.

To achieve compatibility across a wide range of processors, the SPREAD report proposed to create a new task structure for the design process; indeed, the bulk of the report was devoted to explaining how the new design process would be managed. The first step would be to establish a set of rules that all designers would follow. With these so-called architectural ground rules[11] in place, teams at IBM laboratories in New York, California, England, and Germany would work to design specialized pieces of the overall system. At the same time, manufacturing engineers would be designing and building capacity based on new Solid Logic Technology (SLT), described in Chapter 6.[12]

---

[9] Haanstra et al. (1983, p. 12).

[10] Some models of System/360 did achieve compatibility with older IBM machines by emulating the older machines' instruction sets, but emulators were not mentioned, and backward compatibility was explicitly rejected in the initial report.

[11] This is the actual terminology of the report.

[12] SLT planning was already underway and the use of SLT circuits was one of the conditions given the SPREAD task group at the outset of their deliberations. The report in fact says very

In terms of the modular operators defined in chapter 5, the SPREAD task group was recommending a *splitting* of the design and an *inversion* of critical subsets of design information. In addition, the group planned to use microcode to *port* a large instruction set (the criterion of compatibility) to the smaller machines.

Figure 7.2 shows the TSM for System/360, which we reconstructed from the SPREAD report itself. Each box down the main diagonal represents at least one major design project. (A brief description of each project, plus an indication of who would be responsible for it, is given to the left or right of the main diagonal as space allows.) Each box on the main diagonal thus stands for at least one fairly complex and technically challenging project, and some boxes stand for more than one. For example, "Main memories" appears as one box (#16), although in fact three memory projects were needed to span the processor range.

A shaded cell means that the project in that column provided critical design inputs to the project in that row. The lighter the shading, the less interaction. For example, SLT was a set of circuit and manufacturing design rules; these strongly affected every piece of hardware but did not affect software to the same degree. Thus, in the SLT task columns (#1–2), the software rows (#25–32) are lightly shaded.

Consistent with the modular design structure, the report envisioned a three-phase design effort. There was to be a *design rules phase* (tasks #1–6) that would provide coordinating inputs to all the modules; a *parallel work phase* (tasks #7–32) in which work on each module would proceed independently; and an *integration and testing phase* (tasks #33–35) in which all the parts would be brought together.

All the necessary design rules would be established in the first phase, which was projected to last about ninety days. (In fact it took ten months.) Thereafter, work on the individual projects would proceed concurrently. The critical assumption was that if the right design rules were in place, the teams would be able to work independently without close day-to-day coordination. A Central Processor Control (CPC) office would monitor progress and ensure compliance with the design rules, but the details of each design would be left to the individual units.

At the end of three years, the processors and peripherals would be designed, the manufacturing capacity would be built, and the necessary software would be written. The marketing program would then gear up, and the new systems would be tested, shipped, and installed.

---

little about manufacturing; the implicit assumption seems to have been, "if we can design it, they can make it." This says a lot about the capabilities for mass production which IBM had built up in its manufacturing divisions. In fact, producing System/360 machines in volume was not easy, but in the end the group's confidence was not misplaced.

Box 7.1 Compatibility without Modularity

It is worthwhile to consider what the SPREAD committee might have done differently. At the time the group convened, compatibility was a corporate imperative: customers were clamoring for it, and IBM's management committee had shown that they were willing to replace any executive or engineer who did not view compatibility as *the* preeminent corporate goal. Thus the SPREAD task group had no choice but to come up with a recommendation that delivered compatible processors.

But the goal of compatibility did not require a completely new, modular design. A less risky route would have been to select one of the existing processors for further development. Two factors tilted the group (and IBM's top management) in the direction of a totally new design. First, as the SPREAD report went to great lengths to show, IBM's revenues were evenly distributed across all its computer lines. As a result, there was no large concentration of profit in any single set of products.[1] Second, the advent of cheap core memories had created a problem for all existing designs. According to Fred Brooks, an architect of System/360: "[I]t had become evident that the original architectures were running out of gas . . . in a place one can't fix it easily: address space. [All the processors] were all running out of gas in address space."[2]

A case in which similar concerns were weighed differently is that of Intel's x86 family of PC microprocessors. For six generations, ever since the 8088 microprocessor was adopted in the first IBM PC, Intel has faithfully preserved compatibility by imposing legacy design rules—including limits on addressable memory—on each generation. One observer described the x86 family as follows:

Pentium [the fifth generation of these chips] is arguably the fastest 32-bit processor ever to evolve from a custom 8-bit CRT [cathode ray tube] controller. . . . [F]or each new desirable feature the x86 family acquired as it grew there was a vestigial appendage that was faithfully reproduced, and these features—good, bad and ugly—make the x86 processors unique. . . .[3]

There were, in fact, some strong advocates of the legacy approach within IBM. Most vocal was John Haanstra, the president of IBM's General Products Division, whose successful 1401/1410 family was threatened by System/360. Haanstra fought a dogged rear-guard action against the new product line: he continued to fund new 1400-compatible products through 1963. Possibly to hedge the corporation's bets, IBM's Management Committee allowed him to do so for a while. However, in February 1964, on the eve of System/360's announcement, Haanstra was relieved of his responsibilities, and banished from the high councils of the corporation.[4]

---

1. Haanstra et al. (1983, pp. 8–9).
2. Brooks, in Aron et al. (1983, p. 42).
3. John H. Wharton of Applications Research, a participant in the Hot Chips VII symposium, as quoted in *Electronic News,* 14 August 1995.
4. Wise (1966b); Pugh et al. (1991, p. 164).

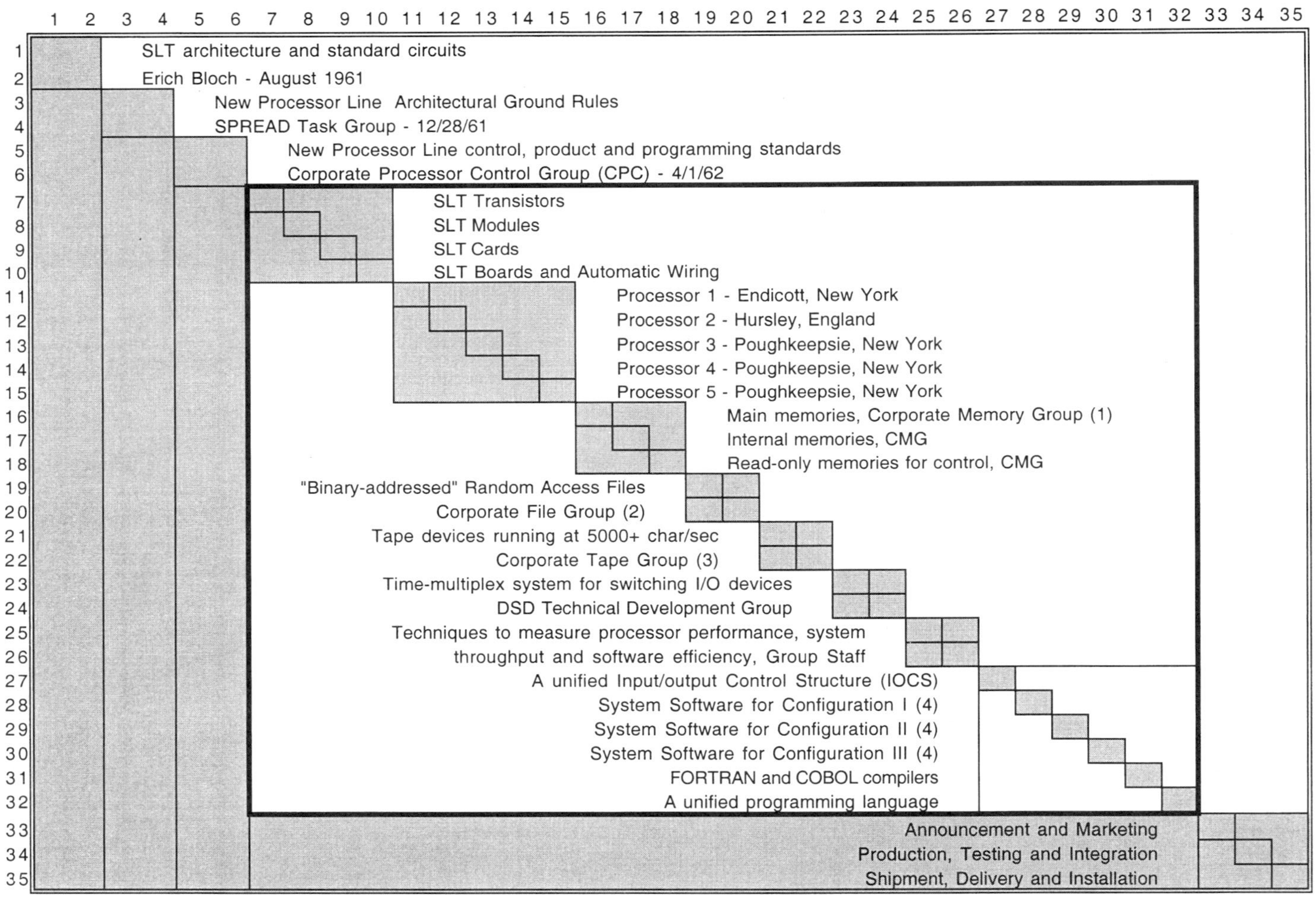

Figure 7.2 Task structure matrix for System/360, early 1962. (1) In 1961, the Corporate Memory Group included 125 engineers located in Poughkeepsie, Endicott, and Kingston, New York, and San Jose, California. The number of people assigned to this group doubled between 1961 and 1965.

The Design Rules Phase

The *architectural ground rules* set forth in the SPREAD report are listed in the appendix to this chapter. They are an interesting combination of goals ("each processor is to be capable of operating correctly all valid machine-language programs . . .") and basic compatibility constraints ("decimal digits and alphanumeric characters will be represented in four-bit and eight-bit bytes . . ."). Since the authors of the report understood very well that these design rules were incomplete, the first set of tasks in the project would be to complete the framework of visible information. In addition, there had to be tests and a way to enforce conformance of modules with the design rules: "[T]he designs of different groups will yield nonconsistent products unless the programs are constantly controlled and measured against a standard."[13]

To keep the separate teams on track, the report recommended that "one architectural-engineering design office" be created. The CPC group would have just three months to come up with logical specifications, complete product definitions, product-control procedures, product standards, and programming standards for the complete line. Thereafter the CPC would monitor the progress of the work, ensure compatibility, and arbitrate conflicts arising among the teams. (It did not escape anyone's notice that the CPC would be a very powerful group if the new product line went forward.)

On January 4, 1962, a week after they submitted their report, the SPREAD task group presented their recommendations to IBM's top fifty executives and their staffs. Many who were there voiced concerns about the scale of the project, but Learson, who had the authority to approve it, heard "no real objections," and decided to proceed.[14] From the historical accounts, we know that the first ten months of the project were spent as follows[15]:

January to mid-March—false start (exploration of stack architecture).

March 15 to April 15—architectural design competition.

Mid-April—sixteen thirty-two-bit general registers, base register addressing, eight-bit character coding were accepted as the foundation of the design. (*Note*: All of these characteristics interact, hence these choices needed to be made simultaneously).[16]

---

[13] Haanstra et al. (1983, p. 19).

[14] Wise (1966a).

[15] Pugh et al. (1991, pp. 144–151).

[16] See Hennessy and Patterson (1990), or Bell and Newell (1971), for discussions of these interactions and the tradeoffs involved.

May–October—discussion and debate about the rest of the design rule choices; five processor teams argued and negotiated.

October—Fred Brooks, director of the CPC, held a two-week "fall festival" to settle all outstanding architectural issues; at these meetings, he heard arguments for and against different design rules, and dispensed "drumhead justice" on the spot.[17] Through a new automated text editing program (one of the first of its kind), these decisions were typed into a manual and printed each night. The design teams went home with the manual in hand.

In effect, the design rules phase of System/360 was the first standards battle in the computer industry. It was a battle because the engineers at IBM did not welcome the discipline that was needed to achieve compatibility across a range of processors. However, thanks to the foresight of the SPREAD committee, there was an institutional mechanism—the CPC—in place to resolve conflicts between the affected parties, and thus the main architectural features of System/360 were decided in relatively short order. By November 1962, most of the rules were in place and the teams could go on to the next phase.

The Design Rules of System/360

What were the design rules of System/360? What did the CPC have to decide and the independent working groups have to know? What were the terms that allowed widely separated design teams to achieve compatibility without day-to-day coordination?

A core set of decisions constitutes the highest level of visible information in a family of computers. These hierarchical design parameters are listed in box 7.2. The composition of this set is determined by the physical and logical structure of computers, and so today textbooks in computer architecture are organized around this set of design decisions. In 1962, however, System/360's architects had no such textbook: they had to know or discover that these were the rules they needed.

Figure 7.3 presents a condensed view of the design hierarchy of System/360, which emerged from the design rules phase. The core of the system is contained within the large box; elements outside the box, like the SMS and SLT standard circuits, were visible to some parts of the system, but were not part of the core design.

---

[17] Pugh et al. (1991, pp. 150–151). Fred Brooks described the critical role he believed documents can play in the management of a large project in two essays: "Why Did the Tower of Babel Fail?" and "The Documentary Hypothesis" (Brooks 1975).

Box 7.2 The Visible Information of System/360:
Decisions Common to All Members of a Compatible Family of Computers

For each processor:

1. the hierarchy of information units: the length of a word (in bits) and the subdivisions of a word;
2. the range of data types accepted (binary integer, floating point, decimal, logical, character) and a format for each;
3. the number of addresses per instruction, the number of registers in the processor, and the uses of the registers;
4. instruction formats: the length of the operator codes and length of operands;
5. the location of operands (in registers or in memory);
6. memory address and access conventions;
7. the codes used for decimals numbers and alphanumeric characters;
8. the instructions (operations) in the instruction set and their codes;

For input-output device:

9. the allocation of I/O control (in the central processor or in specialized processors—System/360 used specialized processors called "channels");
10. the I/O start and stop conventions;
11. the physical characteristics of I/O connections: the number of wires and electrical signal characteristics;
12. I/O commands (these form the instruction set of the I/O processors);

For control:

13. conventions on the sequence of instructions (can an instruction modify itself);
14. methods of branching;
15. provisions for interruption and for restoring the processor state after an interruption;
16. provisions for measuring time;
17. provisions for interaction with real-time input and output;
18. provisions for storage protection, checking of data, and the reporting of errors;
19. provisions for multisystem operation: shared control, interconnected I/O, shared storage, and interprocessor communication.

Note: This list was compiled primarily from Blaauw and Brooks's 1964 article: "The Structure of System/360: Part 1—Outline of the Logical Structure,"[1] with reference to

[1] Blaauw and Brooks (1964).

(continued)

Bell and Newell's characterization of computer structure[2] and Hennessy and Patterson's modern textbook on computer architecture.[3] Supplementary information was taken from the technical history of Pugh et al. of the 360 project and Stevens's discussion of the implementation of System/360.[4]

[2] Bell and Newell (1971, chapters 1–3, pp. 3–88).

[3] Hennessy and Patterson (1990).

[4] Pugh et al. (1991, pp. 144–151); Stevens (1964).

At the highest level of this hierarchy were the global design rules. These included items like the length of a word and its subdivisions; the range of data types; the number of addresses per instruction; in fact, all the items listed in box 7.2 except the instruction set and the I/O commands.

One level down in the hierarchy were three distinct, complementary sets of visible information. First, each processor had its own hardwired, microinstruction set. (Five of the six initial processors used microprogrammed instructions; Model 70, the largest, hardwired the full System/360 instruction set.) A specific processor's instruction set was visible to its designers, who would implement the instructions in hardware; to the designers of the "read-only control stores," who would translate System/360's instructions into processor microinstructions; and to the designers of emulators.

Second were the instructions in the common System/360 instruction set. These were visible, on the one hand, to the writers of processor microcode, and, on the other hand, to machine-language programmers, including those writing compilers for high-level languages like FORTRAN. However, as the figure shows, when these instructions were implemented through a read-only control store (contained in a read-only memory, or ROM), they did not have to be known to the hardware designers. (Hardware designers did have to know enough to choose between hardwired instructions and microcode, but if they elected to use microcode, the details of the instruction set were not a matter of concern to them.)

Third, an IOCS was implemented in special-purpose processors (called "channels"), which in turn communicated with the I/O devices. As a result, the central processors of System/360 needed only four standard I/O commands.[18] The I/O con-

[18] Blaauw and Brooks (1964, reprinted in Bell and Newell, 1971, p. 598).

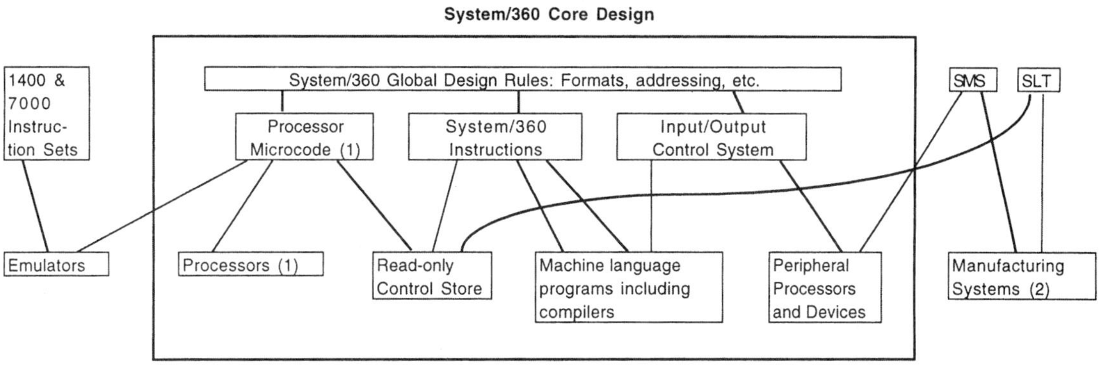

Figure 7.3 Design hierarchy for System/360, circa 1964. (1) Model 70, the largest of the machines announced on April 7, 1964, had hardwired control and hence did not use a read-only control store. System/360's instruction set was directly visible to the designers of that processor. Model 70 did not offer microcode-based emulation. (2) There were two different manufacturing systems—one for the older SMS circuit family (used in all the peripheral devices) and one for the new SLT family (used for the new processors). See chapter 6 for details.

trol system in turn was visible to machine-language programmers and to designers of I/O devices.

Outside the core of System/360, but still visible to hardware designers, were the two sets of standard circuits. The processors in the System/360 family used circuits based on SLT. I/O controllers and peripheral devices were based on the older SMS circuits. These manufacturing standards were visible to hardware designers and to process engineers, but not to programmers.

Also outside the core of the design were the instruction sets of IBM's other processors. The SPREAD report specifically rejected legacy design rules, saying that System/360 would not be "backward compatible" with prior designs. However, during the parallel work phase of the project, two engineers working on the Model 60 (the second-to-largest processor) invented a new type of module called an "emulator," which allowed System/360's processors to execute programs written for earlier machines, like the 7070. Hence the instruction sets of other machines were visible to the designers of emulators. (See box 7.3 for further details.)

Three things are notable about the design rules of System/360. First, by 1962, knowledge about computers had grown to the point where not only could a list of design rules be constructed but designers could argue for and against different choices based on experience with real machines. Therefore, the architectural decisions that underlay System/360 were based on quite detailed knowledge of the

Box 7.3 Microprogramming and Emulation

System 360's basic instruction set—the commands that machine-language programmers would use—had to be large and relatively complex, as befit the high-performance machines planned for the top of the line. The SPREAD report authors realized that hardwiring such an instruction set would be prohibitively expensive for the smaller machines. Therefore the group spent much of its time trying to understand Maurice Wilkes's theory of microprogrammed control (discussed in Chapter 6).

The task group was so impressed by the flexibility inherent in microprogramming that they recommended its adoption even if it appeared to be more expensive than other approaches:

Microprogram controls using a read-only memory shall be employed unless the cost performance of a conventional control system is less than 2/3 that of a microprogram control system.[1]

In the architectural negotiations that the SPREAD group members knew would follow (because the report did not address all of the design rules needed for the new system), a rule in favor of microprogrammed control would make it much easier to argue for a large instruction set. With hardwired control, each instruction has a physical representation in the processor itself, hence is visible and costly to the processor designers. With microprogrammed control, additional instructions do not affect the wiring of the processor itself: they become additional lines in a ROM. As a result, the marginal cost of additional instructions goes down substantially.[2] A second benefit of microprogrammed control was that new instructions could be added to the common repertory late in the design cycle or even after some machines were shipped. Again all that was necessary was to change the ROM that held the instructions.

Five of the six processors announced in 1964, and eight of the first eleven to come to market implemented the common instruction set using microcode and a ROM. Many external critics felt that this dependence on microcode slowed down the machines and led to various processing anomalies. (For example, programs written in a high-level language like FORTRAN might be compiled into the System/360 instruction set and then retranslated into processor microinstructions.)[3]

Serendipitously, however, microprogramming solved another problem—that of "backward compatibility" with IBM's older machines in the 1400 and the 7000 series.

---

1 Haanstra et al. (1983, p. 21).

2 In those days, a large instruction set was equated with programming power and flexibility. It was only much later, after programming switched to high-level languages, and high-performance optimizing compilers became available, that reduced instruction set computing (RISC) designs began to have a following. See Hennessy and Patterson (1990).

3 Bell and Newell (1971, pp. 569, 587).

(continued)

The SPREAD task group did not think it would be feasible for the new product line to be backward compatible. The older lines used a hodgepodge of data types and formats: the 1400s, for example were character-based machines that processed variable-length strings of data. But in 1963, Stewart Tucker and Larry Moss managed to convert the instructions of an IBM 7070 into a combination of microcode and special wiring. When their "emulator" was hooked up, the System/360 Model 60 was able to execute 7070 instructions faster than the original machine. (The speedup was possible because System/360's SLT circuits were inherently faster than the SMS circuits used in the 7070).

Emulation was both an *augmentation* of the basic von Neumann machine,[4] and an application of *the porting operator* in a modular design structure. (See the descriptions of the augmenting and porting operators in chapter 5.) As figure 7.3 indicates, emulators were hidden translator modules, which formed an invisible link between the instruction sets of the new and old product lines. With an emulator in place, programs written for the older machines could run on the new.

Emulation was announced as an option at System/360's introduction (April 7, 1964).[5] Bell and Newell, writing in 1971, said of this feature: "[T]he ability to interpret another computer's instructions at a reasonable performance level . . . [was] undoubtedly the most difficult design constraint to meet . . . and probably the most significant real innovation" in System/360.[6]

In fact, some felt that backward compatibility, and not compatibility within the line, was the true deciding factor in the marketplace's eager acceptance of System/360. If true, this would be an ironic twist, for backward compatibility was the attribute which John Haanstra had championed, and the SPREAD group explicitly rejected. The group chose instead to pursue the vision of a broad compatible product line achieved via a modular design. As it turned out, though, the power and flexibility of the modular design allowed IBM to offer *both* a broad, compatible product line and backward compatibility in the same system. Thus we will never know which attribute was more important to early users.[7]

---

[4] Maurice Wilkes foresaw this possibility and considered it one of the most attractive features of a microprogrammed approach. Wilkes and Stringer (1953).

[5] Emulators were later built for all the microcode-based processors in the product lineup. The smaller machines could emulate the 1401, while the larger machines could emulate both the 1400 and the 7000 series (Pugh et al. 1991).

[6] Bell and Newell (1971, p. 562). Later, when the SPREAD report was published, it became clear that emulation was *not* one of the initial design goals, but was a lucky augmentation.

[7] Indeed, "some [360] users never bothered to convert many of their programs, running them 'under emulation' for years" (Pugh et al. 1991, p. 161).

tradeoffs involved. Although a long period of experimentation with many different computer designs was needed to build that knowledge, we know, with benefit of hindsight, that no drastic errors were made in the 360's design rules.

Second, creating the design rules for System/360 was in itself a highly interconnected set of tasks. At this level of the design, every decision interacted with every other decision. As a result, the stakes involved in setting the rules were high for both individuals and the corporate divisions they represented. A decision made to enhance the performance of one processor could greatly increase the complexity and cost of another.

However—and this is the third point—although the design rules phase took three times as long as planned, the work of setting the rules for System/360 had to be done only once. Once they were set and promulgated, the design rules of System/360 could be used again and again. They were a "sunk cost" of systems design and because System/360 had a long product life, the rules stayed in effect for a very long time.[19]

The Parallel Work Phase

The success of System/360 depended critically on the quality of the new design rules and the corresponding task structure. If the architects did their work well, the modular partition would work, and the design rules would anticipate almost all points of design interdependency. In that case, system integration and testing would be straightforward and painless. If, on the other hand, the architects' knowledge of interdependencies was not up to the task, then integration would fail and the whole effort would collapse back into interconnected cycling and chaos.

Counting the projects already underway for SLT, the new product line envisioned by the SPREAD task group engaged the company in almost one hundred separate design efforts, resulting in upward of fifty new pieces of hardware, plus unknown amounts of new software. When wired together, all these different parts were sup-

---

[19] Robert Evans, the vice chairman of the SPREAD task group, speaking in 1983, called the longevity of System/360's design rules "remarkable":

[W]e were on the cycle of bringing out new systems every three or four years at that time. None of us dreamed that System/360 would last more than 5 or 10 years. If it lasted 10 we thought it would be remarkable. But it has endured and is clearly going to last well into the 1990s, and who knows after that? That is rather remarkable in itself. (Aron et al. 1983, p. 44.)

posed to function as one system. To get these designs done and the products manufactured, the designers figured, would take three years; products could be announced in 1964 and shipped sometime in 1965.

Such was the power of modularity, implemented within the framework of sound technical design rules, a flexible organizational structure, and effective leadership. It was a breathtaking vision. But would the plan work? The amazing thing was that by and large it did. Yet implementation was far from easy.

Consistent with the fact that new task structures must be mapped onto new organizational structures, System/360's modular task structure required many new workgroups to be formed. Most powerful was the CPC office discussed above, but new groups were also needed to design specific processors, to reconfigure peripheral devices, and to manufacture components. These groups were largely anticipated by the SPREAD report. (See figure 7.2, which was constructed from the report.) And, despite the early delay in formulating the design rules, for the first three years the entire project appeared to be going well—so much so that in February 1964, IBM's top management shut off the funding of efforts to extend the 1400 product line (see box 7.1). This action and John Haanstra's removal from the presidency of the General Products Division were the prelude to the company's dramatic public announcement of the new product line.

On April 7, 1964, in simultaneous press conferences at seventy-seven cities around the world, IBM revealed the specifications of six new processors, nineteen processor-memory configurations, and forty-four I/O and storage devices. It also published its plans for a new, unified operating system that would run on all but the smallest processor.

In the product markets, the customers' response did justice to the drama of the announcement. Although all of the machines were at least a year from delivery, IBM booked 1100 orders in the first month after the announcement. (This was equivalent to about 10% of IBM's installed base of machines at that time.) In the months following, orders flowed in. There was ample justification for Thomas Watson, Jr., IBM's CEO, to state: "Within IBM there was a great feeling of celebration because a new era was opening up."[20]

---

[20] Watson (1990, p. 351). The response in the capital markets was much cooler; in parallel with its competitors, IBM declined in market value by about 5% during the month of April 1964. The discrepancy between customers' and investors' reponses to the announcement may have been caused by widespread doubts about the technical feasibility of what IBM was proposing. Customers placed orders in order to obtain machines should they become available, while investors adopted a "wait and see" attitude.

The Integration and Testing Phase

The risks of the modular approach were not apparent until early 1966, almost two years later. Then it became clear that the SPREAD report authors, the Management Committee, and everyone else at IBM had vastly underestimated the job of integrating, testing, and producing System/360, and installing the new systems at customers' sites.

One insidious aspect of first-time modularizations is that design rules flaws are revealed only very late in the process. Once the rules are in place, work on the individual modules proceeds independently and may appear, for a time, to be going very well. It is only when the pieces are brought together for final integration and testing that unforeseen interdependencies are brought to light. If these are serious, major chunks of the system will need to be redesigned. The modular task structure will then dissolve into an interconnected one, often in an unplanned, chaotic way. In addition, much time may have been wasted in parallel efforts that end up having to be scrapped.

In terms of the task structure matrix for System/360 shown in figure 7.2, we may imagine that there were many unknown interdependencies—**x**s—sprinkled outside the boxes, both above and below the main diagonal. These points of interdependency required coordination and consistent decision making, but were not known in advance. Instead they were "discovered" during the testing and integration phase, when some prototype machine or manufacturing process did not work.

In general, a modular design and task structure can withstand the discovery of unforeseen interdependencies as long as they are relatively sparse and remediable. But if they become too dense or are not remediable, the modules must be recombined. In that case, unless the designers start work anew, the resulting artifact will most likely be a hybrid, full of arbitrary parameters and random fixes.

Thus, simply because System/360 was a newly modularized design, the system integration-and-testing phase was bound to be difficult. But to understand the true magnitude of IBM's difficulties in 1966, one has to bear in mind that the company was redesigning its production systems in a modular fashion while it was developing a completely new product line. The changeover to SLT was occurring at the same time as five new processors and many new peripheral devices were being created.

As we described in chapter 6, SLT was a modular production technology, based on four physical levels (circuits, cards, boards, and cross-board wiring). From a process design perspective, each of these levels was a module. Thus, in theory, process design for each level could be conducted by different teams, working in parallel and independently of one another and the product designers. Everything that the product designers needed to know was supposed to be incorporated into a set of visible manufacturing design rules.

In fact, the circuit designs of System/360 *were* standardized: designers used three circuit families (operating at delays of 30, 10, and 6 nss) and thirty-two different circuit types. Hence at this level, the interfaces between product and process were relatively clean.[21] It was a different story, however, at the card, board, and wire-wrapping levels, where the designs were not standardized.[22] As their designs progressed, the designers of processors needed to test their card and board designs. But because the new production systems were not yet in place, physical prototypes of cards and boards often could not be made. The lack of cards and boards then slowed down the testing of the processors, and the completion of their designs. Later, the large number of nonstandard card and board designs vastly complicated the ramp-up of manufacturing to the high volumes needed to satisfy market demand.

Because of these setbacks, which were compounded by raw material shortages, in October 1965 IBM was forced to announce production delays across the line. At this time, systems were being shipped without software and before they were fully tested. Advanced features and even whole devices were quietly "decommitted."[23] The company's internal design, manufacturing, and logistics systems were slipping out of control.

The internal chaos had serious financial consequences. For a short time, IBM's cash flow, which had financed the $4 billion commitment to System/360, dipped into negative territory.[24] To prevent any further erosion of its cash position, IBM issued $371 million of new equity in April 1966.[25]

To rescue System/360 and the company, CEO Watson installed Vincent Learson, then head of IBM sales, as head of the overall System/360 project. Learson seconded four senior IBM executives, and made each responsible for a major chunk of activity.

---

[21] However, for the first time, IBM was manufacturing its circuit components in high volume, and this created another set of problems (Pugh et al. 1991).

[22] Pugh et al. (1991, p.153).

[23] Wise (1966b).

[24] Estimates of the total cost of System/360 range from $4 to $5 billion. Of this amount, $500 million to $1 billion was development cost, while the rest was used to expand manufacturing capacity and produce rental machines. (Most of IBM's machines at this time were leased to customers, not purchased outright.) See Wise (1966b); Evans (1983, p. 44); IBM Annual Reports, 1961–1966.

[25] IBM Annual Reports 1965 and 1966.

Hank Cooley, the executive given charge of System/360 manufacturing at this time, described the situation in early 1966 as follows:

You had your list of 200 to 300 problems on every CPU and every piece of I/O. None of them by themselves were show-stoppers, but when you took that myriad of typical early manufacturing problems on every box and you superimposed that upon all the logistical problems that we were in, due to everything being brand new, it was an absolute nightmare.[26]

Integrating new product designs with a new manufacturing system may have been a nightmare, but it was not a failure. A critical point was that none of the "200 to 300 problems" associated with each new processor was in fact fatal. Although there were more flaws in the modular design than anyone had foreseen, almost all were of the minor, fixable variety, and so, by and large, the initial partition of parameters and tasks held up. (The one exception was in systems software, discussed in box 7.4.) The goal of binary compatibility across the line was maintained, and no major piece of hardware was dropped. One by one the faults of the process were overcome, and by the end of 1966, the company's deliveries were back on schedule.

Conclusion

Thus, by early 1967, IBM had a new product line, backed by an extensive set of new internal manufacturing and design processes. Based on a far-flung, comprehensive modular design, its designers and managers had created

- a new artifact that spanned fifty new pieces of hardware,
- a new set of manufacturing facilities,
- a new supply chain,
- new field service and maintenance operations,
- thousands of new programs,
- a new, decentralized design process.

Then, working on overdrive for a year, they had removed the bugs so that the parts of this system actually worked together in an almost seamless way.

The system to design, make, distribute, and sell System/360 was a new task structure that came into being within IBM between 1962 and 1967. The task structure

[26] Pugh et al. (1991, p. 172).

Box 7.4 Operating Systems of System/360—Where the Modularization Failed

In the mid-1960s, "operating system" was the name given to new concept in software: an integrated set of general-purpose programs, including assemblers, compilers, input, output, utility, and sort programs, all of which were under the management of a control program. Operating systems automated the real-time tasks of managing a computer system, making possible a more efficient utilization of both human programmers and hardware resources.

The SPREAD task group and the CPC envisioned that System/360 would have a compatible set of operating systems that spanned the whole processor range. However, the saga of operating system design for System/360 was a history of false starts, turf wars, personality clashes, and missed deadlines. After three years of interdivisional bickering, systems software development was centralized in December 1963, only four months before announcement of the new product line.

Two issues were then decided. First, the operating system of System/360 would support "multiprogramming," that is, the interleaved processing of several jobs. Second, to keep the systems compatible, the "control program" (the heart of the operating system) would be *modular in use.* In other words, it was envisioned that purchasers would be able to select and combine different parts of the control program to achieve different levels of functionality and to conform to different hardware specifications. These were very ambitious design goals.[1]

In addition, the design managers were under heavy deadline pressure. They attempted to accelerate the project by dividing the design tasks into forty different projects, and adding people to the program.[2] (At the peak of development, more than a thousand people were working on OS/360, as it was called.) But their efforts were to no avail. Deadlines slipped while costs continued to rise, amidst firings and recriminations.[3] In his classic book, *The Mythical Man-Month,* first published in 1974, Fred Brooks captured his experience in what he called Brooks Law: "Adding manpower to a late software project makes it later."[4]

Our theory of how designs become modular, described in chapters 3 and 5, sheds light on IBM's experience with the operating system for System/360. When the SPREAD task group began its work, there was already a protomodular design and task structure in place for hardware and for manufacturing processes. The SPREAD task group partitioned tasks and developed design rules based on their knowledge of this preexisting structure. There was no corresponding structure in place for system software, however. Quite the opposite: operating systems were a newly emerging *architectural module,* and thus System/360's design managers had no prior experience to guide

[1] Pugh, Johnson, and Palmer (1991, pp. 315–318).

[2] Ibid., pp. 328–343.

[3] Wise (1966b); Watson (1990, pp. 353–360).

[4] Brooks (1995).

(continued)

them in establishing sensible design rules or partitioning the tasks of operating system design.

Within operating systems, there were myriad interdependencies (for example, sequencing and scheduling interdependencies) that needed to be worked out by trial and error. Consistent with the fact that operating systems were a new type of software, most of the interdependencies were not known to designers at the time. Thus whenever project managers tried to divide up the design tasks to accelerate delivery, unforeseen interdependencies would crop up. These "bugs" needed to be resolved, and cycling back through the code to fix bugs caused the projects to fall further and further behind. In fact, with benefit of hindsight, we can see that operating systems software was not ready to be modularized: the requisite knowledge did not exist, and would not exist for several years (see the discussion of Unix in chapter 13).

IBM funded a veritable alphabet soup of operating systems for System/360: BOS, TOS, DOS, PCP, MFT, and MVT. These systems did not achieve the initial design goals: they did not have regular, nested modular structures, as was envisioned, nor were they compatible with one another. In the end, DOS and MVT survived, while the rest were discontinued as the product line matured.[5]

[5] Pugh, Johnson, and Palmer (1991, p. 343).

was modular, like the design of the artifact itself.[27] It had to be, because the complexity and breadth of the product line required an efficient division of effort and knowledge within the enterprise, and much parallel work. At the same time, ongoing technical developments in computer science dictated that the product line would have to change, but the nature and timing of specific changes were uncertain. (Recall that in chapter 3 we said that the three reasons to embrace modularity are to manage complexity, to enable parallel work, and to accommodate uncertainty.)

However, despite IBM's success in booking early orders, the system's value was still mostly a potential as of the beginning of 1967. To make money for IBM, and to allow the company to recoup its large, sunk-cost investment (estimated to be around $5 billion in 1966 dollars, equivalent to around $20 billion today), the task structure had to be "packaged" in a contract structure. All the technical and operational successes would be all for naught if the contract structure failed to capture high value for the collective enterprise known as IBM.

[27] Strictly speaking an artifact that is modular in design does not have to be produced, distributed, and serviced via modular processes. But, as we have seen, modular production, distribution, and field service were traditions at IBM that predated System/360.

Appendix: Design Rules for System/360 Contained in the SPREAD Report

1. *Compatibility*

Each processor is to be capable of operating correctly all valid machine-language programs of all processors with the same or smaller I/O and memory configuration.

2. *Formats and Addressing*

a. Address lengths are to be variable so that not all high-order zeros in addresses are expressed.

b. Addressing is to be binary in radix. Efficient use of memory dictates that addressing must be binary or alphanumeric, in preference to decimal. Between these, binary is more flexible, straightforward, and economic.

c. Decimal digits and alphanumeric characters will be represented in four-bit and eight-bit bytes.

d. Variable-length field manipulation, independent of physical memory width, will be standard.

e. Each four-bit byte is to be directly addressable.

f. Move and other streaming operations will operate on fields as short as four bits and as long as memory capacity, although length restrictions may be laid on arithmetic operation.

g. Negative data fields will be repressed in true, not complement, form with the sign, if present, appearing at the low-order end of the field.

h. Address modification through additive indexing is to be standard on all machines.

i. The hardware-software package shall provide automatic translation (at least as late as load time) of symbolic address (indirect addressing), at least for the addresses of similar I/O devices.

j. The hardware-software package shall provide for the automatic and independent relation of program and data, at least as late as load time.

k. A hardware memory protection system shall ensure interprogram protection against any problem-program error.

3. *Operations*

a. No bit combination shall exert any mandatory control function when it occurs in the data stream of a CPU.

b. All fixed-point arithmetic operations shall be provided for radix 10 and radix 2.

c. Floating-point arithmetic shall be available for all CPUs, at least as an option.

d. Compatible sterling arithmetic operations will be available for all CPUs, at least as an option.

e. A uniform subroutine linkage mechanism shall be provided.

f. Program interruptions upon external signal, program invalidity, or machine malfunction shall be provided as standard in all machines.

g. Facilities for the operation of a supervisory program shall be such that the supervisor can retain positive control over any problem program without manual intervention. Nonstop operation shall be possible.
h. A real-time clock and interval timer shall be available for all CPUs, at least as an option.

4. *Input/Output Control*

a. I/O shall be programmed through a sequence of channel-control words, whether a physically separate channel is used or not.
b. I/O operation shall be logically overlappable with processing, but burst operation may be used for high-data-rate devices.
c. Program-controlled cross-channel switching of I/O devices shall be designed for all systems, but not necessarily as standard equipment.
d. Multiplexing and control of low-speed lines and terminals will be accomplished with direct, minimal attachments to standard processors without requiring special-purpose stored-program devices.
e. To any I/O device type, all channels shall appear identical except in data rate.

5. *Reliability and Serviceability*

a. To meet the demands of the new applications, each processor shall attain corporate goals for significant reliability and serviceability improvement.
b. All data paths shall be so completely checked that no single malfunction goes undetected. Controls shall be so checked that the probability of undetected control malfunction is no higher than that of undetected data-path malfunction.
c. Each system portion whose servicing does not prevent system operations shall be furnished with facilities for independent, off-line servicing.
d. Machine-language consistency shall extend to maintenance consoles. Each CPU shall be equipped with appropriate portions of a single full-scale maintenance facility.
e. Ultrareliability shall be achievable by multiple-CPU systems.

6. *Engineering Ground Rules*

The following engineering ground rules must be imposed on all groups working on the proposed processors.

a. Microprogram controls using a ROM all be employed unless the cost performance of a conventional control system is less than two-thirds that of a microprogram control system.
b. A single detailed method for memory-CPU coupling shall be developed and applied to all processors.
c. Timing and priority controls shall be so designed that no processor or channel assumes that it inevitably gets the very next memory cycle after a request.
d. All options announced with a processor shall be field installable.
e. When one processor is substituted for a slower one, the I/O gear shall not need to be changed.
f. Each processor shall be designed to accommodate fifty-cycle power supplies.

8 Enterprise Design: A Task Structure plus a Contract Structure

The announcement of the IBM System/360 in April of 1964 was a landmark event in the history of computers and of the computer industry. With the almost euphoric reaction from customers that followed, one can appreciate the sense of vision and vindication in Tom Watson, Jr.'s comment: "Within IBM there was a great feeling of celebration because a new era was opening up."[1]

As we have seen, turning the vision into reality meant creating a whole new task structure for the corporation, including many new organizational subunits, and a commensurate number of coordinating links. By 1967, however, most inconsistencies had been worked out of the product and process designs, and the idea of a family of compatible, modular processors had become real.

Nevertheless, to bring the new design to competitive life, still more effort was required. IBM's senior managers needed to turn the modular system of artifacts into a broad line of products with a secure, ongoing revenue stream. In other words, they needed to build a profitable enterprise on top of System/360's design and production processes. For that, they needed to wrap a contract structure around the new task structure. The result would be an *enterprise design,* which, they hoped, would be as powerful and effective as the technical design of System/360.

The purpose of this chapter is first, to describe the enterprise design that IBM's managers adopted for System/360, and second, to describe how that enterprise design affected IBM's customers, competitors, employees, and computer architects at other companies. We begin by returning to the concept of contract structure, first introduced in chapter 4. We will compare IBM's prewar contract structure to the one

---

[1] Watson (1990, p. 351).

utilized for System/360, showing that they were in most respects very similar. In effect, IBM's preexisting contract structure was overlaid on its new task structure, and then "tweaked" in response to new issues raised by the new design. We will then show that this enterprise design—a new task structure "wrapped" in a traditional contract structure—was enormously successful by any financial criterion one might choose to apply.

Once we understand how System/360 made money for IBM, we will turn to look at its effect on IBM's competitors. Its immediate impact was to drive most of IBM's major competitors into small niches or out of the computer industry entirely. However, as this was happening, the new design was also causing interesting, almost invisible changes in the "value landscape" of computer artifacts. These changes were related to the fundamental modularity of System/360's design, and in particular, the design's capacity to evolve. The changes in value introduced the possibility of new types of investment (in modules, not whole systems) and new types of enterprises (based on modules, not systems).

As soon as they perceived these possibilities, value-seeking designers began to innovate at the module level of System/360's design. Their actions in turn brought a new dynamic process into the industry: a decentralized, value-seeking process, which we call *design evolution*. Design evolution and its implications for industry structure will be our main topics in the second half of this book. Thus we close this chapter with a description of the significance of System/360 in the history of computer designs, and a preview of what lay ahead.

Contract Structures

Collective enterprises operating in market economies must capture a stream of revenue that is greater than the cost of the collective effort. The organizers of such enterprises secure revenue and the resources needed to sustain it by setting up an interrelated set of contracts. In chapter 4, we defined this "contract structure" as the set of contracts used to form a collective enterprise, and to organize its dealings with its environment, including agreements with suppliers of labor, material, and capital, with customers, with owners and managers, and with governments.

In other words, contracts—agreements and understandings—are what bring resources of all kinds into a collective enterprise and move products out. Contracts establish both the boundaries of an enterprise and the "terms of trade" at those boundaries.[2] (We use the term "contracts" here not in the formal legal sense, but in a looser economic sense, to mean agreements and understandings between two or more parties. In this context, some "contracts" are explicit, some implicit, and some a combination of the two.)

We have said that contract structures have designs, just as all tangible and intangible artifacts have designs. Contract structure designs are chosen on the same basis as any other—for maximum functionality at minimum cost, within the constraints of what is known (about contracting) at a given time and place. A good contract structure design brings capital, people, and supplies into the enterprise, gets products and financial payments out, and does not impose unnecessary costs along the way.

In an advanced market economy, a complete contract structure design also defines the economic surplus (or deficit) created by a particular enterprise, and describes how that surplus will be allocated and divided among claimants. An agreed-upon definition and a plan for dividing the surplus are needed because in such economies, the rights to receive payments from a surplus are recognized as property. As property, contractual rights may be traded in organized markets; moreover, there are sophisticated technologies for estimating the value of a stream of surpluses expected in the future. (These valuation technologies were described in Chapter 4.)

The period-by-period surplus or deficit of a particular enterprise is known as its "net cash flow." As a matter of definition, the net cash flow of an enterprise in time period t equals the sum total of users' willingness to pay for its products, minus the users' surplus, minus all the costs of making the artifacts and getting them into the users' hands:

$$\begin{array}{c}\text{Net cash}\\ \text{flow of a}\\ \text{collective}\\ \text{enterprise}\\ \text{in period } t\end{array} = \begin{array}{c}\text{Users'}\\ \text{willingness}\\ \text{to pay for}\\ \text{the artifact}\\ \text{in period } t\end{array} - \begin{array}{c}\text{Users'}\\ \text{surplus}\\ \text{in period } t\end{array} - \begin{array}{c}\text{Cost of}\\ \text{the task}\\ \text{structure}\\ \text{in period } t\end{array} - \begin{array}{c}\text{Cost of the}\\ \text{contract}\\ \text{structure}\\ \text{in period } t\end{array} \quad (8.1)$$

This equation defines a vector of random variables, corresponding to the net dollar inflows and outflows to the enterprise in each time period.

Each term in the definition arises from a different source. Thus the users' willingness to pay depends on the functionality of the artifact from the users' point of view.

---

[2] There may also be internal contracts, for example, an understanding between two divisions of a corporation, but we are concerned here with contracts at the external boundary. Jensen and Meckling (1976) characterize the firm as "a nexus of contracts."

The users' surplus equals their individual willingness to pay minus the price, summed over all users. The magnitude of this term is determined by the state of competition in the product market for the artifact—highly competitive product markets lead to a higher users' surplus, other things being equal, than monopolistic or oligopolistic product markets.

The cost of the task structure is the cash cost (in a given period) of designing and producing the artifact and getting it into the users' hands. It is important to bear in mind that the timing of cash costs may not match the timing of revenue (which is the algebraic sum of the first two terms). In particular, a tangible artifact must be designed before it can be produced, and produced before it can be used. Hence unless users prepay for use (as they do for magazine subscriptions), the net of the first three terms will be negative for new artifacts for some period of time.

The cost of the contract structure is the cost of assembling the resources, and binding them within a collective enterprise. The enterprise in turn must survive long enough to get the tasks done and collect the revenue. As with the costs of the task structure, some of the costs of the contract structure may precede the receipt of revenue. For example, in order to have a workforce carry out a set of tasks, the enterprise must not only pay people to perform them (a cost of the task structure), but must also set up human resource systems to recruit them, and financial systems to pay them in a timely and efficient way.[3] The latter are costs of the contract structure. (Indeed, methods of recruiting and paying people are important elements of contracting technology, which in a market economy we would expect to change and improve over time.)

Fortunately, there is a set of contracts that reverse the normal cash flow pattern. *Capital contracts* create positive cash flow initially (when the enterprise receives the proceeds from issuing a claim), and negative cash flow later on (when the enterprise repays the claim "with interest"). Thus capital contracts solve the problem of negative cash flow at the start of an enterprise's existence. If the investors' expectations of net cash flow over time are "large enough," then enterprise designers will be able to devise capital contracts that bring money into the enterprise initially, in return for a pledge of repayment or a share of its future surpluses, or both.

The capital market value of the enterprise is then the sum of all its net cash flows, appropriately adjusted for time, risk, and contingencies using the valuation technologies of the capital markets. As we discussed in Chapters 2 and 4, human designers have the capacity to "see and seek" value: this axiom applies as much to the designers

---

[3] The ability to meet a payroll is a fundamental test of the viability of an enterprise.

of enterprises as to the designers of artifacts. Value-seeking enterprise designers generally want to take actions that increase the surplus (net cash flow) in any time period. Given an opportunity to increase cash flow in one period by decreasing it in another, value-seeking designers, to the best of their ability, would look to the valuation methods of the capital markets to see if the change was worth making, i.e., if the impact on the enterprise's value was positive.[4]

Essentially all collective enterprises need a contract structure to provide a framework for their day-to-day activities. The contract structure in turn must take account of the nature of the task structure both in broad outline and in detail. For example, the duration of contracts must match the duration of tasks; nothing is more disruptive (i.e., costly) than to have key resources withdrawn before the tasks are done.[5] Similarly, if input X is essential to the task structure, the contract structure needs to include contractual arrangements with specific suppliers of input X.

There is to our knowledge no general taxonomy, much less a handbook, of contract structures that identifies all the major dimensions and elements of these designs. Instead, it is more common for contracts to be broken out by type, for example, customer contracts, employment contracts, supply contracts, or debt and equity contracts. Each type of generic contract is then studied separately.

Although fragmentation is the norm, there is an important strand in economics and history that focuses on the "fit" among all the contractual elements of an enterprise design. The major intellectual proponent of this point of view has been Alfred D. Chandler, Jr. In three major historical works, Chandler has advanced the theory that the contract structure known as "the modern corporation" arose in response to the possibilities and problems inherent in the technologies of mass production.[6] These technologies, which emerged on the scene in the late nineteenth and early twentieth centuries, promised huge cost savings relative to prior methods of production. However, the new production processes relied on high-volume flows of materials, parts, and products. Managing this flow then required highly synchronized, coordinated efforts by production workers, a salesforce, and managers along a whole chain of supply.

---

[4] For simplicity, we are suppressing the problem of agency, which might cause the enterprise designers to seek value for themselves, and not the enterprise as a whole (Jensen and Meckling 1976).

[5] This is why "ex post opportunism" plays such a large role in the contract theory literature. See, for example, Klein et al. (1978) and Williamson (1985).

[6] Chandler (1962, 1977, 1990).

Spurred by the values implicit in the new technology, managers in a number of industries set about to create contract structures that would support highly synchronized, coordinated, sustained activities by a large number of people. To this end, they constructed vertically integrated corporations with stable workforces; relatively high, but fixed compensation; and implicit employment guarantees. They hired direct salesforces and invested in advertising and R&D to foster customer loyalty and to maintain the quality of their products. They organized the management of these enterprises by function (marketing, production, distribution, finance, etc.), with each functional unit playing a characteristic role in the overall organizational task structure.[7] These, then, became the defining characteristics of "the modern corporation," a type of collective enterprise which has dominated advanced market economies throughout most of the twentieth century.

A number of scholars, including Michael Jensen and William Meckling, and Oliver Williamson, have observed that, in market economies, different types of organizational forms and contract structures will compete with one another to be adopted into enterprise designs.[8] This competition will be especially fierce and effective in economies with advanced capital markets, because there third parties can bid for the rights to control, hence change the contractual designs of specific enterprises.[9] More recently, along the same lines, Bengt Holmstrom and Paul Milgrom have suggested that certain contract types complement one another and are also "well matched" to certain technological attributes.[10] An implication of their approach is that contract structures are likely to evolve into typical patterns and configurations.

The suggestion that certain contract structures may "match" or "be well suited" to certain task structures directs our attention to the contract structure IBM used to encompass the new task structure of System/360. As we will see in the next section, IBM did not invent a radically new contract structure for its new product line. Instead it folded the System/360 task structure into its preexisting contract structure, with only a few minor modifications. We will also see that, measured by the criteria of

---

[7] Later, as the enterprises expanded geographically and across products, the purely functional type of organization gave way to a divisional type of organization, with functions subsumed in each of the divisions (Chandler 1962, 1977, 1990; Williamson 1985).

[8] Jensen and Meckling (1976); Williamson (1985); Alchian and Demsetz (1972); Fama and Jensen (1983a, b); Grossman and Hart (1986); Hart and Moore (1990); Merton and Bodie (1995); Hansmann (1996).

[9] Manne (1965); Jensen and Meckling (1976).

[10] Holmstrom and Milgrom (1991, 1994).

cash flow, return on assets employed, and value enhancement, this was an enormously successful strategy, at least in the short run.

IBM's Contract Structure before the Creation of System/360

Prior to the development of System/360, in fifty-plus years of selling tabulators and other business machines, IBM had instituted a corporate contract structure that satisfied customers, employees, and suppliers, and deterred competitors. This contract structure, coupled with its products, allowed IBM to capture a high, ongoing stream of profits (net cash flows). These profits, in turn, satisfied IBM's investors, and allowed it to finance bold and risky moves like the creation of System/360.

The key elements of IBM's contract structure in its prewar tabulator business were as follows:

- IBM was a single corporation, in the sense defined in Chapter 4.
- It offered a broad product line of compatible tabulators and other business machines.
- The IBM salesforce was in constant contact with the customer's establishment; its members were technically competent and had financial incentives to solve customer problems by reconfiguring their systems.
- IBM owned all its equipment and rented it to users under short-term leases.
- IBM fostered competition among suppliers, shunning joint ventures, strategic alliances, and partnerships.
- IBM offered non-sales employees high, fixed wages and benefits, and opportunities for promotion. By and large, however, it did not tie monetary compensation to short-term, individual performance, nor did it grant stock or stock options to employees. It promised, but did not guarantee, stable employment, including "no layoffs."[11]

The fact that IBM was a corporation, with the property rights of twentieth-century corporations, was fundamental to this contract structure. As a corporation, IBM had the right to hold property, the right to transact and enter into contracts, and the right to issue debt and equity claims against its surplus. Without these rights, none of the other contract structure elements would have been possible.

---

[11] Mills and Friesen (1996).

IBM applied this system of implicit and explicit contracts to its computer business in the 1950s with great success. Indeed, one of the reasons the company managed the product transition from tabulators to electronic computers so well was that it introduced computers into its preexisting system of relationships with established customers.[12] The arrival of System/360 did not change this basic approach—in fact, as we saw earlier in this chapter, the impetus behind the creation of a broad, compatible product line came from IBM's customers in the form of demands for relief from the high cost of machine upgrades and conversions. And those same customers responded to the announcement of the new product line with great enthusiasm.

The Contract Structure Surrounding System/360

The contract structure IBM had developed in its tabulator business was proven and effective.[13] Not much in it needed to change to accommodate System/360. We have listed the key elements of the System/360 contract structure in table 8.1 in a way that shows which elements were old and which were new. The items shown in roman type had their roots in prewar IBM: the ones in italics were either changes in IBM's policies (e.g., its manufacturing policies) or arose in response to the unique demands of the new technology (e.g., nonmodular software, preemptive cannibalization).

For IBM, offering a broad product line of interoperable machines was a return to the company's roots in the tabulator era. In the 1950s with the advent of computers, IBM's offerings came to be dispersed across a number of incompatible product lines, and both the customers' and IBM's internal efficiencies suffered as a result. Hence there were benefits to both associated with the return to a unified line.

However, the stakes for users (hence their willingness to pay) were much higher in the postwar era, because of the technological dynamism and uncertainty surrounding the new artifact class of computers. Particularly for large firms (which formed the core of IBM's customer base), buying a set of computers that filled a range of needs was very expensive. Each system had its own language and required its own applications software. A firm that wanted a small system (such as a 1401) to handle basic accounting but a large system for data processing had to hire people with expertise in two different machine languages, with different peripheral protocols, and so forth. Even when a company used only a single computer system, it still

[12] Christensen and Rosenbloom (1995).

[13] This section draws on Baldwin and Clark (1997b).

Table 8.1 Key elements of the contact structure surrounding System/360

Key elements	Impact
1. One corporation, unified strategy, uniform policies	Credibility and continuity; organizational fluidity
2. Compatible family of computer systems, modular in design, production, and use, covering a wide product range	Users have options to exclude and augment at will; one-stop shopping for customers
3. Direct salesforce, technically competent, focused on solving problems	Strong, enduring customer relationships; deep knowledge of customer's system
4. "Corporate" employment contracts: good salary, small bonus, secure job, excellent benefits, no stock options	A loyal and competent workforce capable of teamwork with little internal competition
5. Own the equipment, rent it to customers	Facilitate product innovation; control over customer configuration of system
6. Conservative financial strategy	Capacity to cope with the unexpected; arsenal to withstand competitive attack
7. Vertical integration, including in-house manufacturing of components, peripherals, and systems	Economies of scale in fabrication of components; control over delivery, quality, and rates of technical change; no vulnerability to holdup by suppliers
8. Proprietary software (operating systems, compilers) and interfaces	High switching costs for users; preempt entry by producers of modules
9. Preemptive cannibalization	Users get system upgrades as technology changes; deters competitors

faced a significant problem when it wanted to upgrade or increase its capacity. Adding more computing power often meant buying another system, with all the attendant headaches of writing new software and setting up new machine operations.

By the late 1950s, therefore, all computer companies were devising ways for customers to migrate from one system to another without rewriting all their software. However, the binary compatibility of System/360 across a broad line was unprecedented, and it offered users very attractive options.[14] Customers could envision moving up the line without the major headaches of the past; moreover, because compatibility was a built-in feature, System/360 established the expectation (an implicit contract) that future system improvements in memory, processors, and peripherals

[14] System/360 conformed to a "write-once, compile-once, run everywhere" standard.

would be compatible with the present system. (Technically, this meant that the visible design rules of System/360 would be maintained for a very long time. They were.)

In addition to being compatible and upgradable, System/360 was also configurable. Customers could initially exclude modules and later add them to their systems; thus a particular installation could be optimized to the user's short-term needs. When those needs changed, the customer could again apply the exclusion and augmentation operators to its own system, terminating some leases and adding others. IBM's contract structure made these marginal changes in the installed machinery very easy and inexpensive.

In this fashion, IBM created a set of implicit agreements with its customers that had considerable "option value." In effect, the modular design of System/360, coupled with IBM's marketing and financial policies (salesforce incentives, leasing), made it cost-effective for customers to take advantage of new hardware technology as soon as it was introduced. This in turn had the effect of changing fundamentally the nature of the customer's purchase decision: rather than waiting (because something better might come along), System/360 gave customers incentives to incorporate new technology right away. [15]

All of these valuable options involved hardware, not software, however. As we described in the previous chapter, systems software development for System/360 had been a major expense and headache for IBM. But the company, almost unaware of what it was doing, turned that headache into a competitive weapon. The power of systems software lay in the fact that, by modular inversion, it became a layer of visible information interposed between programmers and the hardware of the machine. But the design of systems software was not modular—because so little was known, software designs contained many interdependencies.[16] Thus whereas hardware improvements could be rapidly and economically incorporated, operating system changes and software conversions were extremely difficult and expensive.

The complexity of systems software and its visibility to applications programmers had the effect of locking users into their existing installations. There was no compatibility or interoperability across different vendors' machines. What's more, computer operators and programmers could only work productively on the machines they

---

[15] Mason and Merton (1985); Farrell and Shapiro (1988); Beggs and Klemperer (1992); Dixit and Pindyck (1994).

[16] The first operating system to be modular was Unix, which we discuss in chapter 13. Even today, operating systems software generally is not modular by the criterion set forth in chapter 3.

knew. And of course the system with the largest number of programmers and operators was IBM's System/360.[17]

In summary, for System/360, IBM implemented a contract structure that was a direct descendant of its historical contractual design. At the core of the contract structure was a set of implicit and explicit understandings with users: that they would have options (to exclude, augment, and upgrade), low switching costs within the product line, and controlled cannibalization. This combination of promises on the part of IBM created great value for a very large class of users, as evidenced by the immediate, high, and sustained flow of orders for the new machines.

There was one hitch: once a user was committed to the System/360 family of machines, a complex instruction set and an even more complex operating system made it expensive to switch to another vendor's system. This was not a big problem for most users, however, because the breadth of the family, made possible by its modular design, made such switches unnecessary.

Much of what this contract structure did, therefore, was to tie customers happily, but tightly, to IBM. The rest of it served to attract a high-quality workforce, guard against opportunism by suppliers (through backward integration), and deter competition. Competitively, a broad line made it possible to drop prices on selected components in response to competitive entry while maintaining profits on the business as a whole. Proprietary and secret interfaces made it technically difficult and perhaps illegal to attach a non-IBM module to the system.[18] Selective product introductions were an effective deterrent to prospective entrants. Finally, a conservative financial strategy gave IBM the means to cope with unexpected technical developments or competitive moves.[19]

This contract structure—which we label "extensive control"—can be envisioned as a set of contracts that encircle the *entire task structure* of a modular system and its associated products (figure 8.1). In terms of the TSM of System/360 (see figure 7.2), IBM's top managers sought to contain all the modules and all the groups working on any part of the family within the boundaries of the existing corporation.

---

[17] This is a "quasi-network" economy of scale. See, for example, Katz and Shapiro (1986, 1992); Chou and Shy (1990); Church and Gandal (1992, 1993).

[18] These property rights were not well specified, hence the focus of fierce litigation.

[19] Bolton and Scharfstein (1990). The contract structure IBM's managers designed was a combination of many elements which by now are familiar in the economic literature. However, none of the concepts we now use to dissect the contract structure had been published in 1965, when IBM had to design the contract structure of System/360.

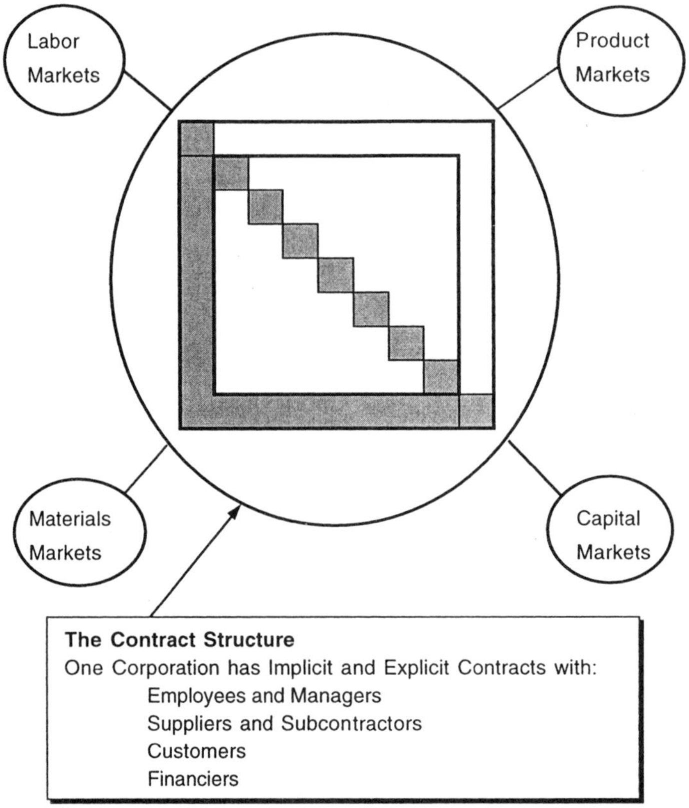

Figure 8.1 Extensive control of a modular design: one set of contracts encircles an entire modular task structure.

Many firms have succeeded in the computer industry by adopting an enterprise design similar to this one. That is, they have combined modular hardware with interdependent, proprietary systems software. They have offered extensive product lines and maintained proprietary interfaces; practiced selective pricing; and controlled the rate of new product introductions. Most have also maintained high levels of financial flexibility. Finally, partly to exercise control over their own system's architecture and key interfaces, and partly as an outlet for their high internal cash flow, these firms have tended to bring more parts of the system within their own boundaries over time.

The Financial Impact of System/360

From the day of announcement, the demand for System/360 far exceeded all forecasts. In 1960, IBM had revenues of $1.4 billion, a quarter of which came from its processors.[20] By 1970, IBM's revenues had grown to $7.5 billion, 85% of which was derived from the processors and peripherals that were part of System/360. In 1980, IBM's revenues reached $26.2 billion, most which was attributable to System/360's modular descendants.

IBM's revenue growth was only part of the story, however. To see how profitable IBM was, we have to look at its economic surplus. Figure 8.2 graphs this surplus[21] from 1960 through 1980 against comparable data for Control Data Corporation, one of IBM's more profitable direct competitors. During this time, IBM's internally generated cash flow went up almost 17 times. In 1970, a staggering 40% of IBM's revenue was a cash surplus available for interest, dividends, taxes, and reinvestment. (One way of calibrating IBM's success in this era is to note that in 1980, IBM's net cash flow after taxes was $6.6 billion: this was four times its *revenue* in 1961, when the SPREAD task group went to work.)

Control Data Corporation experienced good cash flow growth as a result of acquisitions during this period, but its pretax cash flow was less than a tenth of IBM's. It was also substantially less profitable: its cash flow as a percent of revenue hovered in the range of 6% to 18% after 1967, the year that System/360 began to ship in volume. (Before then, Control Data had pretax profit margins in the range of 20% to 25%.)

Another measure of IBM's success was its "return on assets employed," in other words, the surplus divided by the assets needed to generate the surplus. Figure 8.3 graphs IBM's and Control Data's respective return on assets employed for the period

---

[20] These figures were part of the analysis contained in the SPREAD report. (Haanstra et al. 1983.)

[21] IBM's accounting practices tended to push earnings out into the future, and thus obscured the true magnitude and the changes in the company's cash profits from year to year. (For a detailed discussion of these accounting issues, see Fruhan 1979, pp. 180–201.) The best measure of cash flow for our purposes is earnings before interest, taxes, depreciation, and amortization (EBITDA) minus cash taxes paid. This number closely approximates the amount of cash generated by IBM's business in a given year that was available to compensate capital providers and finance growth.

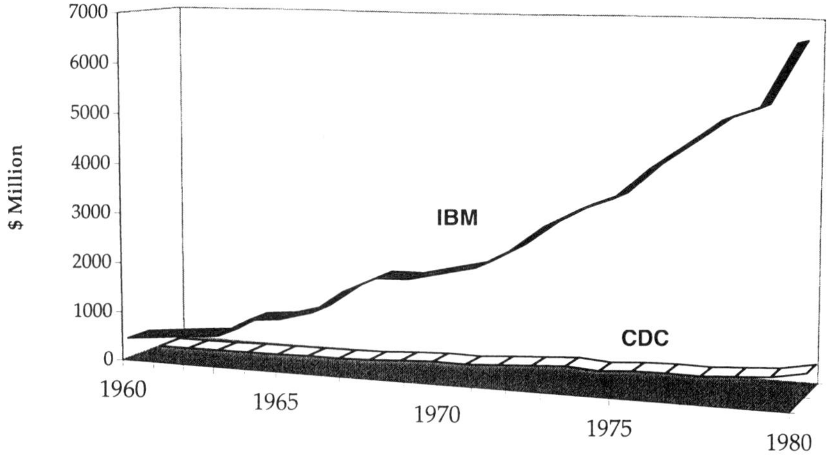

Figure 8.2 Economic surpluses of IBM and Control Data Corporation: 1960–1980.

1960–1980.[22] Figure 8.3 again shows the financial superiority IBM maintained over its nearest competitors. The advantage is even more pronounced when we consider that the relevant cost of capital for companies like IBM and Control Data was in the range of 10% to 12% at the beginning of this time period, and then increased during the 1970s as inflation took its toll. IBM's return on assets employed exceeded the cost of capital easily, whereas Control Data's returns were marginal throughout the whole decade of the 1970s.

The value inherent in the cash flow, growth potential, and return on assets employed of System/360 was not lost on investors. As we indicated in chapter 4, capital

[22] Doing this calculation posed some problems of measurement. One easily solved problem was that IBM's and Control Data's cash balances were not an intrinsic part of running their businesses. Therefore, we excluded cash balances from the asset base when calculating the return on assets employed. However, deeper problems reside in some of the other asset accounts. Most problematic was the fact that throughout the 1960s and 1970s, IBM overdepreciated the machines it put out on lease. Its depreciation charges were consistently too high; therefore its reported assets were too low. To address this problem, we added back to IBM's total assets a number equal to one-half its reserve for accumulated depreciation on rental equipment. We made this adjustment for rental equipment only, because it was here that IBM had the greatest amount of latitude: if it said that a given piece of equipment had an expected life of four years, how could its outside accountants disagree? With respect to other asset categories, IBM was bound by conventional accounting practice: it could err on the side of conservatism, but it could not make up its own rules.

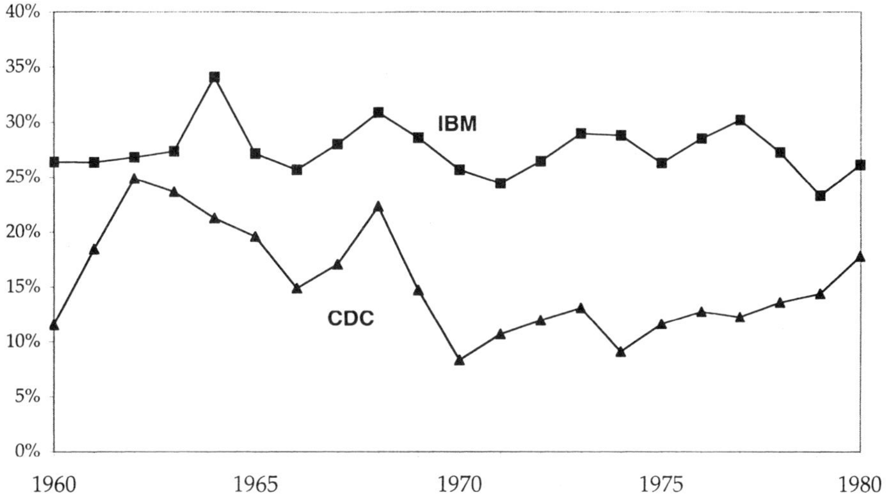

Figure 8.3 Return on assets employed by IBM and Control Data Corporation: 1960–1980.

market valuation technologies are forward-looking, hence market value changes are expected to occur in advance of actual cash flows. Accordingly, it is interesting to see when expectations about System/360's success became "embedded" in IBM's stock price.

Figure 8.4 charts the value of $1 invested in IBM vs. a value-weighted market portfolio and a portfolio made up of seven principal competitors from 1960 through the end of 1969. Significantly, the great fanfare of System/360's announcement in April 1964, and the initial flow of orders, caused barely a blip in IBM's stock. However, as IBM worked through its implementation problems, and began to deliver functioning machines in large volumes, its stock price began to climb.

Essentially all of the change in relative value between IBM and these benchmark portfolios occurred in the years 1966–1967. (The differences in performance between IBM, the market, and its competitors are not significant prior to year-end 1966.) This fact is consistent with the view that investors anticipated high cash flow from System/360, but had to be sure that IBM would resolve its operating problems before they could regard the new system as a definite financial success. By the end of the decade, however, IBM's aggregate market value was hovering around $40 billion (equivalent to about $170 billion in 1996 dollars).[23]

[23] To put this number in perspective, at the start of 1970 the next highest-valued firm after IBM was AT&T, which had a market value of equity of $26 billion ($113 billion in 1996

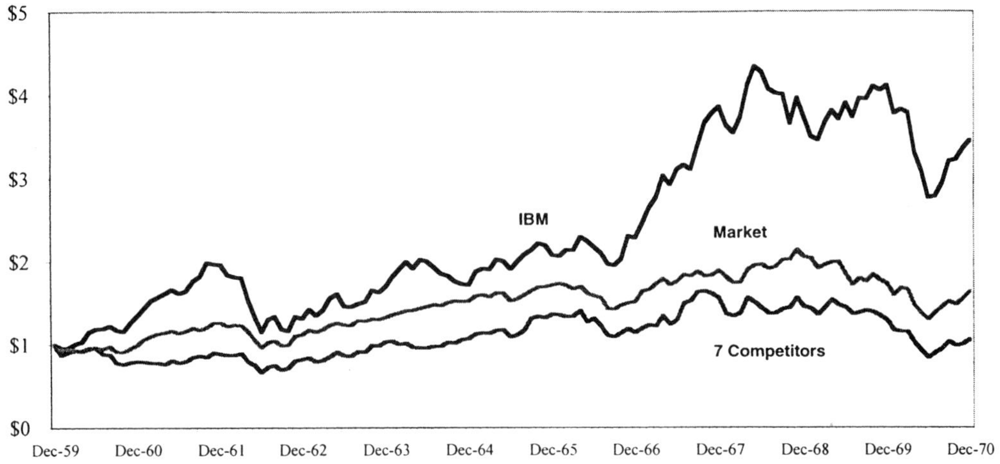

Figure 8.4 The value of $1 invested in IBM, a value-weighted market portfolio, and IBM's seven principal competitors (adjusted for inflation) 1 January 1960–31 December 1970.

The Competitive Impact of System/360

From a competitive standpoint, System/360 established a massive sunk-cost barrier to entry in the market for whole computer systems. Firms seeking to compete head-on with IBM had to contemplate cash expenditures of over $500 million to achieve breakeven scale and to approach comparable breadth in their product lines. (See box 8.1.) Such investments would be directed toward technical objectives that were difficult to achieve; hence there was no guarantee that the money, if spent, would result in technically adequate products.

The magnitude of the necessary investment was driven by two factors: (1) the scope of the System/360 product line, and (2) the need to achieve binary compatibility throughout the line. Even without the compatibility constraint, a project to design and manufacture a set of five processors and the necessary peripheral devices (including printers, terminals, and disk drives) was a massive undertaking. The compati-

dollars). In early 1996, the highest-valued U.S. company according to *Business Week* was General Electric with a market value of $125 billion. Intel Corporation, the then-highest-valued company in the computer industry, was worth $108 billion at the end of 1996. In other words, in absolute terms and relative to its cohort, no company in the mid-1990s was as highly valued as IBM was in the late 1960s.

Box 8.1 The Cost of Competing with System/360

William E. Fruhan, Jr., drawing on court documents submitted in the U.S. v. IBM antitrust case, reported that in 1969, General Electric estimated that it would cost $685 million over five years to develop a broad line of computers capable of achieving a 10% share of the mainframe computer market.[1] The 10% threshold was believed to be a minimum level of sales necessary to achieve breakeven on the indivisible investments in software, product design, and marketing needed to support a broad line of general-purpose computers that could compete with IBM's. The approximately $100 million per year needed amounted to "essentially all" of GE's discretionary cash flow at the time, and would still leave GE a distant number two in the computer industry. GE sold its computer business to Honeywell Corporation soon after completing this analysis.

[1] Fruhan (1979, table 6.3, p. 154).

bility constraint then imposed another degree of purely technical complexity and risk on the effort.

IBM's investment in compatibility thus fits John Sutton's definition of a sunk-cost investment: it was big, difficult, uncertain, and very valuable to customers.[24] The sunk-cost barriers to entry in computer systems were high largely because, at great expense and some risk, IBM had succeeded in achieving binary compatibility throughout a very broad product line. And it was the *principles of modularity*—design rules, independent test blocks, clean interfaces, nested hierarchies, and the separation of hidden and visible information—applied rigorously to both product and process designs, that enabled IBM to achieve that goal. Without the means to divide knowledge and effort among workgroups, while still maintaining necessary levels of coordination and control across them, a broad, compatible product line such as System/360 could not have been created.

Changes in the Value Landscape of Enterprise Designs

In the late 1960s, in the wake of the introduction and rollout of System/360, IBM appeared close to invincible. Its historic competitors were losing money, or at best

[24] Sutton (1991, pp. 2–31).

limping along. Some, like RCA and GE, left the industry entirely; others refocused their businesses and sought to exploit various niches.[25]

However, IBM's success had elements of a Faustian bargain. The modularity of System/360 allowed IBM to introduce a compellingly superior product. However, the modularity of System/360 also weakened substantially the forces that previously had kept whole computer systems within the boundaries of one firm. Figuratively, in the value landscape, while one set of peaks sank "into the sea," along another dimension a new set of peaks broke through sea level. (We are stating this quite loosely; the argument that there were new sources of value inherent in the modular design is made in Chapters 10 through 13.) The new peaks of value corresponded to enterprise designs in which different modules of one system—like System/360—would be provided by different corporations.

These changes occurred in the "space of enterprise designs," which is a space of perceived possibilities. As a result, there was no fanfare surrounding this event—in fact, there was no obvious signal that the underlying structure of the economic system had changed in any way. Nevertheless, if asked to date the emergence of the new peaks in the landscape, we would say they arose as the last "bugs"—interdependencies—were eliminated from System/360's task structure, that is, in early 1967. It was at this point that the operation of splitting the complex task structure was complete, and the resulting design became "truly modular."

This new modular task structure in turn carried within it the seeds of new enterprise designs. Although IBM chose the "one corporation, extensive control" path to capture value, hidden-module designers who knew System/360's design rules could see quite clearly that *they* might capture value under a different "many corporation" contract structure. Moreover, because the overall design was modular, they could act independently of IBM's top management or System/360's central architects.

It did not take System/360's hidden-module designers very long to see these possibilities and act on them. In December 1967, less than a year after the new task structure became a reality, twelve employees left IBM's San Jose laboratory to form a company whose specific aim was to design, make, and sell disk and tape drives that would "plug into" System/360 and therefore compete with products that IBM offered.[26] This was the first of what became a steady stream of defections from IBM's

---

[25] Fishman (1981); DeLamarter (1986); Dorfman (1987).

[26] Pugh et al. (1991, pp. 489–553). There appear to have been isolated defections earlier in 1967.

engineering ranks. The defecting individuals usually joined or founded new corporations that made modules compatible with System/360.

These designers, working independently of one another within System/360's modular framework, set in motion the decentralized process of design evolution. And as the designs of computers changed, the industry began to change in structure as well: it began to evolve into its present form—a highly dispersed *modular cluster* of firms. The nature, the causes, and the consequences of design evolution and the origins of the modular cluster are the central topics of the rest of this book.

Recapitulation: The Role of System/360 in the Evolution of Computer Designs

System/360 was the first modular family of computers to be designed, produced, and offered for sale. A rough idea of the sheer scope of the family can be gleaned from figure 8.5, which shows the principal modules in the family as of 1970. At this time, the product line contained eleven processors, thirty to forty processor-memory configurations, fifteen to twenty emulators, and thirty eight different peripheral devices and an uncounted number of software programs. All of these pieces of equipment and software were compatible, and would work together in many configurations.[27]

We can imagine standing in 1970 and looking backward at the changes in computer designs over the previous quarter century. We would see a gradual splitting apart of the designs year by year. At the time of ENIAC, all tasks of design, production, and use were heavily interconnected. Around 1950, with the construction of EDSAC and the Institute of Advanced Studies machine, programming became separate from design and production. Later, with the institution of the SMS system at IBM, another split took place, this time between the tasks of production and design. But the tasks of design themselves remained highly interconnected.

Nonetheless, throughout this period, designers were accumulating detailed knowledge about the artifact, organizing it, and dividing it up. Dense "knots of structure" and specialized knowledge emerged around the categories defined by von Neumann's functional breakdown.

System/360 was then a radical reconceptualization of a computer's design, based on this accumulated knowledge. The SPREAD task group reasoned that, if a particu-

[27] Our count is based on an overview of the line, compiled by Bell and Newell (1971, pp. 561–587). Our figures are very conservative—the total number of products offered by IBM at this time numbered in the thousands.

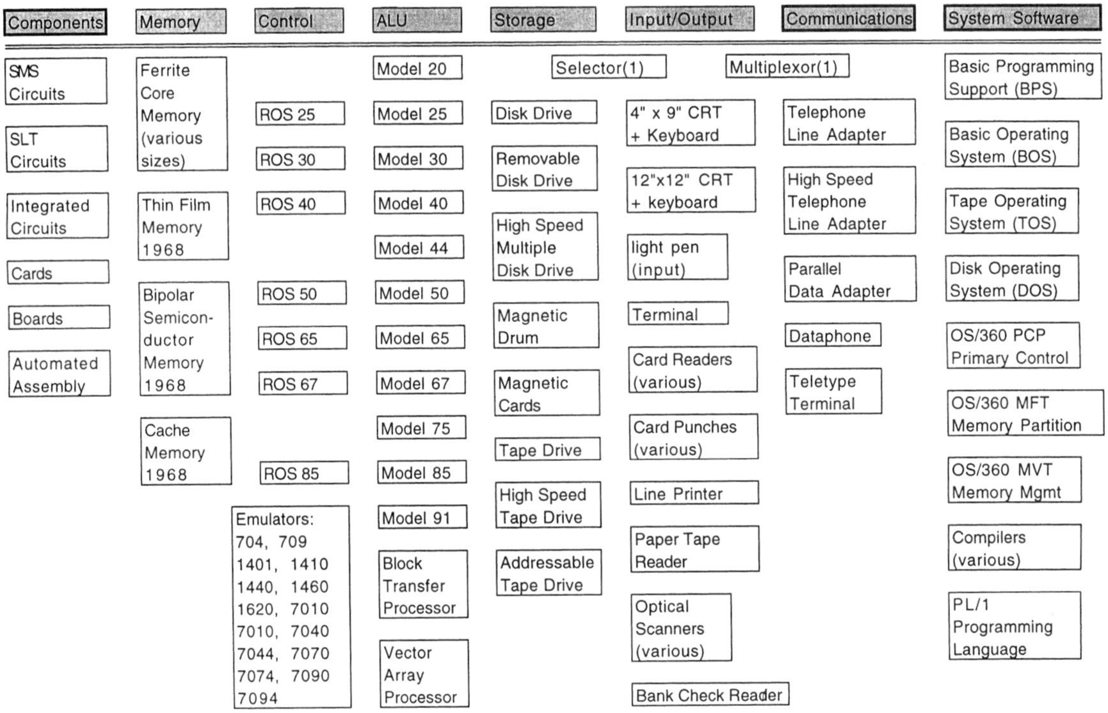

Figure 8.5 The modules of System/360, circa 1970. (1) The selector and multiplexor were special processors that managed the activities of storage, input/output and communication devices. Module categories with dark outlines represent augmentations of the original von Neumann design. Not shown: applications software. (Compiled from Bell and Newell 1971, pp. 561–583 [hardware]; Pugh et al. 1991, pp. 291–345 [software]).

lar set of "privileged" design parameters were chosen first, and communicated broadly, then teams of designers working independently could create a system of interoperable, compatible devices and software. They called the critical set of parameters the "architecture" of the system. (We call this critical body of information the "design rules," which are in turn made up of the "architecture," "interfaces," and "tests"; see box 8.2.)

New design concepts usually do not work perfectly on the first pass, and System/360's modular design was no exception. In fact, an enormous effort was expended within IBM in the years 1965 and 1966 to integrate and debug the system. Many points of interdependency, which had not been fully anticipated by the original designers, were eliminated. But through this process, IBM's designers, engineers, and

Box 8.2 The Changing Concept of Computer Architecture

The concept of architecture is fundamental to all modular systems, not just computers. The word is both an evocative term with a rich set of associations, and a technical term, meant to have a precise meaning in a particular context. However, the meaning of the term "computer architecture" has changed in interesting ways over time.

System/360 was the first computer system to have an "architecture." System/360's architects, Gene Amdahl, Gerrit Blaauw, and Fred Brooks, defined the term in the following way:

The term architecture [means] the attributes of a system as seen by the programmer, i.e., the conceptual structure and functional behavior, as distinct from the organization of the data flow and controls, the logical design and the physical implementation.[1]

Thus for Amdahl, Blaauw, and Brooks, architecture meant "information visible to programmers." This definition envisions a two-level design hierarchy with one body of visible information. Although in its entirety, System/360 was more complicated than this definition allowed (see the design hierarchy diagram in Figure 7.3), this definition of architecture was still a useful concept to take into the design process. Designers could use it to divide the vast number of design parameters that had to be decided into two sets. If a programmer needed to know something (like an instruction or an address format), then that design parameter was part of the "architecture" and needed to be addressed by the CPC.

However, the Amdahl-Blaauw-Brooks definition relied on a concept of "programmer" that was already becoming obsolete. In 1971, when Bell and Newell wrote their book, *Computer Structures* (considered to be the first textbook in the modern field of computer architecture), it was no longer possible to equate "all visible information" with "everything a machine-language programmer needs to know." Multilevel design hierarchies and different apexes of visible information were conceivable, and the concept of architecture needed to expand to encompass these new possibilities. Thus the initial definition of architecture, which served the System/360 designers well, faded in the face of the growing complexity of the artifacts.

---

[1] Amdahl, Blaauw, and Brooks (1964); Brooks (1975).

managers managed to create the first large computer system that was "truly modular" in both product and process designs.

Thus System/360 was not only a significant new artifact, it was also the proof-of-concept for a new approach to designing computers. For the first time in history, a computer system did not have to be created by a close-knit team of designers. Instead architects could designate a set of design rules, and use them as a substitute for ongoing communication and coordination. The rest of the design parameters then became hidden from other parts of the system. Module designers were free to experiment with hidden parameters, confident that they would not trigger unforeseen conflicts in other parts of the system.[28]

A landmark design teaches designers to see the space of design possibilities in new ways. System/360 was such a design. First, it showed that it was possible to specify a modular design with parallel work flow for a large, high-performance computer system. Second, it was an object lesson in how to partition information, set design rules, enforce information hiding, and integrate a system after the fact. Finally, it demonstrated that a modular design approach could be used to create a family of compatible processors and peripheral equipment, which could change over time in response to new technical developments.

The idea of a design with a built-in tolerance for change was a new concept in the computer industry in the mid-1960s. Designers of later systems would adopt and refine the modular, architectural approach pioneered by Amdahl, Blaauw, and Brooks—so much so that, according to John Hennessy and David Patterson, modern authorities on computer architecture, the core ideas "seem obvious to us today."[29]

In addition to being a technical achievement, a system with a market value of $40 billion (equivalent to $170 billion in 1996 dollars) and a built-in tolerance for change was a new thing in the marketplace. A small sliver of System/360 was worth as much as all the rest of the computer industry combined. But the individual parts of System/360 were not especially distinguished.[30] Moreover, the system's design rules were known to a large number of people within IBM, and they were likely to be in place for a very long time.

---

[28] In reality there are always unforeseen interactions. The acid test of modularity is, Are such interactions small enough that they can be fixed locally and unilaterally? This was the case by the end of the System/360 development effort—at least in hardware.

[29] Hennessy and Patterson (1990, p. 128).

[30] Bell and Newell (1971, pp. 561–587).

Modular structures like that of System/360 unlock subsystem designs and release them from interdependencies with the larger system. A hidden-module innovator can then proceed rapidly and unilaterally as long as he or she adheres to the system's design rules. If the innovation works, it can easily be incorporated into the pre-existing modular framework. This in turn creates a whole new set of possibilities for creating value in a large complex system.

The theory we are developing in this book states that modularity in design gives rise to valuable, widely dispersed *design options,* which are spanned by the six modular operators. These options in turn make possible a decentralized, value-seeking process, which we call *design evolution.* We describe the connections between modularity, option values, and evolutionary dynamics in detail in the next chapter.

Then, in chapters 10 through 13, we use financial option valuation techniques to understand the potential market value of each of the modular operators. Following that, in chapters 14 and 15, we show how in the 1970s, entry by new corporations operating in new markets led to the emergence of a modular cluster of firms in the computer industry. Volume 1 ends at that point.

Part III   Design Evolution via Modular Operators

9 Design Options and Design Evolution

Before 1965, as we have said, computers were always whole-system creations. Designs were interdependent, and thus developing a new computer meant creating a new instruction set, peripherals, and applications software from scratch. The new models would reflect what had been learned from previous projects, but each new product line was incompatible, separate, and distinct.

After 1965, the situation was quite different. System/360 proved that, using the principles of modularity (design rules, nested hierarchies, hidden and visible information), it was possible to create a large family of compatible machines. Thereafter, new IBM products were not whole-system creations, but were almost always members of modular families. Further, other companies, like Digital Equipment Corporation (DEC), soon seized on the principles of modularity and used them to build their own compatible and flexible families of products.[1]

What happened to computers and the computer industry after 1965 underscores the power of modularity. In a modular design, adherence to design rules takes the place of direct, ongoing coordination of the groups working on different parts of the system. After a modularization, then, within the limits set by the design rules, module designers can vary designs and seek out more advantageous forms. Within the overall system, the modules are "free to evolve."

In these volumes, moreover, we are building a theory that deals with two sets of evolving entities. On the one hand, the designs of computers have changed—this evolution is apparent in the artifacts themselves. On the other hand, the evolving

---

[1] Designers at other companies were greatly helped by the publication, in 1971, of Gordon Bell and Allen Newell's pathbreaking *Computer Structures,* which explained how to design a group of discrete, yet compatible machines. This book culminated in an analysis of the architecture of System/360.

Box 9.1 The Meaning of "Evolution"

"Evolution" has different meanings in different fields. In common speech it may mean simply any process of incremental change ("the evolution of my thinking. . ."). In economics, evolution may mean "historical patterns of entry and exit," as in "the evolution of an industry."[1] Or it may mean "a process of variation-selection-retention."[2]

Our starting point in defining evolution, however, is the formal mathematical theory of complex adaptive systems, developed by John Holland and his students.[3] According to modern thinking, biological evolution, immune system development, neural path development, language acquisition, and distributed processing and genetic algorithms are subsets of this class of systems, as are missile guidance systems, pattern-recognition algorithms, and games like checkers and chess.

It is useful to divide the large set of all complex adaptive systems into two categories: (1) systems in which the so-called adaptive plan is in the hands of a few agents; and (2) systems in which the adaptive plan is decentralized to many independent agents.

Single-person searches and chess belong in the first category; biological evolution, immune system development, and so on belong in the second. Decentralized complex adaptive systems play host to evolutionary processes, or "evolution" for short.

[1] See, for example, Klepper (1992); Hannan and Freeman (1989); Sutton (1991); Utterback (1994).

[2] See, for example, Nelson and Winter (1982); Burgelman (1994); Tushman and O'Reilly (1996).

[3] Holland (1976).

designs have not remained "within the walls" of one or a few big firms. New firms and new markets have emerged in parallel with new products and new product categories. In 1996, 90% of the firms and 70% of the market value of all computer companies were in industry subclassifications that, practically speaking, did not exist in 1966.

Evolution is a powerful concept to use in understanding these changes. In our language system, evolution is distinguished from other kinds of adaptive processes in that it is decentralized (see box 9.1). In both biological and economic evolution, there is no single entity or group that determines the trajectory of the entire process. Instead, there are many paths of adaptation, and many competing and complementary routes of exploration.[2]

To understand how evolution has worked in the computer industry and to clarify the forces that drive it, we now need a formal theoretical model of adaptation that is applicable to artifacts and their designs, and (subsequently) to firms, markets, and

[2] Holland (1976).

industries. In this chapter we lay the groundwork for that model, building on the language and concepts introduced in previous chapters.

We first describe "design evolution" as a value-seeking process, and compare it with formal theories of biological evolution.[3] Although there are strong parallels between the two theories, the fundamental agents of design evolution are not blind replicators (like genes), but instead are human beings capable of seeing and seeking value in new designs. [4]

We then consider how value is affected by the modularization of a complex design. Modularization permits individuals (or firms) to mix and match alternative designs of the modules of a system. The "rights" to mix and match are options with quantifiable value in the greater economic system. A modularization multiplies design options and at the same time disperses them so that they can be "picked up" by many people, without the permission of any central architect or planner. The pursuit of valuable options by many decentralized actors in turn accelerates the rate of change of the system as a whole.

In fact, a veritable explosion of options occurred after the introduction of System/360. Explaining why this happened, and how the new designs changed the industry's structure between 1965 and 1980, will occupy us through the rest of this volume. Tracing out the evolutionary dynamics of designs and competition in the newly formed "modular cluster" after 1980 will then be the central theme of volume 2.

Design Evolution as a Value-Seeking Process

Metaphorically, one can picture evolution as a process in which *many entities* (designers or genes) are seeking something (value or fitness) in a complex environment. The sources of variation and the mechanisms of selection determine the process by which the seeking takes place. Thus, to build a theory of evolution, one needs four things:

1. units that are selected
2. a selection criterion (a concept of value or fitness)

[3] For expositions of the formal theory of biological evolution, based on recombinant molecular operators and genetic inheritance, see, for example, Holland (1976); Williams (1966); Dawkins (1989, 1996); Edelman (1992); Kauffman (1993); Plotkin (1993).

[4] The fact that evolution can occur in systems where agents have foresight was recognized by Alchian (1950); Simon (1969); and especially by Nelson and Winter (1982). See our discussion of evolutionary processes at the end of this chapter.

3. one or more sources of variation
4. one or more mechanisms of selection

Table 9.1 compares these four elements in different evolutionary settings. We examine each of the elements in turn below.

Units of Selection

A *unit of selection* in an evolutionary process is an entity generated by the process that is indivisible under selection. In other words, units of selection are rated and selected as wholes. By definition, the selection mechanisms of the process cannot "look inside" a unit of selection and pick out and preserve its best components. In evolutionary biology, the units of selection are the combined genotypes and phenotypes of organisms.[5] In economics, the units of selection are, at one level, artifacts, designs, and task structures, and at another level, products, enterprises, and management teams.

The unit of selection greatly influences the outcomes of an evolutionary process. This is because poorly designed subsystems can be propagated if the unit of selection as a whole is successful. Thus in biology, the overall quality ("fitness") of an organism can compensate for the poor quality of some of its parts. The same is true of artifacts. Poorly designed subsystems, like the file-naming protocols of Windows, may survive if they are lodged in a whole system that is "good enough" in other dimensions. Conversely, an excellent subsystem design, like the Macintosh user interface, may disappear if it is embedded in a whole system that loses market acceptance. Similarly, inefficient departments may survive within efficient enterprises, and efficient departments may be disbanded if the enterprise as a whole goes bankrupt.

But with artifacts and organizations, in contrast to organisms, a human designer, armed with appropriate knowledge, can change the unit of selection. Modularization—the conscious splitting apart of a design into independent subunits—has this effect. After a modularization, a user can mix and match modules and construct different combinations. So too can a designer think of preserving the good parts of a modular design, while improving the bad. After modularization, the system as a

---

[5] In biology there is a debate as to whether the true "units of selection" are genes or organisms. The issue hinges on the definition given to the word "selection"—does it mean "passing the test of the environment by surviving and having offspring" or "being replicated in the next generation"? In biology, organisms (phenotypes) are tested, while genes (subsets of genotypes) are replicated. See Plotkin (1993) for a discussion of this issue.

Table 9.1 Basic elements of different theories of evolution

Basic elements of the theory	Genetic evolution	Design evolution	Industry evolution
Units selected	Genes and organisms	Artifacts, designs, and task structures	Enterprises and managements
Selection criterion	Fitness: replication or reproduction	Capital market value	Capital market value
Source of variation	Genetic operators[a]: Crossing over Inversion Mutation Dominance Segregation Translocation Duplication	Modular operators: Splitting Substituting Augmenting Excluding Inverting Porting	Industry operators[b]: Entry Exit Acquisition Divestiture Hiring Firing
Selection mechanisms	Survival and reproduction	Realization, purchase, and use	Survival; selection for inclusion in a portfolio

[a]This list of genetic operators was compiled from Holland (1976).
[b]Although thcrc arc many cxtant thcorics of industry cvolution (cf. Nclson 1995 and citations in the text), to our knowledge, none has been formalized in terms of operators comparable to the genetic or modular operators shown here. This is our candidate list of actions that can be taken with respect to enterprises and their managements.

whole is no longer the only unit of selection: some of the selective action shifts to modules at a lower level of the design hierarchy.

For example, figure 9.1 shows a generic three-level, four-module design hierarchy. At the lowest level are what we have called the *hidden modules* of the system. These units "look to" visible information conveyed in the form of design rules. But they also incorporate hidden-design parameters, which may be chosen independently of other modules as long as the design rules are obeyed. At the next level are units we have called *architectural modules.* These convey visible information to some, but not all hidden modules. Finally, at the highest level of the design hierarchy are the *global design rules.* These rules establish the architecture, interfaces, and test protocols for all modules, hence are visible (directly or indirectly) to all modules in the system.

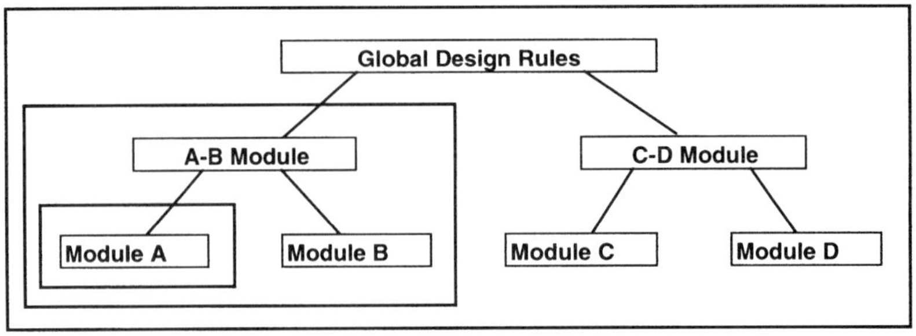

Figure 9.1 A three-level design hierarchy has three levels of selection. The hidden modules—A,B,C,D—are the first and lowest level of selection. The architectural subsystems are the second (intermediate) level of selection. The whole system is the third and highest level of selection.

In this system, each of the hidden modules is a unit of selection; a group consisting of an architectural module and the modules that "look up to it" is a unit of selection; and the whole system is a unit of selection. Designers can work on any level. They can strive to create better hidden modules, to improve architectural modules, and to create whole new systems.

The capacity to be modularized—split apart so that separate units can be worked on independently while remaining part of the whole—is to our knowledge a property unique to manmade designs. In biological evolution, there is no genetic operator that splits organisms and allows natural selection to operate on organs.[6] In contrast, as we saw in the case of VLSI chips in chapter 3 and in the case of System/360 in chapter 7, complex manmade designs (and their corresponding task structures) *can* be split apart via modularization, as long as the architects know enough about the underlying design structure to transform design parameter interdependencies into appropriate design rules.

A Selection Criterion (a Concept of Value or Fitness)

In biology, the quality of an organism is measured using a metric called "fitness." In economics, the concept analogous to fitness is "value," or, more precisely, "value

[6] The process of sexual reproduction can be viewed as a special, highly constrained type of modularization. Mixing and matching of alleles within the framework of an organism's design does occur when genes recombine in sexual reproduction. But how alleles (the hidden mod-

added." In chapter 4 we described at some length the different kinds of value that coexist in an advanced market economy. These included expected value and realized value, personal value and market value, and product market value and capital market value. We also discussed "guidance technologies" that could "point" people and enterprises toward value, or bring different concepts of value into agreement.

In this book we assume that value established in the capital markets is the primary selection criterion affecting the evolution of computer designs. As we discussed in Chapter 4, the capital market value of a design is the sum of the willingness of every member of society to pay for the artifact less the cost of supplying it. Future and risky values can be discounted to the present using appropriate prices of time and risk.[7] We recognize that other kinds of value (e.g., personal value) exist and are important. However, to an overwhelming degree, changes in the computer industry over the last fifty years have been (and continue to be) driven by values set in the capital markets, and so we will adopt that criterion here.

Sources of Variation

The wonder of evolution, in both biology and economics, lies in the fact that complex structures, which are both robust and beautifully specialized, can emerge from the repeated application of a small number of simple operators. Charles Darwin expressed his sense of wonder at this process in his famous conclusion to *On the Origin of Species:*

There is grandeur in this view of life . . . that, whilst this planet has gone cycling on according to the fixed laws of gravity . . . endless forms most beautiful and most wonderful have been, and are being, evolved.[8]

In evolutionary processes, new forms—that is, new designs—are generated by modifying old ones in structured ways. In biology, for example, the agents of change

---

ules) arise and what determines the "breakpoints" of the chromosomes (the architecture) is not yet known. Some theorists, e.g., Kauffman (1993, 1995), have argued that natural selection will favor designs that are "evolvable," but at this point such statements are only conjectures.

[7] Here we are relying on the capital markets to provide a valuation technology for measuring the worth of designs. The value captured by the creators and owners of designs and reflected in their capital market values (their stock prices, for example) will depend on definitions of property rights in the greater economic system. Thus the value of an enterprise may be far less than the value of the designs that it creates.

[8] Darwin (1859).

are genes, and new organisms are generated by a process of genetic recombination. Over the years, genetic theorists have identified a number of simple changes that perform the work of recombination: these are the so-called genetic operators listed in table 9.1.[9]

Similarly, design evolution occurs when designers make local value-seeking changes to a design. If the design is modular, the admissible "moves" are the six operators described in Chapter 5:

1. *splitting* an interconnected task structure into modules;
2. *substituting* one version of a module (or set of modules) for another;
3. *augmenting* the design by adding a module with new functions at a particular interface;
4. *excluding* a module from a system;
5. *inverting* a recurring design element into an architectural module;
6. *porting* a module to a new system.

(*Note:* splitting and substitution can be applied to interdependent design structures, whereas the other operators can be applied only to modular structures.)

As in genetic evolution, the basic role of the operators is to create new structures on which the forces of selection can act. Thus the operators are the fundamental source of variation in modular designs.

Mechanisms of Selection

In an evolutionary process, a *mechanism of selection* is something that chooses units of selection for transmission to the next generation or round of the process. For organisms, survival and reproduction are the mechanisms of selection.[10] For designs, the mechanisms of selection are (1) to be made into a real artifact; and (2) to be

---

[9] As we said in Chapter 5 (see Box 5.1), the specific genetic operators that biologists believed were worth investigating were those that reflected the *physical structure* of genetic material. Chromosomes, which store and transfer genetic material, are known to be linear strings of alleles (specific instances of genes), hence the genetic operators are all basically ways of cutting and pasting pairs of strings. More generally, the structure of an object is an important source of insight as to its probable behavior, its ontogeny, and the forces acting on it. See Thompson (1942); Edelman (1992); and Kauffman (1993) on this point.

[10] For genes, survival and replication are the mechanisms of selection: successful genes must do both.

Box 9.2 Coevolution Makes Variation Responsive to Selection and Vice Versa

Evolutionary biologists point out that, over many generations, genetic variation and selection are not strictly separated. With no conscious foresight whatsoever, genes do in fact "learn" about their environment, and the environment also "learns" about the genes. Given a long enough time span, genes will adapt their patterns of variation to their environment's patterns of selection. Those patterns of selection, in turn, are often patterns of variation generated by other sets of genes. This interactive adaptation of selected units and selective mechanisms is called "coevolution."

Coevolution is often the result of competitive interaction between different sets of genes. For example, in his theory of how the capacity to learn may have evolved via natural selection, W. Ross Ashby uses the example of a kitten learning to hunt. The kitten "selects" actions to catch a mouse. As she learns, her selections get better; her actions become more sophisticated and coordinated. But the mouse (or mice generally) may simultaneously develop more sophisticated evasive maneuvers. In this fashion, "[An organism's gene-pattern] does not specify in detail how a kitten shall catch a mouse, but provides a learning mechanism and a tendency to play, so that it is the mouse which teaches the kitten the finer points of how to catch mice."[1]

For this kind of coevolution to occur, the critical requirement is that the kitten's and the mouse's "rules" for selecting actions themselves be subject to selection, hence endogenous to the adaptive process.

---

[1] W. Ross Ashby (1960). Ashby's work had a profound effect on both Herbert Simon's and Christopher Alexander's theories of design (see Box 3.1). We thank our colleague Charles Christenson for locating this quote, which appears in the second edition of *Design for a Brain,* but not the first.

purchased by a user. To succeed in the economic system, a design must pass over both these hurdles.

In biological evolution, variation and selection are strictly separated. Genes lack foresight: they cannot choose their strategy based on what the environment is likely do. (Genes and their environment do "coevolve"; see box 9.2.) The separation of variation and selection is one of the fundamental differences between "natural selection" and "design."[11]

In contrast to biological processes, in human design processes there is no strict separation of variation and selection. Human designers working with foresight and knowledge can decide both what to try (variation), and what to select. Moreover,

---

[11] Dawkins (1989, 1996); Maynard-Smith (1989).

multiple rounds of variation and selection can go on in the mind of a single individual with no outward manifestation (other than perhaps a wrinkled brow).

Indeed, human designers can do even more: Using their capacity for imaginary manipulation,[12] they can develop models of their design processes, and imagine changes in them, and then improve those processes. Hence, not only are variation and selection not separate in human design processes, the relationship between variation and selection itself may be endogenous, adaptive, and the result of conscious design.

Therefore, patterns of variation and selection in design processes are not "hard-wired": they can and do vary across contexts and over time. It is as if the kitten mentioned in box 9.2, besides being able to learn how to hunt by trial and error (a variation and selection process within the kitten's mind), from time to time could reflect on the activity of hunting, and then change and improve it.

Recursive Design Processes

In Chapters 2 and 5 we said that human design processes are *recursive.* By this we meant that human designers, as part of their higher-order consciousness, have the ability to operate on structures at many levels of a complex system.[13] Using their powers of conscious foresight and other mental capacities, human designers can partition complex problems and create abstractions that simplify the relationships among parts. They can also "move outside" of a process they are part of, construct a mental representation of the process, and then consider how to make the process better.

What does recursiveness have to do with modularity? Simply this: the modularization of a design is essentially a recursive operation—it is a change in the design of a process that changes designs. As we have seen, designers may choose to modularize designs and their corresponding task structures to make the design process more efficient and flexible. Efficiency and flexibility are valuable properties for any process to have, but designers who do not operate recursively—for example, beavers—would not necessarily hit on modularity as a strategy for improving their own processes.

To see this more clearly, recall the laptop computer example of chapter 3 (see figure 3.1). The possibility of a four-block modularization is suggested by the DSM, which is a view from "outside" the system. However, in the absence of that view, the probability of eliminating all the out-of-block **x**s, and thus arriving at a modular

[12] Edelman (1992); Bucciarelli (1994).

[13] Edelman (1992).

structure (see figure 3.4) via a series of "random, directionless changes" is essentially zero.

Human designers can and do operate recursively in their imaginations. Thus they can envision hierarchical systems, nested structures, and modular designs, even when those things do not yet exist. As a result of these powers of imagination, the idea of making a process more efficient and flexible via a modularization is a strategy that human designers can conceive of, and one that they sometimes select.

This fact in turn presents us with a challenge in modeling design evolution as an economic phenomenon. There is a longstanding debate in economics about the relative importance of human foresight vs. "blind" variation and "natural" selection. Alchian (1950) opened up the debate, describing two extreme modeling approaches:

[One can] start with complete uncertainty and nonmotivation and then . . . add elements of foresight and motivation in the process of building the model. The opposite approach . . . starts with certainty and unique motivation, [and] must abandon its basic principles as soon as uncertainty and mixed motivations are recognized.[14]

The "evolutionary strand" of economics theory often emphasizes the "blindness" or "myopia" of the actors who make economic choices.[15] The emphasis on bounded rationality and blindness has arisen partly in reaction to neoclassical and game theoretic models, which often assume the actors have unrealistically precise knowledge of the structure of the "games" they play.

However, the emphasis on bounded rationality and blindness in evolutionary economics has had the unintended consequence of limiting investigations into "smarter ways" of dealing with evolutionary dynamics. Modularity is one of these "smarter strategies"—it works (as we will show), but to select this strategy, designers have to have a certain amount of insight and command over the design process itself.

Our approach to economic evolution is strikingly similar to the one Alchian recommended fifty years ago. We take the world to be fundamentally uncertain, but have endowed the actors (designers) with both foresight ("to see value") and intent ("to seek value"). And as we begin to model the evolutionary processes that result from seeing and seeking value, we will make use of a powerful metaphor that economists and biologists have in common. This is the metaphor of the "value landscape" in economics or the "fitness landscape" in biology.

---

14 Alchian (1950, p. 221).

15 See, for example, Simon (1957, 1969) on "bounded rationality"; Nelson and Winter (1982) on the "tacit" (i.e., unconscious) origins of human skills and of organizational "routines." Nelson (1995) provides an excellent overview of this literature.

The Value Landscape

In his seminal article on economic evolution, Armen Alchian offers the following evocative analogy: "A near-sighted grasshopper on a mound of rocks can crawl to the top of a particular rock. But there is no assurance that he can also get to the top of the mound, for he might have to descend for a while, or hop to new rocks."[16]

To our knowledge, this is the earliest use of the value landscape metaphor in economics. (Note that Alchian did not say that the grasshopper would not get to the top, but only that he *might* not. Whether the grasshopper lands on the top of the rocks or in a crevice depends on how well he sees, and how accurately he jumps. Exactly the same principles apply to design processes.)

In chapter 2 we introduced the concept of the "space of designs" and the value associated with each point in the space. The mapping of designs to value, we said, can be pictured as a landscape. Each distinct point in the landscape is a unit of selection—a design that can be an outcome of the design process. The altitude of the point equals the value of that design; thus the most valuable designs are the "high peaks" in the landscape.

Moving around in complex physical spaces is an experience universally shared by all human beings, which, like design, involves both imagination and reality. Hence, this metaphor is an extremely useful one to keep in mind as we begin to create mathematical models of design evolution driven by designers who "see and seek value." Because we rely extensively on the landscape metaphor, before we begin to lay out our theory of design evolution in earnest, we would like to draw our readers' attention to certain correspondences between real landscapes and "value landscapes." (Readers familiar with the metaphor may skip the next two sections.)

Similarities between a Real Landscape and a Value Landscape

One or more designers may be thought of as moving over the landscape (like grasshoppers or mountain goats) in search of higher value.[17] Points close to one another in the landscape are similar.[18] By changing parameters, designers can "walk" (or

---

[16] Alchian (1950, p. 219).

[17] Some models of physical systems invert the metric and have entities searching for low points, which are equilibriums or points of rest in the system. However, in economics, we are used to thinking of higher values as better—we will not attempt to invert that intuition here.

[18] The notion of distance can be formalized in various ways, none of which is entirely satisfactory.

jump or fly) from one point in the space—one particular design—to another. Given adequate knowledge of where to look and the ability to move about in the "space," value-seeking behavior will drive them "upward" toward higher altitudes.

New designs correspond to newly discovered points in this space. Thus the process of creating new designs is an exploration and mapping of some part of the value landscape.

Cliffs and crevices in the landscape correspond to sharp, sometimes unexpected, drops in value. These discontinuities can occur when a particular design parameter moves out of its appropriate range. More difficult to anticipate, however, are the cliffs and crevices caused by design parameter interdependencies. For example, figure 2.1 showed how the "fit interdependency" between the vessel and cap of the mug design appears as a steep ridge of value in the subspace of the cap and vessel diameters.

An issue of some interest in biological models of evolution is, Are the high-value designs connected? In other words, is there a route of continuous, positive value from one form of an organism to other, better forms?[19] Or is it sometimes true that "you can't get there from here?"[20]

The same question can be asked about the designs of artifacts. Groups of good designs that are not connected to others are like "islands in a sea." (Sea level is the point where value equals zero—negative values are below sea level.) Islands do seem to be present in the space of artifact designs, but they are not as problematic for human designers as for genes. Specifically, human designers are clearly willing and able to explore "seas" wherein the designs have negative value, in hopes of encountering one or more new islands of positive value.

Within a subspace of positively valued, connected designs (a landmass), the difference in value between any two neighboring points corresponds to a *gradient* in the landscape. Positive gradients point in the direction of higher value and are good; negative gradients point in the direction of lower value and are bad. A very important subset of the guidance technologies, which we described in Chapter 4, are in effect gradient measurement technologies.

Differences between a Real Landscape and a Value Landscape

There are also differences between real landscapes and value landscapes that can lead human intuition astray. The first and foremost of these is the extremely high dimensionality of the space of designs. Real landscapes have three dimensions.

[19] Dawkins (1996).

[20] Thompson (1942).

Design spaces and consequently value landscapes have thousands and sometimes millions of dimensions.

Human beings are actually quite good at moving around in high-dimensional spaces in their imaginations.[21] For example, a typical living room contains a couple of hundred objects, which in aggregate have tens of thousands of design dimensions. We as human beings deal with the hyperdimensionality of the space of living room designs using standard techniques: we create hierarchies, form abstractions, develop representations, define interfaces, hide information, define tests, make prototypes, and modularize and sequence decisions. As long as we can use these techniques, we can manage to move around in the imaginary space of living room designs with ease.

Another difference between value landscapes and real landscapes is the fact, already mentioned, that several different kinds of value may exist at the same time. Different concepts of value include expected value, realized value, personal value, product market value, and capital market value. In principle, there is a different value landscape corresponding to each one of these concepts.[22]

Finally, in contrast to most real landscapes, value landscapes can shift over time.[23] Something like plate tectonics can occur, with "the ground rising or falling" (figuratively) under the designers' feet. We will see an example of this in the next chapter, when we discuss how new knowledge and new tests may evolve in conjunction with a modular design and then affect its value.

Modularity and Design Options

The larger economic system is a context—a very large value landscape, in effect—which rewards the creation of new artifacts, new enterprises, and ultimately new knowledge. Within this landscape, value is not only "the height" or "altitude" of a

[21] See Edelman (1992) and McClelland and Rumelhart (1995) on the neurological and cognitive structures of human beings, which permit us to navigate in hyper-dimensional imaginary spaces and landscapes.

[22] It is a constant challenge for value-seeking enterprises to convince their employees to substitute a collective value measure for their personal value measures. A great deal of work on motivation and incentives is aimed at understanding how such substitutions can be effected and maintained.

[23] Stuart Kauffman has coined the evocative phrase "dancing landscapes" to describe this effect. (Kauffman 1993, 1995.)

design, it is also a force that pulls both designers and firms "upward" in the landscape. For example, in the computer industry, shared perceptions of value are what cause entrepreneurs and contract designers to create firms and artifact designers to join them in seeking to improve existing computer designs or create new ones.

If design outcomes were predictable, the "force" of market value would pull designers unambiguously in one direction. Design processes would then be straightforward deductive exercises focused on implementation. But we have already said that design outcomes are never wholly predictable. Figuratively, the locations of the "high peaks" in a value landscape are uncertain. The steps that need to be taken and the time needed to reach a given peak are unpredictable as well. In some cases, even the "direction" of the nearest peak may be unknown.

This fundamental uncertainty, combined with the inherent complexity of designs (the high dimensionality of design spaces), makes certain approaches to design processes much more efficient and effective than others. Specifically, in the presence of uncertainty and complexity, design options become very valuable. *Modularity creates design options and in so doing can radically change the market value of a given set of designs.*

Real Options

Formally, an option is the "the right but not the obligation" to purchase or sell something—a security, a contract, or a design—in the future. Options provide a "nonlinear" payoff to a gamble. Rather than facing symmetric gains and losses, the option holder can wait to see how the gamble turns out. If the outcome is good (high value), he or she takes the gain; if it is bad, the option holder can walk away and thereby avoid the loss. Loosely speaking, then, options are structured so that "heads, I win; tails, I stay the same."

The loss-limitation feature of options is very valuable, and thus one generally has to pay for options purchased and will be paid for options sold.[24] Based on the pathbreaking work of Fischer Black, Myron Scholes, and Robert Merton, there are now well-established market valuation technologies for options. These are an important subset of the financial sector's valuation technologies, which we discussed in Chapter 4.[25] We use these valuation methods in our models below.

---

[24] Selling an option is like writing an insurance policy. The seller assumes the consequences (low or negative value) of bad outcomes.

[25] Black and Scholes (1973); Merton (1973a, 1990).

The best-known kinds of options are financial options, which are created through trade. For example, a company may purchase an option (to borrow money at a fixed interest rate, for example) from a bank. By buying the option, the company transfers some of the risk of high interest rates to the bank, which may be better positioned to handle it. In these transactions, risk gets moved around, but, in a very precise sense, the total amount of risk in the system remains the same. Hence for every purchaser of a financial option there is a seller, who must stand ready to absorb the consequences of the bad outcome if it occurs.

In contrast to financial options, *real options* are embedded in designs, technologies, and production processes. For example, a company may choose a flexible production process that allows it to change inputs and outputs in various ways. This flexibility constitutes a set of options on the value of underlying products.[26] But in contrast to financial options, these real options come from selecting one technology over another. If a bad outcome arises, the option holder will not look to transfer the consequences to the seller of the option, but will look instead to change its operating plans.

How Modularity Creates Design Options

As a general rule, an interconnected design process will deliver one option—to take the output of the process or leave it. In contrast, a modular design process creates many options.[27] It isn't necessary to take an all-or-nothing approach to a modular design: one can mix and match modules as long as they conform to the system's overall design rules. Thus, a modular design process creates at least as many options as there are modules.[28] In other words, modularity *multiplies* the options inherent in a design.

Figure 9.2 depicts the option structure of a design process before and after modularization. As it shows, modularizing a design process changes both the "size" and "location" of design options. The options become "smaller" in the sense that they involve parts of the system, not the whole. In addition, individual hidden-module

---

[26] Mason and Merton (1985); Hodder and Triantis (1992); Dixit and Pindyck (1994); Trigeorgis (1996); Amram and Kulatilaka (1998).

[27] Independent in this context means that the modules can be worked on by separate groups. They may not have strictly independent probability distributions of market value.

[28] The substitution operator may lead to there being more than one option per module. We discuss this in chapter 10.

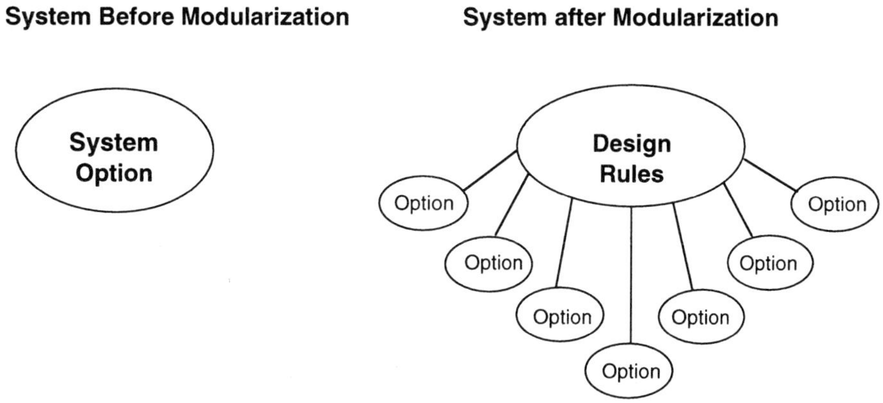

Figure 9.2 Modularity creates design options.

designers can make choices (including the choice to offer new designs) without consulting the designers of other modules or the system's architects. Metaphorically, a modularization shifts design options from the "center" to the "periphery" of the system. Hence, modularity *decentralizes* the options inherent in a design.

Through modularity then, the value associated with the different parts of a new design can be broken up. Each component that is a module can succeed or fail, survive or be eliminated, on its own: its value is no longer hostage to that of distant elements of the system. As a result, each module becomes an independent unit of selection and a point of potential value creation within the overall system.

The Dynamics of Modular Systems—Design Evolution

A shift from interdependency to modularity, because it changes the number, value, and location of design options, has the power to accelerate dramatically the rate of change of the system as a whole.[29] As we will show, if there is uncertainty about outcomes of a design process, then a modularization causes the market value of design options to go up by a lot. The number of people who can unilaterally exercise those options goes up by a lot as well. (We will formalize what we mean by "a lot" in the next chapter.) As a result, many individuals and firms will have strong economic

[29] It is almost as if the "laws of motion" governed the system change. In fact the most fundamental law—that designers see and seek value—does not change at all.

incentives to offer new module designs. And with the advent of so many new module designs, the system itself will begin to change in new ways.

If we look at what happened in the computer industry in the aftermath of the introduction of System/360, we can actually see the dynamics of modularity take hold. By 1970, six years after its introduction, System/360 housed over sixty different modular units (see figure 8.4). In figure 9.3, we have grouped these sixty modules into twenty-five categories by putting closely related modules (e.g., all disk drives) in the same category.

The modules within each of these major groups were individual units of selection, which could be improved independently of one another as long as they conformed to System/360's design rules. System/360's modular design created these new units of selection, and thus replaced all-or-nothing system-level options with options to improve modules. What had been a single design "trajectory"—a single path of design improvement—was split into something on the order of twenty-five different module trajectories.[30]

The effects of modularity did not stop there, however. If we zoomed in on one of the modules in figure 9.3, and looked forward into the 1970s and 1980s, we would see further branching of design trajectories at the submodule level. The branching took two forms. First, designers often split complex modules into their component parts. Thus disk storage systems, for example, were modularized into heads, arms, motors, media, controllers, and so on.[31]

Equally important was the fact that designers soon began to *augment* the basic module designs by adding new module architectures. For example, within the disk storage module, designers worked with smaller and smaller "form factors."[32] The initial System/360 disk drive used a 14-in. disk. Drives with 8-in. disks were introduced in 1978, and became popular in minicomputers. Between 1980 and 1983, smaller drives with 5.25-in. disks became an integral part of the first personal computers. Still smaller 3.5-in. drives became widespread in the late 1980s, fueled by the boom in portable computers.[33]

Figure 9.4 shows these five distinct new categories of disk drives branching off from the initial module design. The work of Clay Christensen demonstrates that the

---

[30] On technological trajectories, see, for example, Marples (1961); Rosenberg (1982); Nelson and Winter (1982); Clark (1985); Dosi (1986 a,b); Sutton (1998).

[31] Sahlman and Stevenson (1985); Christensen (1993).

[32] Christensen (1993).

[33] Christensen (1993, 1997); Christensen and Rosenbloom (1995); Lerner (1995b, 1997).

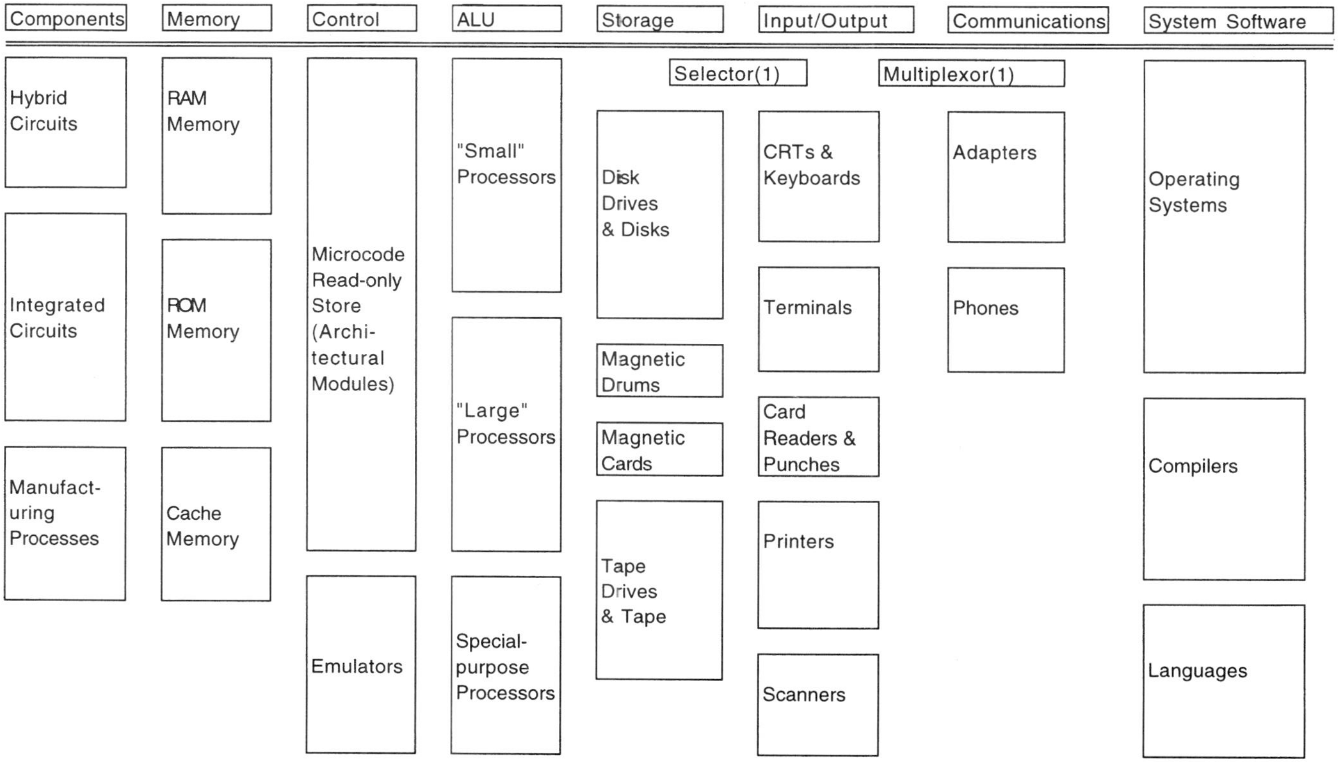

Figure 9.3 The modules of System/360 grouped by major category. (1) The selector and multiplexor were special I/O processors that managed the activities of storage, input/output, and communication devices. (2) Shaded boxes indicate architectural modules, whose designs had to remain relatively fixed. Not shown: application software. (Compiled from Bell and Newell 1971, pp. 561–583 [hardware]; Pugh et al. 1991, pp. 291–345 [software]).

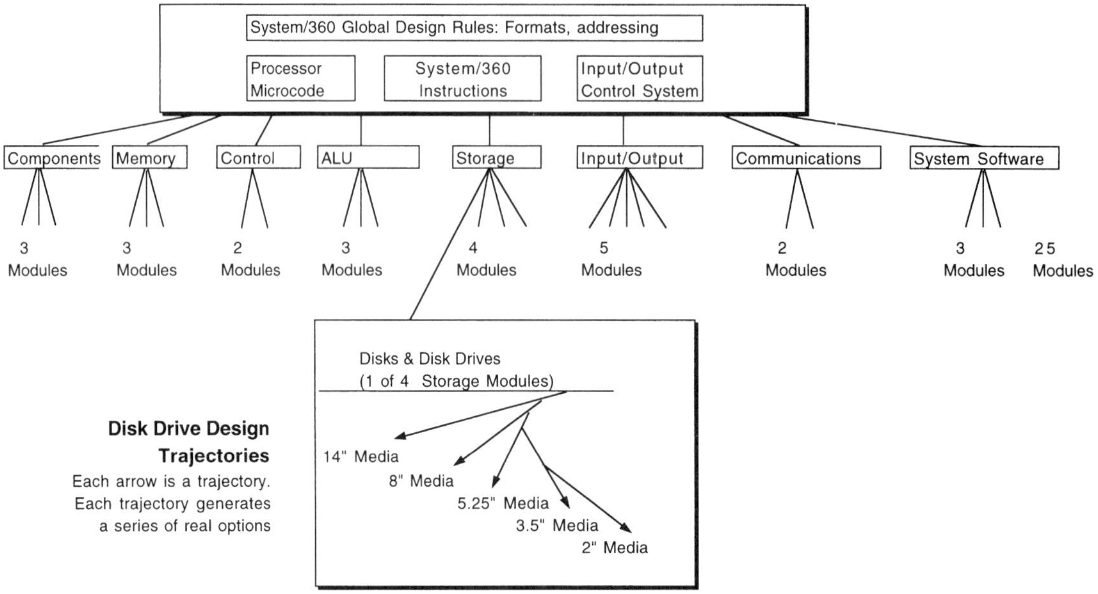

Figure 9.4 Design trajectories for System/360, circa 1970.

new disk drive designs "descended" from older ones in the sense we defined in Chapter 5.[34] A common pattern was that a designer in an established firm would propose a new design and see it rejected. At that point, he would leave to found a new firm with venture capital financing. (We explore the economic logic behind such "defections" in volume 2.)

Often the new disk drive designs were inferior to the old ones when measured by traditional standards. But the new architectures had features that the older ones did not, which made them attractive to the designers of new systems.[35] Thus by a complicated process of decentralized bootstrapping, new module designs made it possible to create new system designs, and the new systems in turn created demand for the new modules.

In this fashion, the working out of the potential of an initial modular design (System/360) led to the further splitting of modules, as well as to the augmentation,

[34] Christensen (1993; see especially figure 7 on p. 575).

[35] Christensen and Rosenbloom (1995).

exclusion, inversion, and porting of modules. Ultimately, new systems were created from the augmented, inverted, and ported modular parts. The end result was an incredibly diverse, constantly changing set of module and system designs. Moreover, as we indicated in chapter 1, these changing designs did not stay within the control of one or a few firms. Instead, over time they came to be dispersed across a modular cluster of more than one thousand firms in sixteen industry subcategories. (Refer back to table 1.1 and figures 1.1 and 1.2. We discuss the emergence of this modular cluster in chapter 14.)

When design options are no longer closely controlled by a few enterprises, but are instead widely dispersed throughout a modular cluster of independent firms, there will be many points where individuals and firms can respond to local incentives to capture or create value. Each of these points marks a locally determined trajectory; from that point, the module design may continue to develop along its preestablished route, or it may branch onto new paths (as disk drives did).

We must emphasize, however, that the locally independent modules are not *wholly* independent of one another. Designers of modules must always take account of their system's visible information, and adhere to the design rules established for the system as a whole. Nevertheless, as control of the design passes from a single, closely coordinated workgroup to many independent workgroups lodged in different enterprises, centralized design development gives way to what we call "design evolution." At the level of modules, individuals and firms (like the disk-drive designers and their firms) can pursue local options and opportunities.

The trajectory of the whole system then becomes an aggregation of all these individual, locally determined trajectories. This decentralized, complex adaptive process is in turn highly reminiscent of biological and other evolutionary processes, for example, the formation of vertebrate immune systems, the development pathways of neurons, and some models of language acquisition.[36] That is why we think it is appropriate to call this process "design evolution."

Even though the agents of design evolution are not blind replicators, in our judgment the similarities between design evolution and other forms of evolution overshadow the differences. Fundamentally it is more useful to think of design evolution as belonging to a larger class of decentralized complex adaptive processes than it is think of it in a category by itself. (Box 9.3 describes "memetic evolution," another form of cultural evolution that more closely resembles biological evolution.)

---

[36] Edelman (1992); Plotkin (1993); Pinker (1994); Deacon (1997).

Box 9.3 Design Evolution vs. Memetic Evolution

Design evolution—the decentralized adaptive process by which designs are changed and artifacts improved—is a feature of human culture. But there is another way in which evolution can occur in human societies. In his book *The Selfish Gene,* Richard Dawkins named the units of selection in this other process "memes."[1] Memetic evolution and design evolution are different processes, although both are important to the development of human artifacts and economies. We will explain how they differ.

Memes are memorable units of knowledge—patterns that can be stored in some form of memory and transmitted from one human being to another. Songs, stories, and routine practices are all memes. According to Dawkins's theory, memes operate in a manner very similar to genes in biological evolution: they replicate and compete with other memes for a larger share of human consciousness and memory. Some memes get copied because they are useful (like recipes), others because they are addictive (like jingles). The more copies of a meme there are, roughly speaking, the more "fit" it is, and the more likely it is to survive.[2]

In theory, a bunch of memes can be assembled into a complex mental artifact, for example (Dawkins would say), a book, a science, or a religion. If the artifact has coherence and satisfies a human need, this "super-meme" will be transmitted, copied, and will survive as part of the culture.

Memetic evolution differs from design evolution in two important respects. First, the agents of memetic evolution (the memes themselves) are *blind replicators.* Memes may have important functions and roles in society, but their functions are incidental to their success, which is measured by the number of "true copies" of a meme there are. In contrast, designs are created by human beings for particular purposes, hence the success of a design is measured by how well the corresponding artifact performs its functions and how important those functions are.

The second difference is that design evolution is destructive of memes. Human beings actively work to improve designs—their goal is to destroy previous patterns and replace previous designs (existing memes) with new designs that are better. (If we were to push the biological analogy to an extreme point, one could think of this "creative destruction" as simply a form of competition between memes— a blind, quasi-natural process.[3] However, if humans consciously set up memetic competitions to improve their artifacts, then, from a human perspective, those competitions become part of an even larger design process. We do not know whether human designers or the memes are at the top level of this recursive hierarchy.)

[1] Dawkins (1989).

[2] Dawkins (1989, 1996). For elaborations of the theory, see Lumsden and Wilson (1981); Wilson (1998); Hayek (1988); Plotkin (1993); Kelly (1994); Dyson (1997).

[3] Schumpeter (1976).

(continued)

> Design evolution and memetic evolution are different, but they coexist in human societies, and in fact often complement one another. A designer may consciously try to create a meme that will replicate itself (an advertising slogan or a melody, for example). And memes in the form of routines, traditions, and rules of thumb may in turn be used to create new designs. As social scientists, our challenge is to understand how these two modes of cultural evolution differ from one another and how they interact. As designers, our challenge is to use both processes constructively to achieve our ends.

Preview of the Next Chapters

Our purpose in the rest of this volume and in volume 2 is to understand how modular design evolution—a process that occurs at the level of artifacts and designs—shaped the development of the computer industry—a set of firms and markets—from 1965 to the present. We have already seen how modularity emerged as a design principle for computers, and how the first modular product family was created. Now we want to understand how and why computer designs evolved as they did, and how their evolution influenced (and continues to influence) the structure of the computer industry.

We contend that design evolution in the computer industry is fundamentally an economic phenomenon. That means it can be understood in terms of value-seeking moves by economically motivated agents operating on designs in a complex technological and industrial environment. However, because the complexity of the overall system is so great, we must break the phenomenon down into parts, and understand each part in its own terms. Thus in chapters 10 to 13, we look at the economic value of the six modular operators that were defined in chapter 5. Then in chapters 14 and 15 we turn our attention to firms and markets and show how a "modular cluster" of computer companies came into being in the 1970s. Volume 2 will pick up the story in 1980, and follow the evolution of computer designs and the competitive interactions of computer firms through the 1980s and 1990s.

10 The Value of Modularity—Splitting and Substitution

To understand the drivers of the evolutionary process and the patterns of technological change and competition that grew out of it, it is not enough simply to establish the fact that computer systems became modular, that a modular task structure allowed modules to change at different rates, and that new module concepts were introduced by designers trying to create and capture economic value. We need to understand *how* the modular operators create value, *why* designers choose one set of operators rather than another, and *why* some modules evolve at very different rates and come to play very different competitive roles.

Therefore in the next four chapters, we will attempt to understand how economic value is created in a modular design by building mathematical models of option value for each of the six basic modular operators. We begin by laying out the assumptions found in all chapters, and then look at the basic splitting and substitution operators. Following that, we look at the four remaining operators, which can only be applied to designs that have already been modularized.

Two fundamental ideas lie at the core of our models. The first is that designers' actions have both benefits and costs. The second is that designers have access to the valuation technologies of the capital markets, and thus at each stage of the design process, they will seek to increase the *net option value* of their designs.[1] Responding to the motivating force of value, designers will create new modules and new groupings of modules. Their actions in turn will cause the overall system to evolve.

Thus, our theory of design evolution can be boiled down to the following propositions:

---

[1] Technically, net option value is a calculation of probabilistic rewards in the future, appropriately valued at prices set in the present. In an advanced capital market, some portion of this amount can be monetized and consumed in the present by the creators of the option value.

1. Design evolution is a value-seeking process.
2. There are six basic modular operators, each of which has an associated option value.
3. In a modular design, the six basic operators can be applied at many points and in different combinations.
4. Over time, this value-seeking process gives rise to an ever more complex, diverse, and dispersed modular system.

Let us begin by imagining a designer or a team of designers faced with the challenge of designing a complex product. To specify our models, we must make some assumptions about his, her, or their situation. We have divided these assumptions into three groups: basic framework, tasks and modules, and uncertainty. Later we shall also make assumptions about the costs of different actions.

Basic Framework

Our assumptions about the basic framework flow from the fundamental task structure of modular designs (see chapter 3). Recall that a modular design process has three basic stages: (1) the formulation of *design rules;* (2) *parallel work* on hidden modules; and (3) *testing and integration.*

Figure 10.1, which is based on figure 3.4, shows how the divisions of the task structure matrix correspond to the three different stages of the design process. It then lays out the basic framework of the model in terms of these stages. (The division of the time line does not correspond to the percentage of time spent in each stage, which will vary from case to case.)

Actions and Events in Each Stage

In the first stage, we assume, designers create a design structure and corresponding task structure (recall the two are isomorphic). The number, boundaries, and interfaces of the modules in the system are determined at this point. If the existing design is interdependent, designers may apply the splitting operator to make it modular. (In chapters 3 and 7, we saw how this was done for Mead and Conway's multiproject chips and, on a much larger scale, for System/360.) Or, if a modular structure already exists, the designers may apply one or more of the six basic operators to come up with a new modular design that is "descended" from the old in the sense we described in chapter 5.

Mathematically, in our model the choice of a design and task structure corresponds to the choice of a payoff function for the design. Because the features of the ultimate design are unknown and because consumers' reactions to it are not perfectly predictable, the payoff function is uncertain when the design process begins. In mathematical language, it is a random variable. This means that the function does not have a specific value in stage 1—it has instead a set of potential values and an associated probability distribution. The characteristics of the design and task structure—the number of modules, the number of experiments, and the number of tasks—affect the probability distribution of the payoff function in ways that we describe below.

In the second stage, the task structure is implemented. Figuratively, one can imagine that "a wheel spins" and determines the value of the design. Mathematically, one can model the "spinning wheel" as a draw from the probability distribution associated with the random-variable payoff function.

In the third stage, the value of the new design is revealed. At this point, the option structure of the design becomes important. As we discussed in the previous chapter, sequential and interconnected task structures do not create many options—one either takes the outcome of the process or rejects it. In contrast, a modular design creates many options, hence many individual accept-or-reject decisions. Because there are so many options, modular task structures require the explicit testing and integration of modules in the final stage.

Costs at Each Stage

Each of the three stages has its own set of costs. In stage 1 are the costs of formulating the design rules. Intuitively, these costs should go up as the number of modules goes up. However, as we saw in the case of System/360, design rules once established can last a very long time, and in later rounds of the design process, they are "sunk." Thus the cost of formulating design rules is an investment, whose returns may be garnered over several rounds of work on the hidden modules.

The costs of stage 2 are those of selecting design parameters and running experiments. Other things being equal, these should increase as the number of design parameters (hence the complexity of the design) goes up, and as the number of parallel experiments increases. Unlike the costs of formulating the design rules, these costs are incurred each time a module is redesigned.

Finally, in stage 3, designers incur costs of testing modules and of doing the incremental "fixes" needed to get the modules to work together. In comparison to design rule costs, which are sunk, and module experimentation costs, which are recurring, these costs tend to change over the lifetime of the system. Early on, when

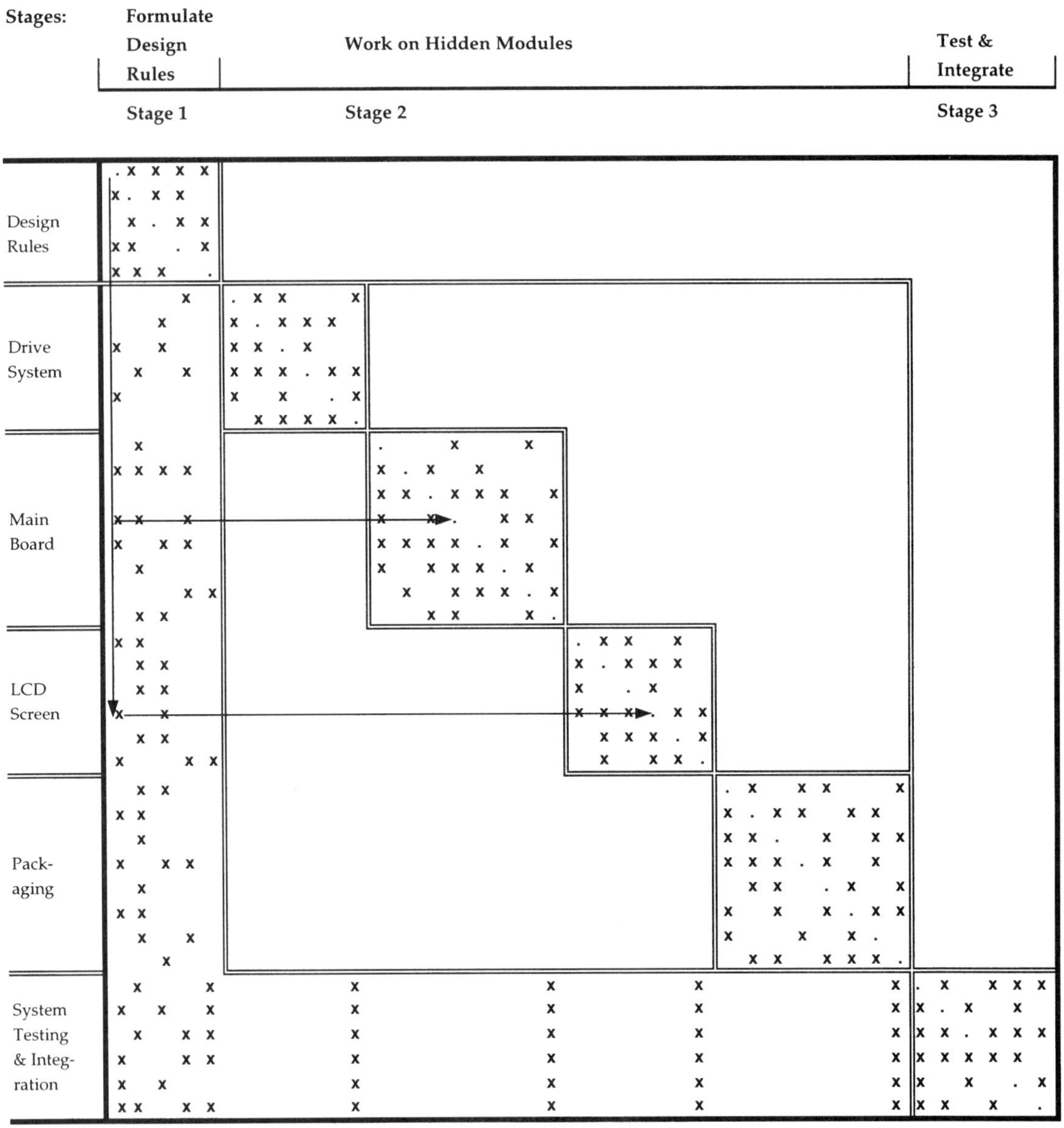

Figure 10.1 The basic framework of our model of a modular design process.

Stages	Stage 1	Stage 2	Stage 3
Time Line	**Create Task Structure & Design Rules**	**Implement Task Structure for Modules**	**Test, Integrate, Evaluate System**

What Actually Happens

Actions	Choose operators	Carry out tasks	Test results & Exercise options
Events	Splitting Substituting Augmenting Excluding Inverting Porting	*"The Wheel Spins"* 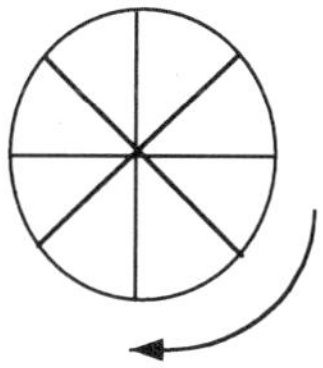	Economic value is revealed; Best outcomes are selected

Mathematical Representation

Benefits	A payoff in the form of a random variable is chosen. $\tilde{X}$	An outcome is drawn from the distribution of the random variable. $\tilde{X} \overset{\bigcirc}{\rightarrow} \tilde{X}$	The value corresponding to the outcome of the random variable is revealed; where options exist, the best outcomes are selected. $\max(\tilde{X}, 0)$
Costs	Cost of designing task structure	Cost of implementing task structure	Cost of testing and integration
Basis of Choice	Highest Net Option Value	Highest Net Option Value given task structure	Highest Value given outcomes and tests

Figure 10.1 (continued)

the modularization is new and imperfect, the testing process will be inefficient and stage 3 costs will be high. Later, as knowledge about the system accumulates, interactions requiring "fixes" will be codified in the design rules. As a result, the testing and integration process will become more efficient and may even disappear. The ideal is "plug and play" compatibility, where the consumer can do his or her own integration of modules. (Later in this chapter, we discuss the economic implications of achieving that ideal.)

Designers "Select" on the Basis of Capital Market Value

Thus far, we have spoken of what happens as the design process unfolds, but not of the value of different paths. However, as we have already noted (cf. chapter 4), an extremely important property of advanced market economies is that their capital markets can translate complex random-variable payoffs into single-point market values. Thus, when confronted with a choice between actions with probabilistic outcomes, value-seeking actors may ask, What is the *capital market value* of choosing action A over action B?

Modern valuation technologies, especially the theory of option value, provide a rigorous way to answer this question.[2] Stripped to their essence, these theories say that the value of a random-variable payoff equals its mathematical expectation, appropriately translated for risk, and discounted to reflect the passage of time. In other words, from the probability distribution of the payoff function, one forms a mathematical expectation.[3] This reduces the many-valued random variable to a single number, which is then discounted to the present at the prevailing interest rate.

As we indicated in Chapter 4, the resulting number is the present value of the payoff function—its equivalent in terms of wealth today. If an option is involved, we will call this number the *net option value* (NOV) of the probabilistic payoff function. Given a menu of possible actions, each of which generates a complex random payoff sometime in the future, value-seeking actors can calculate the net option value of each action, and choose the one with the highest value at the present moment.

---

[2] Mason and Merton (1985); Dixit and Pindyck (1994).

[3] A mathematical expectation is the sum of endpoint values weighted by their associated probabilities. Intuitively, it is a "mean" or "average." In the evaluation of options, the probability distribution applied to the endpoint values must be translated into its "risk-neutral" equivalent. This is a technical detail—the important thing is that it can be done, and the result is a single number, not a random variable. (See Merton 1973b; Cox and Ross 1976.)

Box 10.1 Tasks

Tasks. The experimentation phase (stage 2) of the design process consists of a set of tasks, numbered 1 to *N*. Each task requires a design decision and results in the specification of a design parameter. The design parameters in turn affect the performance of the system (as viewed by users), but the precise impact of a particular decision on performance is uncertain.
Complexity. The complexity of the system is proportional to the total number of tasks, that is, proportional to *N*. In our analysis, we will hold *N* fixed: in reality, complexity is endogenous and *N* may vary.
Performance. The "total value" of the system, denoted *X*, reflects the consumers' willingness to pay. We assume that the designers are seeking to deliver higher *X*. The net option value on which the designers base their decisions is fundamentally an expectation of *X*, adjusted for time and risk, minus the costs incurred along the way.
Benchmark. As a theoretical benchmark, we assume there is a "current model" of the product which represents a feasible combination of the *N* design parameters. Without loss of generality, the value of the "current model" is set to zero.

Then, as they move through the design process, the actors will confront new choices. However, at each decision point, they can look forward in time, assess the payoff function for each action, and calculate the corresponding net option value. We assume that when they have a choice, the designers will want to select the action with the highest value, calculated this way. In many circumstances, economically-motivated actors can do no better than to follow this rule.[4]

Tasks and Modules

We now turn to the definition of tasks and modules. As we showed in chapter 2, no matter how complex a product is, it can be described in terms of a finite set of design parameters. These parameters in turn correspond to tasks, each of which fixes a parameter in the final design. A task structure matrix (TSM) shows which choices depend on which other choices. Box 10.1 summarizes our assumptions about tasks.

[4] We have defined "subgame perfect" or "time-consistent" strategy. In interactive games, there are cases in which these strategies lead to suboptimal outcomes for all players. It is then an open question whether the players can break out of their suboptimal game and realize a higher-valued joint outcome. We take up these issues in volume 2.

We assume that N tasks are divided among j modules. The modules may be symmetric (each having the same number of tasks) or asymmetric (each having a different number of tasks). In reality, of course, modular partitions are determined by opportunities embedded within the design itself, and thus it is far more likely that modules will be asymmetric than symmetric. Notwithstanding this fact, it is helpful to look at symmetric modularizations first, for they can be visualized in two or three dimensions, whereas asymmetric modularizations cannot. Box 10.2 summarizes our assumptions about modules.

The essential aspect of modularity lies in the fact that subsets of the design are broken out at the beginning of the design process. Tasks within different modules can then proceed independently, and the resulting designs can be combined in a variety of ways. Thus in a one-module design process, only two outcomes are possible: either the new design will be superior to the old and will replace it, or the new design will fall short and the old design will continue to be used.

However, if the design has been split into two modules, then, when the module designs are finished, the designers will have four choices. They can introduce an entirely new system, keep the old system, or combine one old module with one new one.[5] Which of these options is most attractive depends on the value of the new modules (denoted X_1 and X_2, respectively) relative to the corresponding old ones.

Similarly, a three-module task structure generates eight options, a four-module structure generates sixteen, and so on. This expansion in the number of options is the direct result of changing the *units of selection* in the design. It is what we meant in chapter 9, when we said that modularization multiplies options, and thereby initiates the process of design evolution.

Uncertainty

The world of design is fraught with uncertainty. Uncertainty arises because the designs themselves are complex and not well understood, and because consumer pref-

---

[5] Implicitly we are assuming that the act of modularization partitions the design of both the old and the new product. This assumption is most likely to be true when the older product's design has been partially rationalized (cf. Shirley 1990). However, in some cases, the preexisting product design cannot be partitioned: for example, software engineers speak of the impossibility of breaking apart "spaghetti code" (Korson and Vaishnavi 1986). The designers must then either take on two projects per module (see the discussion of experimentation below), or wait a generation before realizing the option values inherent in the modularization. (There may still be nonoption values derived from the modularization, for example, manageable complexity, parallel work; see the discussion in chapter 3.)

Box 10.2 Modules

Modules. At the outset of the design process, tasks are partitioned, that is, divided up among modules. Tasks are highly interconnected within modules and independent across modules.

Number of modules. The total number of modules, denoted j, is a choice variable in stage 1, and fixed in stages 2 and 3.

Size of modules. The "size" or "complexity" of each module is equivalent to the number of tasks needed to design that module. If a modular partition is symmetric, then all modules have the same number of tasks, equal to N/j. If the modules are not symmetric, their "sizes" are given by a αN, βN, . . ., where α, β . . . are fractions that sum to one.

Module performance. The overall performance of the system, X, can be partitioned into a system-level value measure, S_0, and j module value measures ($X_1, \ldots, X_j$), one for each module. We assume that the total value of the system in the eyes of a user can be expressed as the sum of the total system value plus the incremental value added by the performance of each module:

$$X = S_0 + \sum_{i=1}^{j} X_i .$$

Because of S_0, the value of the system is not just the sum of the value of its parts.[1] However, for now, we will assume that new module designs do not change this system-level value; thus without loss of generality, we normalize the system-level value, S_0 to zero.

Note: In our model, the value of the system-as-a-whole or any module may be set to zero as a matter of mathematical convenience. This does not mean that the system is unimportant, or that the module somehow "isn't there." *There must be a system with a potential module in each position for the design to have any value at all.* However, within that preexisting system, we can examine the incremental contributions associated with better module designs. That is what we are doing in this chapter.

Modular selection. If the design was initially partitioned into modules, the designers can observe X_i for each of the j modules, and compare it with the performance of the corresponding module in the current model (normalized to zero). Subject to costs, which we discuss below, the designers can mix and match modules of the new design with complementary modules of the old one.

[1] There are some very subtle system definition and preference issues subsumed in this assumption. We are aware of them. Worried readers should know that the qualitative results in this and later chapters are robust to many alternative specifications of the value function. For example, the basic results hold if the value function is supermodular in the role attributes of the modules. (See Milgrom and Roberts 1995; Schaefer 1997 for the definition and discussions of supermodularity.) The results do not hold, however, if the value function is subject to hidden interdependencies, including complex, epistatic effects à la Kauffman (1993). The purpose of design rules, of course, is to eliminate subregions of the design space, thereby "smoothing out" and "ordering" the landscape of the actual search.

erences are both complex and unknown. For example, each of us knows around a thousand facts about the computer that sits on our desktop. Each of the things we know about the system affects our willingness to pay for it in some fashion.[6] And underlying the things we know are many thousands and even millions of additional things that we don't know. Although they are hidden from us (because they are not essential to our use of the computer), these parameters define the system's structure. The system occupies only a cubic yard of physical space, but is of mind-boggling complexity.

Given such a high degree of complexity, it simply is not possible for designers to know enough about the system to eliminate all uncertainty. Thus each new design is fundamentally an experiment. Its outcome may be guessed, but it cannot be known ahead of time.

Box 10.3 summarizes our assumptions about uncertainty. In choosing these assumptions, our aim was to simplify the results as far as possible, without violating any known facts. Thus we assumed that at the beginning of the design process, the value of the basic system, S_0, is known and equal to zero. (We will relax this assumption when we discuss the augmenting and excluding operators in chapter 12.) In addition, each module's value has a normal distribution with a mean of zero and a variance proportional to the number of tasks in the module.

The sums of normally distributed random variables are also normally distributed. Since we intend to add up module values, this is a convenient property for our probability distributions to have. Normal distributions are also consistent with what we know empirically about value distributions—that very high and very low outcomes are rare, while outcomes in the middle are more common.

Setting the mean of each module's value equal to zero is another convenient assumption. It signifies that we expect some outcomes of design processes to be better than what was available, and other outcomes to be worse. Anyone who has seen a favorite feature disappear when a new version of hardware or software is introduced knows that design changes can destroy value in the eyes of consumers, as well as enhance it. However, the fact that the "raw" expectation is zero does not mean that the expectation of the ultimate outcome is zero. The designers have the option to reject any module if it fails to clear the benchmark of the established design. Because

[6] Remember that one of the earliest definitions of computer architecture was "those things the user must know" (Amdahl et al. 1964). This concept of architecture is still powerful in software engineering, where user interfaces (the things the user knows) are difficult to do well but are also very important to the success of the design.

Box 10.3 Uncertainty

Normal distribution. Every modularization results in a normal distribution of value for each module design. In other words, every random variable X_i, corresponding to a module value, has a normal distribution.
Expectation = 0. The expectation of value for each module design, denoted $E(X_i)$, is zero.
Variance proportional to complexity. For a module design of n tasks, the variance of its value, denoted σ_n^2, increases linearly with the number of tasks: $\sigma_n^2 = \sigma^2 n$. The variance factor σ^2 is a parameter common to all modules.
Conservation of performance value. If the number of tasks is held constant, the probability distribution of value is unchanged by modularization.

these options are built into the design process, raw expectations in every instance can be zero and yet ultimate outcomes can be positive. This sounds mysterious, but really it is not—it merely means that designers have the ability to reject a new design if it is no better than what is already available.

Finally, we assume that the variance of outcomes depends on the number of tasks in each module. Here we are saying that larger task structures, which correspond to more complex designs, have more upside and downside potential than their smaller and simpler counterparts. However, large, complex modules are also more expensive to create, test, and integrate than small ones, and one must balance their economic potential against these costs and the risk of failure.

The Value of an Interconnected (One-Module) Design Process

Given these assumptions, we can calculate the value of an interconnected task structure with only one module. Let V_1 denote the value of a one-module design with N tasks. Under the assumptions listed above, the final outcome of the design process, X, is a random variable that is normally distributed with mean zero and variance $\sigma^2 N$. Consistent with this being a one-module design, once the designers implement the task structure, they can only take the new design or leave it. The value of the new design *if it is superior to the old one* (which was normalized to zero), is

$$V_1 = S_0 + E(X_N^+), \tag{10.1}$$

where E indicates "expected value" and X has a normal distribution with mean zero and variance $\sigma^2 N$. The superscript + on X means that the expectation applies only

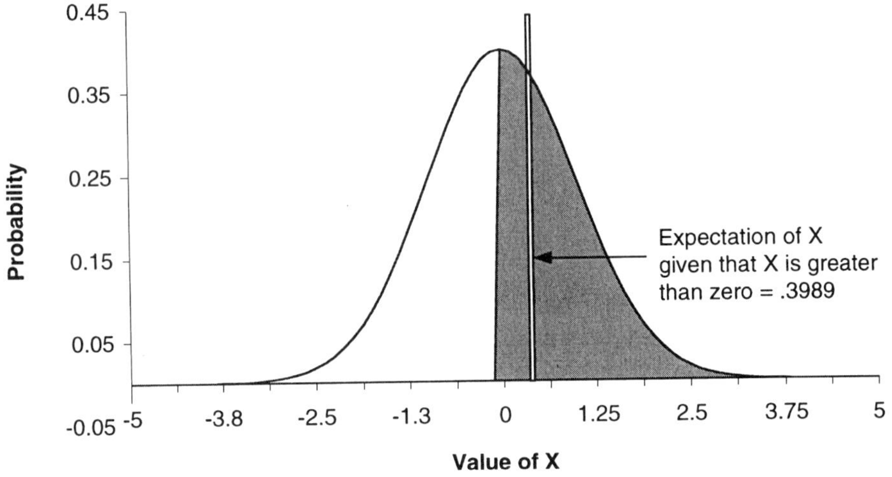

Figure 10.2 A normal probability distribution.

to outcomes above zero.[7] (*Note:* S_0 equals zero by assumption. However, we have left it in the expression to remind readers that the system provides a platform for the modules. This will be important when we look at the value of whole systems in the next chapter.)

Figure 10.2 depicts the standard bell curve of a normal probability distribution. The shaded region is the range wherein the random variable X is greater than zero. $E(X^+)$ is the mathematical expectation of X taken over this range; technically, it equals the area of the shaded region. The relationship of $E(X^+)$ to the range of the random variable is indicated by the vertical column in the figure.

For a standard normal distribution, $E(X^+)$ equals .3989.[8] For any normal distribution, with mean zero and variance $\sigma^2 n$, $E(X_n^+)$ equals .3989 times the standard devia-

[7] More formally, expected performance is calculated by weighing each potential outcome by its probability and summing up as in:

$$E(X^+) = \int_0^\infty X f(X) dX,$$

where $f(X)$ is the density of a normal distribution. We shall use the simpler notation throughout the text. In the interest of simplicity and clarity, we are also suppressing adjustments for time and risk. ($E(X^+)$ is not exactly a so-called conditional expectation though it is closely related—the expectation of X conditional on its being greater than zero equals $E(X^+)/.5$.)

[8] A standard normal distribution has a mean zero and a variance and standard deviation of one. Any normally distributed random variable can be expressed as a linear transformation of a standard normal random variable.

tion of that distribution: $E(X_n^+) = .3989\ \sigma n^{1/2}$. This is a convenient fact, which we make use of below.

Most of the fundamental building blocks needed for our investigation are in place. We can now begin to use the mathematical apparatus to understand the values implicit in different actions. Our goal in this and the next three chapters is to develop a formula for the NOV of each of the six fundamental modular operators—splitting, substituting, augmenting, excluding, inverting, and porting. Of these, the most basic operators are *splitting* and *substituting,* and it is to their valuation that we now turn.

The Value of Splitting

Every attempt at a new design is a foray into unknown territory. As a case in point, the choice to modularize is a bet that the state of knowledge will support a partitioning of tasks, and that the resulting options will justify the added costs. At the beginning of the process, the state of knowledge and the costs involved are uncertain, hence the choice is never clear-cut. (Box 10.4 describes the kinds of debates that swirl around a decision to split a design.)

Early in the history of a design, designers will have limited knowledge of critical interdependencies. Then a functional design can be achieved only through an interconnected task structure, which allows the interdependencies to be addressed during the design process itself. But the design that emerges from this process almost always lacks flexibility.

A different approach, we have seen, is to create design rules and divide up the design tasks among independent modules. This requires greater knowledge and more investment at the beginning of the process. Moreover, if the design rules are incomplete, the resulting systems will underperform other systems that were designed using an interconnected process. But over time, the modular systems that result from this second approach will come to equal their interdependent counterparts. Furthermore, modular systems can accommodate change and improvement in the underlying module designs, and thus have greater flexibility.

Clearly then, the choice between an interconnected and a modular approach to a particular design has economic consequences. Just as clearly, there is no economic "right answer" that works in all cases. Our task here is to develop a model that will allow us to compare the values of interdependent and modular alternatives, in order to illuminate the tradeoffs that will appear in individual situations.

To continue, let us suppose a design process is partitioned into j independent modules, while the number of tasks, N, remains the same. The expected value of the modular design, denoted V_j, is then

Box 10.4 A Debate over Splitting

Most of the time, the costs and risks of different design structures are debated privately. (Recall the deliberations of the SPREAD committee, which were only revealed long after the fact.) But every once in a while, disagreements among designers lead to public bets on opposite strategies. One such case arose in mid-1995 in the realm of graphics-enhanced personal computer chipsets.[1] The choice was between a "loosely coupled" vs. a "tightly coupled" unified memory architecture (UMA).

Both approaches sought to have the central processing unit and the graphics controller share memory (cf. the graphics controller example in chapter 3). In "loosely coupled" architectures, however, the graphics engine and the core logic were modular, connected by a communications protocol to avoid memory conflicts. In "tightly coupled" architectures, core logic and graphics capabilities were inextricably linked in a single chipset. In the former case, tasks could be spread over two firms; in the latter case, the chipset had to be designed as a single module at one firm.[2]

The benefits claimed for the interdependent design were functionality and low cost in relation to modular alternatives. Proponents of this approach claimed

[The] loosely coupled approach has yet to prove that it measures up to a fully integrated solution. With separate graphics and separate core logic, you will end up shutting the system down a lot. If someone delivered a loosely-coupled architecture, the challenge is to do it with good performance.[3]

A tightly coupled UMA chipset was introduced by Weitek Corporation in June 1995. However, the chipset lacked flexibility, as Weitek's competitors were quick to point out:

[The] issue is whether you are combining the best core logic with the best graphics controller. . . . Weitek's . . . graphics controller [may not be] as good as somebody else's.[4]

These quotes are representative of the kinds of debates that constantly recur among computer designers of both hardware and software. The debates can be heated because neither modular designs nor interdependent designs are *inherently* superior: the costs and benefits of each approach vary by case and over time. Gaining a better understanding of the structure of those costs and benefits is the purpose of this chapter.

---

[1] Cataldo, A., "Weitek Debuts 'Unified Memory.'" *Electronic News,* 5 June 1995.

[2] The chipset was a single module, but was designed to "contain" other modules, specifically a central processing unit and memory chips. This is a case where the physical hierarchy and the design hierarchy are different (see chapter 5).

[3] Doug Green, Graphics Products Manager, Weitek, quoted in *Electronic News,* 5 June 1995, p. 66.

[4] Prem Talreja, Director of Marketing for Core Logic, OPTi, quoted in *Electronic News,* 5 June 1995, p. 66. At the time, OPTi was working with four graphics controller companies to develop a common protocol for communication between core logic and graphics controllers.

$$V_j = S_0 + E(X_1^+) + E(X_2^+) + \ldots + E(X_j^+), \tag{10.2}$$

where X_i is the contribution to overall system value of the ith module. Equation (10.2) indicates that each module's value can be compared with the benchmark established for that module. If the new module design has value greater than zero, meaning that its performance in the eyes of consumers is superior to the existing one's, the new design will be incorporated into the system. Otherwise the existing design will continue to be utilized.

As long as the distribution of the sum of module values remains the same as before the split, the modular approach is bound to yield a higher total value than the unmodularized approach. For one thing, the modular designers could tie their own hands and commit to take all or none of the new designs. If they did so, they could expect their design to perform as well as a corresponding one-module design.

However, the modular designers can also consider module-level improvements. These options only add to the value of the whole. Mathematically, the option values are reflected in the fact that each expectation in equation (10.2) ranges over the positive half of a set of possible outcomes; realizations that are negative (i.e., fall short of the existing design) will be culled out.

A Modular Design Is a Portfolio of Options

Thus, holding the distribution of aggregate value fixed, higher degrees of modularity increase the value of a complex design. This result is a special case of a well-known theorem, first stated by Robert Merton in 1973.[9] For general probability distributions, assuming aggregate value is conserved, Merton showed that for any distribution of underlying value, a "portfolio of options" is more valuable than an "option on a portfolio."

The components of a complex system may be likened to a portfolio of stocks. Holding options on a number of different stocks permits the best outcomes of the risky stocks and a riskless asset to be combined. In a similar fashion, the options built into a modular design permit the best outcomes of old and new component designs to be combined. As long as the component designs are not perfectly correlated, modularity will exploit the lack of correlation, just as options on stocks exploit the lack of correlation of individual securities.

Merton's argument works for any distribution of outcomes, as long as the probability distribution of aggregate value is unchanged by modularization. If modularization makes the distribution of outcomes better, then the conclusion holds a fortiori. If, on

---

[9] Merton (1973b).

the other hand, modularization makes the distribution of outcomes worse, then there may not be a net gain to splitting the parameters and tasks of design.

We have already seen that designers must know about parameter interdependencies to formulate sensible design rules. If the requisite knowledge isn't there, and designers attempt to modularize anyway, the resulting systems will miss the "high peaks of value," and, in the end, may not work at all. Thus a premature splitting of the design will make the distribution of outcomes worse in some cases.[10] Nevertheless, as knowledge increases, the "value penalty" associated with splitting a design and its corresponding task structure will go down. Designers will come to know where and how to split without harming the system. Thus the interesting economic questions are: When does it make sense to switch from an interdependent to a modular design? And how much knowledge is needed to transform a blind gamble on modularity into a good bet?

Equation (10.2) says that if designers wait until splitting is value-neutral (i.e., until the distribution of outcomes is unaffected by the modularization), they may have waited too long. In economic terms, the option value inherent in the modular approach will offset some deterioration in the distribution of outcomes. Although the *average outcome* in a modular design may be worse than in an interdependent design, a system that incorporates the best module outcomes may still outperform an interdependent system that is a bundle of good and bad outcomes.

We should remember, however, that these option-related gains do not reflect the costs of achieving modularity, which include the cost of specifying design rules, the cost of experimentation, and the cost of testing and system integration. Taken as a whole, these costs may substantially reduce or even wipe out the option value of a modular design.

A Square-Root Relationship between Modules and Value

We will now specialize the assumptions to look at symmetric modules with normally distributed outcomes. Our purpose in looking at this special case is to see how much option value is created by a modularization. Thus let N tasks be partitioned symmetrically into j modules. Each module contains N/j tasks, and is independent of all the other modules. Under the assumptions, module values will be distributed normally, with mean zero and variance equal to $\sigma^2 N/j$.

Here we can use the general expression for $E(X^+)$, which was noted above. Each module's value is

[10] Recall IBM's experience with the systems software of System/360 (see chapter 7).

$$E(X_1^+) = .3989\sigma(N/j)^{1/2}.$$

We can substitute this expression for each term in equation (10.2), divide by $E(X_N^+) = .3989\ \sigma N^{1/2}$, and collect terms to get

$$V_j = j^{1/2} V_1. \tag{10.3}$$

(Here it is convenient that S_0 has been normalized to zero. If the system value were not normalized in this fashion, the expression would be more complicated, but the intuition would be the same.)

This result can be interpreted as follows. Suppose a change in architecture converts a one-module product into a twenty-five-module system. (This is approximately what happened when System/360 was modularized.) The overall quality of the design effort remains the same, but designers now have options to mix and match new solutions with old ones in different parts of the system. Equation (10.3) shows that there is a square-root law at work. Ignoring the cost of the new architecture, the twenty-five-module design can be expected to perform five times as well, and be five times as valuable as the one-module design.

Box 10.5 generalizes this square-root law to the case of asymmetric modules. The results are basically the same. Holding the total number of design parameters constant, value increases as the sum of the square roots of the fraction of parameters allocated to each module. Thus successive splitting increases value but at a decreasing rate.

Even though equation (10.3) neglects costs, it still gives us new ways to think about the economic choices designers make and the tradeoffs they confront when designing a complex system. Take the example of System/360, which we discussed in chapter 7. The SPREAD report engineers thought that in the space of *only twelve weeks* they could specify design rules that would allow the new product line to be split into between twenty and thirty independent modules. They did not anticipate any significant degradation in system performance from this modularization.[11]

According to equation (10.3), that modularization should have increased the value of the system by a factor of about five in the first generation after introduction.[12] By

---

[11] Haanstra et. al. (1983); Aron et al. (1983); Pugh et al. (1991).

[12] To be precise, assume that the pure performance of the first System/360 was identical to the unobserved, nonmodular alternatives. (This was true with respect to the enhanced 1400s, which, under Haanstra's auspices, almost made it to market.) In the next generation, System/360 could be upgraded in a modular fashion (in fact this is what happened). In contrast, the nonmodular alternatives would need to be redesigned from scratch. Our model says that,

anyone's accounting, this was a large reward for what initially looked like a mere ninety days of work. In this sense, computer designs were "ripe for modularization" in 1960. The designers saw that by recasting the design structure in seemingly minor ways, they could make the machines much more flexible, as well as compatible, across a wide range of performance and cost.[13]

However, the IBM architects greatly underestimated the costs of delivering a modular computer family on the scale of System/360. The formulation of design rules alone took ten months, and even then they were incomplete. Producing the software and integrating the manufacturing subtasks later absorbed many hundreds of man-years of effort. Luckily, the designers underestimated System/360's value to users by at least as much as they underestimated the cost of designing and building the new system.

The Value of Substitution

The *substitution operator* allows a designer (or user) to swap one module of the system for a better version of the same module. As users, every time we upgrade a piece of software, replace an old hard drive, or buy a faster modem, we are applying the substitution operator to our own systems.[14] And as consumers, we look more favorably on loosely-coupled designs if we know that we will want to upgrade and mix and match components in the future. Thus, intuitively, it seems that the values of splitting and substitution should be related—more of one should enhance the value of the other. Indeed this turns out to be the case.

---

holding the "quality" of the design efforts constant, the ability to mix and match the best of old and new design solutions on twenty-five symmetric modules increases the value of the system by a factor of five.

[13] Haanstra et al. (1983). This perception was fairly widespread in the late 1950s and early 1960s, and other companies besides IBM were experimenting with modular architectures. Thus had IBM adopted a different design strategy (Haanstra's strategy of extending the 1400 architecture, for example), modular designs would still have arrived on the scene, albeit by a different route.

[14] Garud and Kumaraswamy (1994) credit Sun Microsystems with inventing "modular upgradability," which we would call modular substitution. However, the roots of this operator go back to the very first computer designs. An appreciation for the value of substitution can be traced back to the early 1950s. It was one of the things that drove Maurice Wilkes and others to make computers systems more modular rather than less (Wilkes 1985).

Box 10.5 Splitting a Design into Asymmetric Modules

Let X_α be the performance of a module of size αN. By assumption, X_α is a random variable, normally distributed with mean zero and variance $\sigma^2(\alpha N)$. We define z_α as follows:

$$z_\alpha = \frac{X_\alpha}{\sigma(\alpha N)^{1/2}},$$

z_α is normally distributed with mean zero and variance one.

Substituting standard normal variates in equation (2), dropping S_0, and collecting terms, we have

$$V_\alpha = \sigma N^{1/2}(\alpha_1^{1/2} + \alpha_2^{1/2} \dots \alpha_j^{1/2}) E(z^+),$$

where $E(z^+)$ is the expectation of the right tail of a standard normal distribution and equals .3989. (See figure 10.2.)

Thus, under the assumptions of our model, we can easily compare the expected performance of a modular partition of the design to the performance of the corresponding unmodularized effort. We summarize the relationship that comes out of our model in the following proposition:

Proposition 1. Under the assumptions of our model, let a design problem of complexity N be partitioned into j independent modules of complexity $(\alpha_1 N, \alpha_2 N, \dots, \alpha_j N)$ as defined above. The modular design has value

$$V_\alpha = (\alpha_1^{1/2} + \alpha_2^{1/2} \dots \alpha_j^{1/2}) V_1,$$

relative to the corresponding one module design effort.

Proof. By definition, a one-module design has both j and α_j equal to one. Thus $V_1 = \sigma N^{1/2} E(z^+)$. Collecting terms and substituting in the expression for V_α yields the result.

From the fact that $< \alpha_1, \alpha_2, \dots, \alpha_j >$ are fractions that sum to one, it follows that the sum of their square roots is greater than one. Thus, as expected, if we ignore any change in system value and the costs of achieving modularity, a modular design is always more valuable than the corresponding nonmodular design.

Moreover, additional modularization increases value: if a module of size α is split into submodules of size β and γ, such that $\beta + \gamma = \alpha$, then the subdivided module's contribution to overall value will rise because $\beta^{1/2} + \gamma^{1/2} > \alpha^{1/2}$.

We can use our model to quantify the interaction of the two operators. Previously we assumed that designers would create only *one* new design per module. However, they could as well decide to run several parallel experiments on each module and select the best of these outcomes. The value of substitution is the answer to the question, How much better is the first-best from the second- or third-best experimental design? Embedded in this question is another: How many separate, parallel design processes does it make sense to run?

Parallel design processes are in fact experiments that map the "value landscape" of possible outcomes. Thus the value of experiments goes up if designers are ignorant of the possibilities, and if the experiments can be structured to explore a wide range of potential solutions. Intuitively, modularity decouples experimental outcomes and allows the designers to select the best solution for each subcomponent. (In the computer industry, this is known as a "best of breed" approach to a solution.)

To quantify the value of this approach, let us suppose that for a module comprising n tasks, the designers initiate k parallel, independent design efforts. They then have the option to select the best of k outcomes for the final design.[15] Let $Q(\tilde{X}; k)$ denote the "value of the best of k designs," as long as it is better than zero, for a random payoff function $\tilde{X}$.[16] The value of a design process with j modules and k_i experiments in the ith module is then

$$V(\tilde{X}_1 \ldots \tilde{X}_j; k_1 \ldots k_j) = S_0 + Q(\tilde{X}_1; k_1) + Q(\tilde{X}_2; k_2) + \ldots + Q(\tilde{X}_j; k_j) \qquad (10.4)$$

Like equation (10.2), equation (10.4) applies to any set of probability distributions, as long as the system value is unchanged by modularization and module values are additive. But, again, it is too general to be of much use. We can gain additional insight by focusing on normal distributions and symmetric modules.

When referring to standard normal distributions, we can suppress $\tilde{X}$ and denote the value of the best of k designs simply as $Q(k)$.[17] Figure 10.3 shows $Q(k)$ for values of k ranging from 1 to 10. Note that $Q(1) = E(X^+) = .3989$.

15 Stulz (1982) provides a general analysis of parallel options in his valuation of the option to select the maximum of two risky assets. Sanchez (1991) later applied Stulz's framework to the special case of parallel R&D.

16 This distribution of the best of k designs is well-known in statistics: it is the distribution of the "maximum order statistic of a sample of size k" (Lindgren 1968).

17 Formally

$$Q(k) = k \int_0^{\infty} z[N(z)]^{k-1} n(z) dz,$$

where $N(z)$ and $n(z)$ are, respectively, the standard normal distribution and density functions (Lindgren 1968).

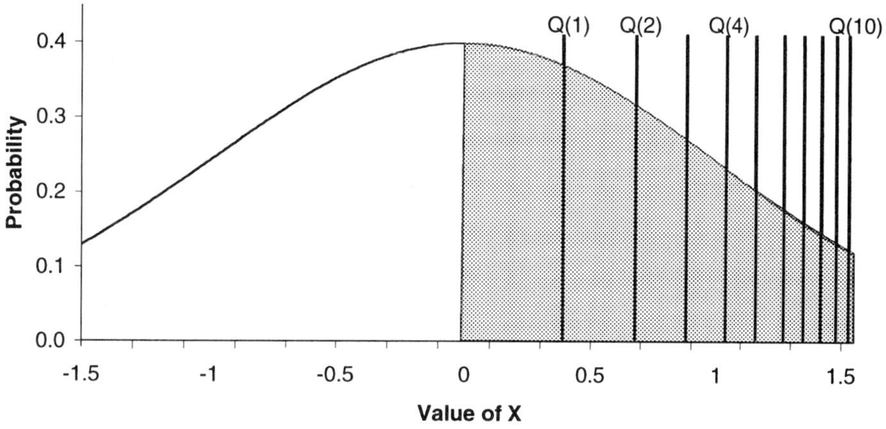

Figure 10.3 $Q(k)$—the expected "best draws" from k parallel experiments.

In general, the appropriate number of experiments to run on a module depends on that module's technical potential, complexity, and visibility to other modules.[18] We discuss these characteristics in the next chapter. For now, however, we continue to assume that modules are symmetric, in which case it is optimal to run the same number of experiments on each module. The $2j$ arguments in equation (10.4) then collapse to two, and the value of the design process as a whole, denoted $V(j,k)$, is

$$V(j,k) = S_0 + \sigma(Nj)^{1/2}Q(k). \qquad (10.5)$$

Plate 10.1 graphs this function in three dimensions. In addition to normalizing S_0 to zero, entries have been scaled so that the value of one experiment in a one-module process is one: $V(1,1) = 1$. Throughout the rest of the chapter, all design processes will be scaled to this value, which we label "the value of the basic process."

Along both dimensions in the figure, value increases at a decreasing rate. On the right edge, we see the square-root relationship discussed above [$V(j,1) = j^{1/2}V(1,1)$]. On the left edge, the function maps out $Q(k)$: [$V(1,k) = Q(k)\,V(1,1)$]. However, the highest rate of increase in the function is along the 45-degree line. Twenty-five

---

[18] Implicitly, we are assuming that it is possible to formulate any number of independent approaches to a given design problem, subject only to cost constraints. This is in essence an assumption about the technical potential inherent in a given module. In reality, modules differ in terms of their technical potential. We treat these and other differences among modules in chapter 11.

experiments on each of twenty-five modules increases the value of the basic program almost twenty-five times![19]

The Economic Consequences of Splitting and Substitution

The economic message of the figure is at once compelling and—for the creators of the design—surprising and dangerous.

It is compelling because the value created by combining the operators of splitting and substitution is large. In the case of System/360, we surmised that splitting alone may have led to a fivefold increase in value over previous designs. But plate 10.1 indicates that this fivefold increase was only the tip of the iceberg. System/360's modular structure permitted "good" experimental outcomes on modules like disk drives and memory to be accommodated within the basic architecture of the system. Because of this fact, the design's true potential lay in combining its modular structure with much higher levels of experimentation than had been contemplated before. By our calculations, modularity plus experimentation could lead to a *twenty-five-fold* increase in the system's value.

The message of the figure is surprising because it shows the splitting operator interacting with the substitution operator in a very powerful way. This interactive effect is hard to perceive without the help of a model—after all, why should a company reevaluate all of its experimental projects just because it adopts a modular product architecture? But our theory indicates that such a revaluation is called for—in fact, it is imperative. The effects of a modularization on the value of experimentation are too strong to be ignored.

To see how this revaluation works in practice, let us look at what happens to the value of "secondary" research efforts when the system architecture goes from one to twenty-five modules. We can relate this example to the actual case of disk drives as

---

[19] It is visually obvious and easy to verify that this function exhibits the property of "supermodularity" with respect to j and k. Formally, let $\mathbf{x}$ and $\mathbf{y}$ be vectors. A function $f(\mathbf{x},\mathbf{y})$ is supermodular on a sublattice $(\mathbf{x},\mathbf{y})$ if $f(\mathbf{x}\vee\mathbf{y} + f(\mathbf{x}\wedge\mathbf{y}) \geq f(\mathbf{x}) + f(\mathbf{y})$ where $\mathbf{x}\vee\mathbf{y}$ is the component-wise maximum and $\mathbf{x} \wedge \mathbf{y}$ the component-wise minimum of the vectors $\mathbf{x}$ and $\mathbf{y}$. If f is smooth, supermodularity is equivalent to the condition: $\partial f/\partial \mathbf{x}_i \partial \mathbf{y}_j \geq 0$ for all $i \neq j$. Intuitively, supermodularity means that increasing investments in one dimension increases the returns to investment in some other dimension. In other words, more modules increase the returns to more experiments. Supermodularity in turn has profound implications for evolutionary dynamics. (See Milgrom and Roberts 1990, 1995; Holmstrom and Milgrom 1994.)

IBM made the transition from its various interconnected architectures to System/360's modular architecture.

At the time System/360 was introduced, IBM was funding the development of new disk drive designs at its R&D center in San Jose. We can calibrate the value of these efforts by assigning the most promising disk drive design in the laboratory a value of one. Our model indicates that in a one-module design, the *second most promising design* would most likely be worth only about 70% of the first.[20] The costs of the second project would have to be set against that value, and it is quite likely that the second project would not be worth funding. For the sake of argument, let us assume the second project was *not* funded. (In this example, the exact number of projects being funded and the value of each are unimportant details; what matters is that there is a funding cutoff. In economic language, we are defining the second project as "marginal" for a one-module design.)

When the system went from one module to twenty-five modules, our model suggests that the value of the first disk drive project went up by a factor of five. That was cause for celebration, but since we assumed the project was already being funded, no change in strategy was called for. However, the value of the *second* project also went up by a factor of five—from 0.7 to 3.5 times the value of the original first disk drive project. Assuming costs stayed the same, it is quite likely that, after the modularization, the second project should have been funded. A third, fourth, fifth, and sixth disk drive project might have merited funding as well!

In effect, our model says, a modularization on the scale of System/360 would call for a massive revaluation of R&D efforts, a large increase in R&D investment, and (probably) wholesale recontracting with the participants in the R&D process. Neither the architects of System/360 nor the managers at IBM who set R&D budgets understood these facts, however.[21] At the time, modularity was viewed through the lens of the status quo (the set of R&D projects currently being funded). The effect of the modularization on the value of additional experiments went unnoticed, and the called-for changes in R&D investment levels and strategy were not made—at least not at IBM.

That brings us to the third part of the economic message: the value inherent in a modular design is potentially dangerous to its creators. The danger arises because

---

[20] This follows from the fact that $V(1,1) = Q(1) = .3989$ and $V(1,2) = Q(2) = .681$. The ratio is 1.707.

[21] The SPREAD report did not discuss this consequence, nor did IBM's R&D/sales ratio change in the 1960s (Haanstra et al. 1983; IBM Annual Reports).

modularity, in addition to creating options and option value, also "moves" the options from the center of the system to the modules. The initial designers of the system thereby "lose control" of their creation. In this respect, it is misleading to view modularity as simply a scheme for multiplying option values. Modularity is also a powerful organizational tool that makes possible decentralized decision making while at the same time maintaining those forms of coordination that are essential to the system's functioning.

What this means in practice is that system designers (or top managers) do not have to "grant permission" for experimentation to take place at the module level. The fact that modules exist means that experiments (new R&D projects) can be started without the architects' even being aware of them. The experimenters simply have to adhere to the design rules for their products to function within the system. Thus a major tool of control—the right to determine what is and what is not part of the whole—passes from system architects to module designers when the former adopt a modular design structure.

System-level architects can retain de facto control over the modules if substantial amounts of system-level testing and integration are needed to finish the design and bring it to market. In effect, if a lot of work remains to be done in stage 3 of the design process, the delegation implicit in the design rules is imperfect. From a purely technical perspective then, module designers can initiate experiments, but their actions must be ratified by the system testers and integrators.[22] In these circumstances, "rogue" modules will have difficulty gaining admission to the system.

However, we have already seen that the rationalization of a modular design and task structure over time will cause more and more out-of-module interdependencies to become codified in design rules. Thus, as knowledge about the design accumulates, the design rules will become more complete, and the number of system integration ("bug-fixing") tasks will shrink. Moreover, in the next section we show that once a set of modules comes into existence, designers will have large incentives to develop decentralized, module-level testing technologies.

These two facts imply that the system integrator's control over a modular design—like the architect's—may be short-lived. Improved design rules and new tests will tend to diminish the need for after-the-fact integration. In the limit of pure "plug-and-play" compatibility, module designers can take over the tasks of testing, and they or the users can perform the tasks of integration.

At that point, the system architects and integrators will have no further role: they will have modularized themselves out of existence. Ironically, then, the better the

---

[22] Fama and Jensen (1983a).

architects of a modular design do their work, the sooner they may find themselves out of a job. That is what we mean when we say that the combined effects of splitting and substitution are potentially dangerous to the creators of such designs.

The Net Option Value of Splitting and Substitution

Of course, modularity is not free. Modular designs require investments in architecture and design rules. Experiments are costly to run, and their results must be tested and integrated to arrive at the final design. Thus, in deciding how much modularity to pursue, the option value of splitting and substitution needs to be compared to the relevant costs.

As a first approximation, it is reasonable to assume that the stage 1 costs of formulating design rules will be roughly proportional to the number of modules, j, and that the stage 2 costs of experimentation will be proportional to the number of experiments, k. Define c_j as the design cost per module, c_k as the cost per experiment, and $T(j,k)$ as the cost of testing j modules and k experiments per module. The cost $C(j,k)$ of a modular, multiexperiment process is then

$$C(j,k) = c_j j + c_k k + T(j,k). \tag{10.6}$$

We are now in a position to write down the expression for the *net option value* of the combined operators splitting and substitution:

$$\text{NOV}(j,k) = \sigma(Nj)^{1/2}Q(k) - c_j j - c_k k - T(j,k). \tag{10.7}$$

This is the economic value designers will "see" in stage 1, as they make decisions about the fundamental design structure of the artifact.

Although they appear symmetrically in equation (10.7), there is an important difference between the cost of modularizing and the cost of experimenting on a system. If a particular architecture and the associated interface designs survive for more than one generation, these elements do not have to be designed again. At that point, the stage 1 costs, $c_j j$, are sunk—they will disappear from value calculations from that time forward. In contrast, the stage 2 costs are incurred each time a different experiment is run, and thus are variable, recurring costs.

The cost of testing depends on whether tests are conducted at the module or the system level. Testing is the main focus of the next section, but for now we will set that term equal to zero. Although this limiting case is unrealistic, it is a useful upper bound on the net option value of the modular operators.

Plate 10.2 shows how the net option value function behaves as the number of modules and the number of experiments per module increase. To complete the numerical specification, we set the value of the basic process, $V(1,1)$, equal to 1, and $c_j = c_k = 1/2$. These assumptions cause the basic process to just break even [NOV(1,1) =0].

In the figure, net option value appears as a single-peaked function, which achieves its maximum at fourteen modules and nine experiments (per module). The complementarity between modules and experiments, which we said could lead to "surprising and dangerous" developments for the architects of a modular design, is quite pronounced here. Because of this complementarity, the value-maximizing strategy cannot be found by looking at the number of modules or experiments independently of one another.[23]

For example, in the figure, if we started with a one-module, one-experiment design and varied only the number of modules, investments in modularity would not appear to be profitable. Each new module "costs" half the value of the basic process. The first modularization increases value by only 40%, and subsequent modularizations increase it even less. In effect, square-root increases in value are just not powerful enough to offset the linear costs of defining additional modules.

Along the other axis, if we again started with the one-module design, it appears that three experiments are optimal. From that new baseline, it is worthwhile to invest in five modules. And with five modules in place, three more experiments are justified (for a total of six). Repeating this process, designers might arrive at the value-maximizing combination of fourteen modules and nine experiments through a series of incremental value-increasing actions. In effect, they can reach the top of the "hill" by a series of small upward steps.

Unfortunately, although it happens to work in this case, such "hill-climbing" does not always work in the sense of leading to the best outcome (the "global maximum" in the value landscape). When actions are complementary, as they are here, plausible cost structures may give rise to multipeaked value functions. Plate 10.3, for example, shows the effect of subtracting a fixed cost of 1/2 $V(1,1)$ from all designs having more than one module. This assumption is reasonable if there is a fixed cost to creating a modular design as well as a per-module cost.

The net option value function in this figure has a local maximum at the point where a one-module design is combined with three experiments ($j=1$, $k=3$). The *global* maximum, however, occurs at the same point as in plate 10.3—fourteen mod-

---

[23] Milgrom and Roberts (1995).

ules and nine experiments. What is noteworthy about this figure is that, if the designers start at $V(1,1)$, there is no one-step move that both increases value and puts their design on the higher-valued slope (from whence "hill-climbing" can reach the global optimum). Moreover, large investments involving *modularization alone* or *experimentation alone* have negative values.[24] It is only when splitting and substitution are combined that the modular design takes on positive net option value.

Thus designers with imperfect foresight or limited control over the design process can easily get trapped in a low-value state. In the example shown here, getting from the original design to higher-valued opportunities requires a discontinuous change in strategy and a coordinated set of investments in both new modules and additional experiments.

For an enterprise to make this leap successfully, three conditions must be met. First, its designers must see that the value is there: they must perceive that splitting and substitution are complementary, and that the value of the combination justifies an investment. Second, those who perceive the value must have the resources to seek it by making the necessary investments. They must be able to change the design strategy so as to "move" in the value landscape along two dimensions at once. Given this pattern of benefits and costs, a big move on one dimension but not the other may be far worse than no move at all. Lastly, the enterprise must have a way to capture the value it creates, and to reward its designers (assuming they are value-seekers, too) for the value that was created and captured by their efforts.

If any of these conditions fails to hold, the design may get "stuck" at an inferior, local maximum.[25] Thus even when every individual actor is motivated to see and seek economic value, ignorance, poor enterprise designs, and inappropriate guidance mechanisms can be substantial barriers to the effective, coordinated use of the splitting and substitution operators.

The Value of Good Tests

Earlier in this chapter, we saw that a design strategy combining modularity with a judicious amount of parallel experimentation created options that could increase

[24] Rather than attempting to plot negative numbers, which leads to confusing graphs, the figure sets negative values equal to zero. The flat surfaces along each axis have negative net option value.

[25] Levinthal (1994a)

substantially the value of a complex design. In formulating the model, we assumed that the designers could find the best modules wherever and whenever they arose. In reality, however, to determine which modules are the best, one must test them.

Testing costs are the Achilles' heel of modular designs. A combinatorial explosion results when many experiments on several modules must all be tested, because even a small number of modules and experiments creates an astronomical number of candidate solutions. The cost of sorting through the different combinations can easily swamp the option values achieved through mix-and-match flexibility.

For this reason, design evolution can proceed only as fast as the designers can devise "good tests" for the modules of the system. Testing technologies act as brakes on the splitting and substitution operators and therefore on the rate of change in the system itself.

To frame this issue, it is useful to define two extreme forms of testing technologies. *System-level tests* embed modules in prototypes of the whole system and measure the performance of those prototypes. *Module-level tests* rely on *principles* that describe the module's contribution to the system-as-a-whole. Such tests measure performance of the module using *metrics* that are appropriate to the module's role in the system. (Recall the discussion of principles and metrics as guidance technologies in chapter 4.) Thus in general, module-level testing requires greater depth of knowledge than system-level testing.[26]

System and module testing are not mutually exclusive; generally both will be used to measure the performance of different elements of a complex system. However, design processes differ substantially in the weightings given to these two forms of testing. For example, Microsoft's "synchronize and stabilize" process for designing large software programs[27] and Intel's "copy exactly" procedures for designing chip fabrication facilities[28] make heavy use of system-level tests. In contrast, Sun Microsystems's[29] methods of designing hardware using off-the-shelf components make comparatively greater use of module-level tests.

---

[26] Depth of knowledge should not be confused with the cost of testing a system or a module. Modules are (by definition) smaller and simpler than the systems in which they function, and thus it is usually cheaper to construct and carry out module-level tests than system-level tests. However, in order to conduct a module-level test, designers must have fairly detailed concepts of how the module contributes to the system's overall functionality.

[27] Zachary (1994); Cusumano and Selby (1995).

[28] Iansiti and West (1997).

[29] Baldwin and Clark (1997b).

PACKING SLIP

DATE	ORDER #
7/19/2000	207845

BILL TO

IBM CORPORATION
ACCOUNTS PAYABLE
PO BOX 9005
ENDICOTT, NY 13761-9005

P.O. NO.	TERMS	DUE DATE	REP	SHIP DATE	SHIP VIA	WWW#	REQ #	STATUS
GOYAL		7/19/2000		8/2/2000	Airborne	PH: NLK		COMPLETE

LINE	QTY	ISBN #	TITLE/DESCRIPTION	SHIPPED	STATUS
1	1	0262024667	DESIGN RULES: THE POWER OF MODULARITY (BALDWICN, C.)	1	8/2

System-Level Tests

When little is known about a complex system and the way customers value it, the only viable testing method may be to build the system, and offer it to customers in the marketplace. Customer purchase decisions then determine value. Customer purchase is, by definition, a system-level test.

As knowledge and experience accumulate, however, the system's designers will develop ways of testing the system before it goes to market. Initially, these tests will be mappings of system-level features onto market value. It will be extremely difficult to form or test hypotheses that address the contributions of individual elements to the whole.[30] Even so, a vaguely understood component or module can be placed in a prototype system, and designers can then test the system to see if it indeed "runs better." This approach does not require much knowledge about what the component or module does, but it may entail costly cycling, rework, and debugging after the fact. The costs of cycling, rework, and debugging in turn can be avoided only through deeper knowledge of the system, its external functions, and its internal structure.[31]

To see how system-level testing constrains the value of modular designs, consider a complex artifact made up of ten components. Assume that users value the artifact for its speed, but that its designers have no idea about what made different versions go faster or slower. Despite this lack of understanding, suppose an architect hit upon a way to modularize the design, so that different versions of the ten components could be mixed and matched at will. Further suppose that there were ten variants of each component available. How would designers then know which specific components to combine to make the best possible system?

Lest the reader think that this is a totally unrealistic example, consider the example of food and food recipes. The analog of speed is flavor. The exact flavor of a finished dish comprising ten ingredients is very difficult to predict, and there are a very large number of variants to be tested. In the course of developing a new dish, even the best cooks have to try many different combinations. Moreover, there are always untested combinations; hence it is impossible to know if a given recipe is indeed "the best possible" for the dish in question.

If the designers of a system lack a valid theory of how each component affects the performance of the whole, outwardly every combination of modules will appear equally good. Then, to find the *best* combination among all possibilities, the

---

[30] John Holland calls this the "credit assignment" problem (Holland 1992, chapters 4 and 5).

[31] Alexander (1964); Clark (1985).

designers will have to measure the performance of every single combination against every other combination. The number of tests in this case would equal 10^{10}, that is, 10 billion tests!

We can generalize this result as follows. If all testing takes place at the system level, and the cost per test is c_t, then the overall cost $T(j,k)$ of testing j modules and k experiments per module to find the best combination is

$$T(j,k) = [(k + 1)^j - 1]c_t. \tag{10.8}$$

Obviously this number increases very rapidly as j or k goes up.[32] For example, recall that in plate 10.3, the optimal combination of modules and experiments assuming no testing costs was fourteen and nine. However, nine experiments on each of fourteen modules results in an astronomical number—10^{14}—of candidate systems to be tested. Plate 10.4a shows what a heavy burden this is. It graphs the net option value of splitting and substitution assuming that a system-level test costs 10% of the value of the basic program [c_t = .1 $V(1,1)$]. (This cost is roughly in line with the cost of building a physical prototype of a complex device like a computer or an automobile.[33]) The figure shows that the option value created by splitting and substitution is completely swamped by the combinatorial cost of testing. The maximum potential option value is almost twenty-five times $V(1,1)$, the value of the basic process in this domain (see plate 10.1), yet the highest *net* option value realizable with system-level tests is only about 12.5% of $V(1,1)$, or one two-hundredth of the potential value! And this value is obtained by running only three experiments on a one-module system.[34]

These numerical results are driven in part by the assumption that the basic design process just breaks even. However, the picture changes very little if we assume that

---

[32] Faced with this problem in reality, designers would not attempt to test all combinations exhaustively, but would settle instead for something less than the best system design. This is "satisficing" by Herbert Simon's definition (Simon 1969). Our point is that the ability to run experiments, which is conferred by having a modular design, is worth very little if the results of the experiments cannot be tabulated at a reasonable cost.

[33] To complete the specifications, we assume that the variable costs of creating design rules and running experiments are each equal to 45% of the value of the benchmark program: [cj = ck =.45 $V(1,1)$]. Thus, as in the previous section, the net option value of the basic process is zero.

[34] In fact, holding the other costs fixed, testing costs must fall to less than 5% of the value of the basic process before it is worthwhile to take the first step toward modularity, and split the tasks from one module to two.

the basic process is very profitable. For example, we might set the variable costs of design and of experiments to 5% of the value of the basic process, while maintaining the cost of testing at 10%. The basic one-module, one-experiment design process would then have a 500% return on investment. The value-maximizing configuration is then to run nine experiments on a one-module system. In other words, a massive change in profitability calls for more experimentation, but no more modularity.

In summary, the combinatorial explosion of system variants that follows a modularization means that, as long as system-level testing is the dominant mode of evaluating trial designs, designers will have no choice but to pursue unmodularized designs with low levels of experimentation. Therefore, heavy reliance on system-level testing will inhibit the development of modular designs. Unless designers can break through the system-testing cost barrier, the option values derived from splitting and substitution might as well not exist.

Very Low System-Level Testing Costs Do Not Solve the Problem

Over the last decade, as a result of advances in computer designs, a major breakthrough in system-level testing costs *has* taken place. Computer-aided design (CAD), manufacturing (CAM), and software engineering (CASE) have reduced the cost of evaluating trial designs significantly. Where once it was necessary to construct a physical prototype to test the behavior of a system, emulators and simulators now provide virtual prototypes at a fraction of the former cost. Design changes that once required costly handwork or rewiring, can now be coded in a few keystrokes. This drop in the cost and time required for design evaluation amounts to a profound change in the guidance technologies available to support design processes.[35]

A triad of IBM designers, working on the AS/400 computer family in 1987, described the difference between debugging a physical prototype and using a sophisticated simulation package to test circuit designs for the AS/400:

> *[I]n the past, the back of a debugged board would look like a plate of spaghetti, with 300 or more wires streching to and fro. But when we debugged the [computer]-tested board, there were only six—*just six *— wires on the back. . . . In the past, debugging a system in its entirety usually took us as long as a whole year. But, thanks to [the simulation], we had a prototype . . . up and running just* six *weeks after debugging began.*[36]

---

[35] Gilder (1989); Thomke et al. (1998); Thomke (1998).

[36] Bauer, Collar, and Tang (1992, pp. 105–106); italics in original.

As automated design tools have spread throughout the computer industry, such stories have become common. For example, in 1996, speaking of the implementation of a PowerPC 604-compatible microprocessor, Gordon Campbell, chairman of Exponential Technology, observed: "We've always anticipated having a turn or two after we got first silicon. It's all built into the schedule. The wonderful surprise is that [the first chip] came back and it was really functional."[37] Campbell attributed this rapid turnaround to "an inordinate amount of simulation" in the early stages of the design process.

Cheap and rapid tests—design verification, validation, and simulation—have indeed wrought great changes in the computer industry over the last ten years. But very low system-level testing costs do not do away with the combinatorial nightmare of modular designs. Plate 10.4b shows the net option value of splitting and substituting when the cost per test is one ten-millionth of the value of the basic program. This is roughly in line with the variable costs of automated testing using emulators or simulators: for a design worth $10 million, the cost per test would be $1.[38]

What is noteworthy in the figure is the rate at which the net option value drops off for higher levels of modularity and experimentation. Six modules and seven experiments are optimal, but it is not worthwhile to pursue any more of either. The option values are still there, but with system-level tests it is too difficult to cull the best modules out from the myriad combinations. In fact, if by some chance the system happened to have more "natural" modules than the six shown here, designers would have to lump them together in some fashion to bring testing costs within bounds.

The reason for the dramatic falling off in value can be found in equation (10.6), wherein the number of modules enters as an exponent of the number of experiments. As long as modularity affects costs in this way, almost no reduction in the cost per test will suffice to make high levels of modularity economically attractive. In some design arenas, testing costs have come down by six or seven orders of magnitude in the last decade. This means that a given test may cost one millionth to one ten-millionth of what it did ten years ago. However, with only system-level tests, costs would have to come down another million times to make a fourteen-module design economical.

Faced with a combinatorial explosion of alternatives, the designers may abandon the goal of finding the absolute best design and turn to a less perfect, but more cost-

[37] Gordon Campbell, quoted in *Electronic News,* 6 May 1996, p. 1.

[38] Our assumption ignores the investment in tools, models, and training needed to use these technologies. However, changing the cost by an order of magnitude does not affect the figure in any important way.

effective form of testing. One commonsensical approach is to rely on the designers' intuition and experience. Another approach is to define subsystems (groups of modules) and test combinatorially at that level. Still another approach is to use genetic algorithms to "evolve toward" better system designs. However, all of these approaches tend to break down as the designs become more complex.

Module-Level Tests

Thus where system-level testing is the rule, lowering the cost per test, even by many orders of magnitude, will not in itself be enough to justify investments in high levels of modularity. However, the fact that testing costs can foreclose a set of valuable modular options provides strong incentives for designers to create better ways to test their designs.

A different way to attack the problem is to design tests so that the best module can be selected "at its own level." To evaluate a module without embedding it in a prototype system requires detailed knowledge about what the module contributes to the whole, as well as how different modules interact. In particular, dysfunctional interactions (like one module transferring heat to its neighbor, or a subroutine failing to return to the calling program) must be understood, so they can be avoided.

Defining the role of each module (i.e., how it functions within the system) and measuring how much each contributes to the performance of the system-as-a-whole requires knowledge.[39] Such knowledge does not come free or ready-made. However, it is the same kind of knowledge as that needed to effect a successful modularization. Hence designers who have mentally split a system into components and are working to modularize the design and corresponding task structure are well on their way to being able to test modules "at their own level."

Knowledge at the module level gets built up in the following way. First, as individual designers gain experience with a specific module, it is natural for them to develop a better understanding of that particular part of the system. They will be able to "see" characteristics of the module, and can then frame and test hypotheses relating those characteristics to system performance and value to the user. The result will be new principles and metrics that define and calibrate the quality of a module.

When these principles and metrics are in place and the hypotheses about how modules affect system performance have been validated, the designers can approach testing in a new way. Instead of employing exhaustive, combinatorial system-level

[39] Clark (1987); Henderson and Clark (1989).

tests, they can test module implementations against module metrics. Practice will then shift from "build and test the system" to "model the system and test the modules." In this fashion, testing technology progresses from a more primitive to a more advanced state.

If module-level testing replaces system-level testing across the board, the cost of testing will fall precipitously for higher levels of modularity and experimentation. Specifically, the cost of testing j modules and k experiments per module becomes

$$T(j,k) \;=\; jkc_t, \qquad (10.9)$$

where c_t again denotes the cost of one test. In this expression, the number of modules, j, now enters as a multiplicative factor and not as an exponent. For a system with fourteen modules and nine experiments on each module, this means that the cost of discovering the best combination drops from 10^{14} times the cost of one test to merely 126 times the cost of one test.[40]

All of this implies that design processes that can rely on module-level tests will have very different structures from those that must depend on system-level tests. Where system-level tests predominate, there is very little to be gained from higher levels of modularity. Module-level tests, in contrast, make high levels of modularity both feasible and profitable.

The impact of a change from system-level to module-level testing can be seen in table 10.1. Shown are the points of highest net option value for testing costs ranging from one-tenth to one-billionth of the value of the basic process. (For these calculations, the unit costs of modularization and experimentation were set equal to .45, so that the basic process combined with the highest testing cost just broke even.)

The figure shows how the levels of modularity and experimentation are endogenously determined by the interaction of option values and testing costs and testing technologies. For example, if designers face testing costs of 10% of $V(1,1)$ and system testing is the norm, then there is little to be gained from modularization. If system-testing costs fall to 1%, designers will have an incentive to split the system into two modules and run three experiments on each. With module-level tests, however, a 1% testing cost calls for eleven modules and seven experiments.

Comparing the first and second panels of the table, it is clear that high levels of modularity are attainable, even with relatively high testing costs, if designers can test

---

40 For this comparison, we assumed that it costs the same to test a module as a system. In reality, modules are less complex than systems, and thus module tests may be cheaper than system tests.

Table 10.1 Value-maximizing combinations of modules and experiments for different testing costs (Cj=Ck=.45)

Cost per test—fraction of V (1,1)	Modules	System testing experiments	Value	Modules	Module testing experiments	Value
0.1	1	3	0.13	2	3	0.30
0.01	2	3	0.75	11	7	2.38
0.001	3	5	1.26	19	11	3.63
0.0001	4	5	1.69	21	13	3.83
0.00001	5	5	1.98	22	13	3.86
0.000001	6	6	2.20	22	13	3.86
0.0000001	7	7	2.47	22	13	3.86
0	22	13	3.86	22	13	3.86

at the module level. However, the question then arises, How does the knowledge needed to test eleven modules emerge before the modules themselves come into existence?

Designers can develop module-testing capabilities by means of a bootstrapping process. That is, they can first use whatever knowledge they have to split the task structure, and then experiment on the split system. With data from these experiments, they can develop module-level tests and then proceed to split again. The result will be a steady increase in the number of modules and in the value of the system. If other costs stay fixed, our hypothetical system, which started out with two modules, can be transformed into one with eleven modules by means of this process.

In general, however, costs will not stay fixed while this process runs its course. The costs of design, experimentation, and testing will all tend to decline over time because of the accumulation of system-specific knowledge, and ongoing improvements in design tools and technologies. A decline in any of the three cost elements of a modular design—architecture, experimentation, testing—will in turn increase the number of modules that the system can sustain.[41]

Figuratively speaking, at each point in time, there will be a "hill" in the value landscape, whose highest peak represents the best combination of modularity and

---

[41] The impact of cost declines on the optimal number of experiments is more ambiguous. For example, when system testing is the norm, a decline in the cost of testing, holding other costs fixed, leads to an *increase* in the optimal number of modules and a *decrease* in the optimal number of experiments. When module testing is the norm, any cost decline increases the optimal number of modules and experiments.

experimentation for the current set of costs (including testing costs). To the best of their ability, value-seeking designers will attempt to identify the value-maximizing combination and implement it. Metaphorically, they will try to climb to the top of the hill of the moment.

Over time, however, the costs that determine the locus of the peak will change, causing the "ground" underneath the design to shift. If we could track the highest-valued peaks over time, we would see them moving steadily "outward" toward higher levels of modularity, more experimentation, and better, cheaper tests.[42] As long as costs keep declining and tests keep improving, investments in splitting and substitution will continue to create value, and the overall design will continue to evolve along these pathways.

Design Evolution

Exploiting the option values inherent in splitting and substitution requires three types of investments:

1. in design rules to establish the architecture and interfaces;
2. in independent experiments to explore the possibilities inherent in the design;
3. in module-level tests to identify superior combinations efficiently.

These investments are highly complementary. Investing in one or even two creates much less value than investing in all three together.

These investments also occur in a fixed order. The definition of design rules precedes experimentation, and some experimentation usually precedes the development of module-level tests. Figure 10.4 shows how this dynamic pattern of investment is related to the evolutionary framework, which we laid out at the beginning of Chapter 9 (refer back to table 9.1). In this process, a "generation" begins with the formulation of design rules that define a modular system. These modules become the *units of selection* in the evolutionary process. If they are new, then the only *mechanisms of selection* available will be system-level tests. Value-seeking behavior then justifies a set of experiments (potential substitutions), which are the *sources of variation* in the evolutionary process. The experiments will generate a variety of designs, which can be evaluated and selected using the existing testing technology. They also provide

[42] This follows from the supermodularity of the underlying functions (Milgrom and Roberts 1995).

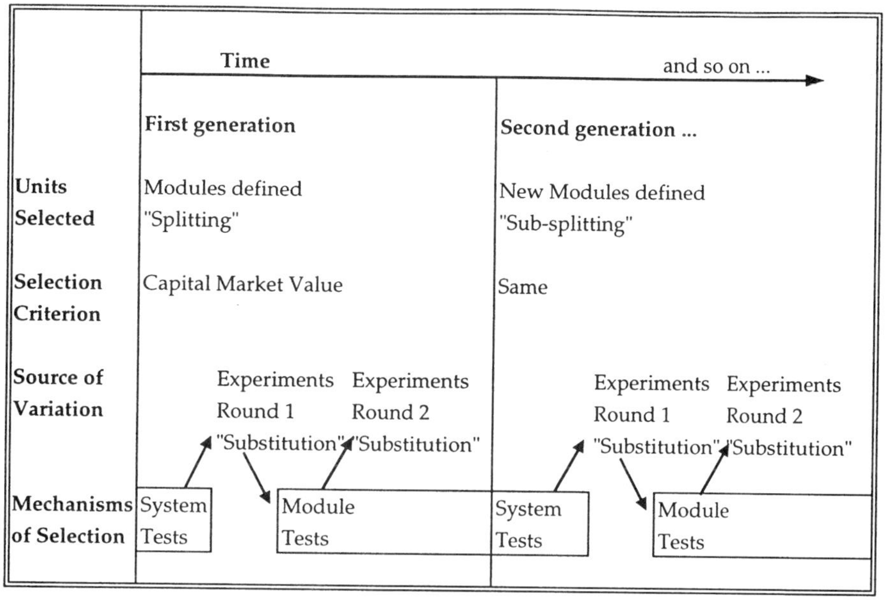

Figure 10.4 Splitting and substitution in an evolutionary framework.

data needed for the framing of module-level tests. And the module tests in turn will provide the impetus for another round of experiments.

At the end of this "second wave" of experiments, the full option value of splitting and substitution for this degree of modularity may have been exploited. However, from their experiments and tests, designers will have learned even more about the underlying design structure. In particular they may know which interdependencies in the existing structure offer opportunities for further splitting. This knowledge may then set the stage for another round of splitting and substitution.

In this process, the units selected and the mechanisms of selection evolve in tandem. It is only when these two elements coevolve that sustained improvement in the design is possible.

This process of design evolution is "path dependent," meaning that each generation's design is inextricably tied to knowledge gained in the preceding generation. The net option value that designers "see" in the present depends in fundamental ways on investments made in the past. However, if we were to step back and look at the system as a whole, we would see the same basic process repeated across the system. Stripped to its essence, the basic pattern is splitting, followed by substitution, followed by splitting again.

This sequence may occur over and over again in various contexts throughout the system. It is a pattern that applies to the system as a whole and transcends the histories of the individual components and modules. Thus the overall *pattern* of design evolution is independent of the path taken by any particular module design.

Clearly, over time, successive rounds of splitting and substitution can cause the value of a system to be dispersed over a large set of modules. Broadly speaking, this is what has happened in the computer industry over the last thirty years: the industry split into sixteen subindustries, and approximately one thousand firms came to be participants in an extended "modular cluster."

Without design rules providing strict partitions of information and tasks, a modular cluster type of industry structure would not have been possible. But, there are still many unanswered questions and issues. First, our assumptions about symmetry are unrealistic: complex systems do not divide nicely into symmetric modules and experiments. Second, there are four other modular operators that may also influence the path of development of a complex set of designs. Finally, design evolution via modular operators does not require a cluster-type industry structure; in theory, it could take place within one or a few big firms.

In the next three chapters, we consider the net option value of a system made up of asymmetric modules, and then go on to discuss the values of the other four operators. Then, in chapters 14 and 15, we look at how modular designs made possible the formation of a "modular cluster" of firms and markets in the computer industry during the 1970s. Volume 2 will then consider how modular computer designs and this modular cluster evolved in tandem during the 1980s and 1990s.

11 All Modules Are Not Created Equal

The defining characteristic of modules is that they are independent of one another, constrained only by their adherence to a common set of design rules. In the early stages of a modularization, this degree of independence may be more of an ideal than an accomplished fact. Nevertheless, as we saw in the case of Mead and Conway's multiproject chip in chapter 3, and in the creation of System/360 in chapter 7, as the process of modularization moves forward, the lingering conflicts do tend to be worked out, so that eventually "true" modular independence is achieved. When that time arrives, the future of a given module can be decided by its own designers. This is an important milestone in the economic history of the design.

If incremental module values are additive,[1] the net option value of the system can be expressed as the sum of the net option values of its individual parts. For a system of *j* modules we write

$$\text{System value} = S_0 + \text{NOV}_1 + \text{NOV}_2 + \ldots + \text{NOV}_j. \tag{11.1}$$

If we focus on one specific module, we can expand this expression and thereby gain insight into the factors that drive investment in, hence the evolution of, a single module.

To fix ideas, we adopt the assumptions set forth at the beginning of chapter 10, but now allow modules to vary on the following three dimensions:

1. the number of tasks needed to redesign the module (denoted n);
2. the standard deviation of potential value (denoted σ);
3. the "visibility" of the module to other parts of the system.

---

[1] If system and module values are not additive, modular designs will still generally have option value, but the expressions will be more complicated. See note 6 in chapter 10.

We can then expand any one of the net option values of equation (11.1) as follows:

$$\mathrm{NOV}_i = \max_{\substack{k_i,\ \text{other} \\ \text{decision} \\ \text{variables}}} \underbrace{\sigma_i(n_i)^{1/2}Q(k_i)}_{\text{Option value}} - \underbrace{C_i(n_i)k_I}_{\text{Cost of experiments}} - \underbrace{Z_i}_{\text{Cost of visibility}} . \qquad (11.2)$$

This is a generic expression for valuing the option to redesign a module, and so we will look at it in detail.

The maximization operator (max) at the beginning of the expression indicates that decisions must be made: the parameters chosen will then fix the value of the function. The decision criterion is the maximization of capital market value (refer back to chapters 4 and 9).

At this point in the design process, the modules have already been defined, hence the number of modules, j, is not a decision variable. However, decisions to run further experiments are subject to review, hence the number of experiments, k_i, *is* a decision variable. In addition, there may be other decisions—affecting the number of tasks, technical risks, and so on—that need to be considered at this point. Rather than try to enumerate them, we lump these under the heading "other decision variables."

The Option Value of Hidden-Module Design Efforts

The first term in the expression represents the option-value benefit of experimenting with alternative designs of the hidden module in question. The number of experiments k_i may now differ from module to module, hence the subscript i. In addition, modules may differ in terms of their "size" and their "technical potential." We parameterize "size" in terms of the number of tasks, n_j, and "technical potential" in terms of standard deviation of outcomes, σ_i.

Given the assumptions about uncertainty discussed at the beginning of Chapter 10 (normal distributions, zero means, variance proportional to the number of tasks), the option-value benefit of running k_i experiments on module i's design is the product of

1. the standard deviation coefficient, σ_i;
2. the square root of the number of tasks, $n_i^{1/2}$; and
3. the expectation of the best of k_i trials drawn from a standard normal distribution, $Q(k_i)$.

This term encapsulates a set of hypotheses about which modules afford the best targets for design efforts. Ignoring costs for now, our model says that the economic value of module experimentation goes up as the module's technical risk goes up, as its complexity (measured by the number of tasks) goes up, and as the number of experiments goes up.

At first glance, some of these relationships (e.g., the best modules to redesign are those with the most technical risk) may seem surprising. However, these counterintuitive findings are the direct result of the options embedded in the modular architecture. Recall that when designers initiate the redesign of a particular module, they are not committing themselves to accept the outcome. The modular structure of the system gives the designers the option to reject any new module design that is inferior to those already in existence.

If the downside risk of "bad draws" in the design effort can be controlled by rejecting bad outcomes, technical risk and complexity may be good things. The reason is that modules with more technical risk or complexity may have wider distributions of outcomes than other modules. If the downside risk can be controlled, that leaves only the "upside risk"—the possibility that the experiments will uncover very good designs (high peaks in the value landscape). The designers only have to take the best, and it is the expectation of the best (of k_i trials) that enters the valuation expression.

The World Turned Upside Down—Implications for the Management of Design Processes

What this means is that switching from an interconnected to a modular design and task structure will have a dramatic impact not only on overall system value (as we saw in Chapter 10) but on the relative values of different components of the design. If the overarching system is interdependent, then the designers will be committed to accept the outcome of whatever approach they initially select for each component. Precommitment is implied by all the interdependencies in the design, which appear as **x**s, sprinkled throughout an interconnected task structure matrix (TSM). After the fact, each subsystem's design will be "wired in" by hundreds of coordinating links and fixes taken along the way.

In this context, narrowly predictable outcomes are valuable, for if a component design is poor, it can compromise the whole system. The designers must live with the consequences of their earliest decisions, and thus they will rationally be inclined to shun component designs involving a lot of risk.

All this changes in a modular design, however. There is no longer a need to place a premium on predictability of outcomes. Instead, the most valuable components (now separate modules) are those offering many avenues of independent exploration. Pursuing those avenues may lead to positive surprises. Negative surprises can be rejected, and will never enter the system.

The reranking of design projects that follows a modularization can pose a subtle trap for designers whose experience is derived from interdependent systems, and who imperfectly understand the relationship between modularity and option value. With interdependent systems, designers may learn through trial and error to favor tried-and-true solutions, to seek incremental improvements, and to avoid approaches with highly variable outcomes.[2]

These intuitions are the right ones to apply to interdependent designs, but the wrong ones to bring to modular systems. A large-scale modularization thus requires not only an upward revaluation of R&D projects (as we suggested in chapter 10) but also a reassessment of the relative merits of existing and potential projects. Other things being equal, this reassessment will cause resources to be withdrawn from low-risk, low-payoff projects, and redirected toward higher-risk, higher-payoff projects. However, to exploit the potential of a new modular design, the designers must understand the implications of delayed commitment and of the options that modularity creates. Intuition gained in the old setting will not serve them very well in this respect. A modular design will overturn the standard relationships: old rules of thumb that may have worked well in an older interdependent design process will be counterproductive in the new context.

Costs of Experimentation and Visibility

The second and third terms in equation (11.2) summarize the costs of module experimentation. The costs of creating the modular architecture are sunk and hence do not appear. We have divided the remaining costs of redesign into the module's "own cost," and the costs incurred in the rest of the system. With respect to own cost, we have lumped the costs of experimentation and testing together and made the total a function of the module's complexity (measured by the number of tasks, n_i). This formulation assumes that testing takes place at the module level, and that it is more expensive to experiment on and test modules that are "larger" in the sense of having a greater number of design parameters and tasks.

The last term in the equation represents the costs incurred in the rest of the system. If the module's design parameters are "hidden" from the rest of the system, then this cost is zero. However, certain modules embody visible design rules that must be followed by the "hidden modules" if they are to function as part of an integrated

---

2 This is a form of organizational rigidity. See the discussion of "frozen organizations" in chapter 2.

whole. For example, computer instruction sets are often hardwired in the control lines of the central processing unit, and thus redesigning a CPU means redesigning many other parts of the system. The "costs of visibility" are positive in such cases.

The measure of a module's visibility is the number of *other* modules that "see" the visible information contained in it.[3] For any module, this measure can be ascertained from the system's design hierarchy diagram: the modules that would need to be redesigned are those that "look up" (directly or indirectly) to the module in question. (Refer back to figures 3.5 and 9.1.)

Which Modules Are Most Valuable?

Let us now look at how net option value varies for modules that differ in terms of their size and visibility. We focus on two values (large, small; hidden, visible) in each dimension. Figure 11.1 gives examples of modules that fall into each of the four resulting categories, and graphs the net option value for each combination of size and visibility. (The detailed assumptions used to construct the graph are given in box 11.1.)

Figure 11.1 shows that differences across modules on these two dimensions give rise to dramatically different incentives for investment. The optimal course of action is to fund *nine or ten* projects aimed at improving small, hidden modules, *three* projects for the large hidden modules, and *no* projects for the visible modules.

The particulars of these optimizations are not important, for if we varied our assumptions we could generate a large number of different scenarios. The key point is that different parts of the system present investors and designers with radically different incentives. As a result, in an open and informed capital market, different modules would be expected to attract different amounts of investment. Some would attract none; others would attract enough to fund many parallel projects. Hidden modules, especially, are an attractive focus of investment, while visible modules are not so attractive.

Different rates of investment across modules in turn will lead to different rates of improvement in the underlying designs. Because they are the focus of considerable experimental effort, hidden-module designs should evolve very rapidly. There will be turnover among the best designs, which may be reflected in the market shares of

---

[3] This definition is carefully stated because it is possible—in fact common—to modify architectural modules without changing their "appearance" to other modules. In such cases, the preexisting visible information becomes a "legacy design rule," and a constraint on the next round of redesign. We discuss this possibility in the next section.

Size	Visibility	Examples
large	hidden	disk drive; large application program
large	visible	microprocessor; operating system
small	hidden	cache memory; small application program
small	visible	instruction set; internal bus

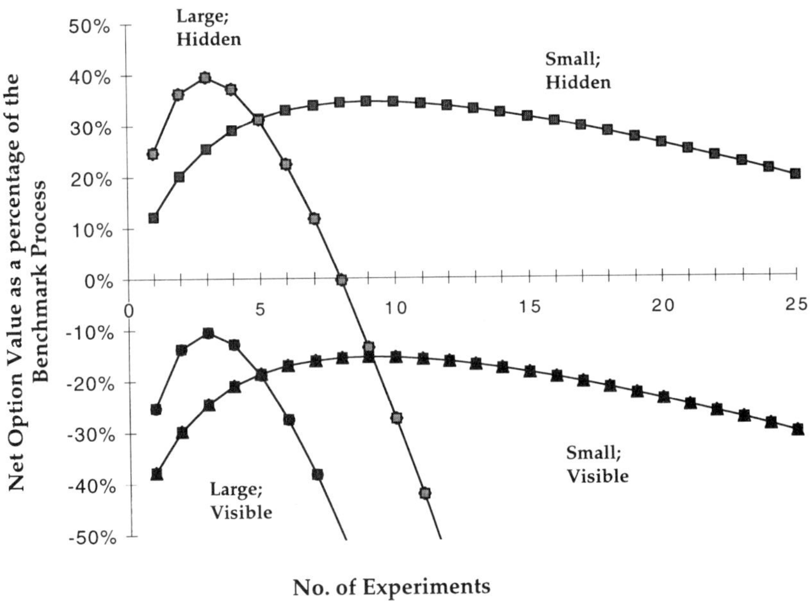

Figure 11.1 Module net option value for different combinations of size and visibility.

Box 11.1 The Effect of Module Size and Visibility on Incentive to Experiment: Numerical Assumptions

The "system" comprises N tasks.

The option value of one experiment on an unmodularized design process is one:

$$\sigma(N)^{1/2}Q(1) = 1$$

Costs of redesigning a module are proportional to the number of tasks in the module. Costs per task, c, are the same for all modules:

$$C(n_i) = cn_i$$

One experiment on an unmodularized system just breaks even:

$$\sigma_N(N)^{1/2}Q(1) - cN = 0.$$

Module size (number of tasks) and visibility are measured in relation to the whole system.

- A "large" module constitutes 20% of the system; a "small" module constitutes 2%.
- A highly visible module sets design rules for 50% of the system; a hidden module sets design rules for none.

The cost of visibility is proportional to the number of modules that "see" the visible module, and the number of tasks in each of those modules. Costs of redesign are the same whether the effort is aimed at discovering superior designs, or accommodating changes in visible information:

$$Z_i = \sum_{j \text{ ``sees'' } i} cn_j.$$

firms in the industry. In contrast, architectural modules that embody visible information will tend to change more slowly.

Legacy Design Rules in Architectural Modules

Our calculations do not take account of the opportunity that may exist to change an architectural module without changing its visible information. As we indicated in Chapter 7,[4] this type of change involves imposing a "legacy design rule" on the design process for the architectural module. The attractiveness of doing so can be easily

[4] Refer to chapter 7, Legacy Design Rules, and box 7.1.

understood in terms of equation (11.2). With legacy design rules ensuring backward compatibility, the costs of visibility, Z_i in equation (11.2), will go down. Counterbalancing this effect, the cost of experiments may increase, and the overall technical potential of the design effort may decline as well. However, the drop in the cost of redesigning other parts of the system (captured by the decline in Z_i) can easily outweigh the other effects. In that case, a legacy-constrained redesign of an architectural module may be worthwhile, when other types of experimentation on that module are not.

An interesting case arises when the net option value of the design effort is negative for low values of k_i, positive for intermediate values, and negative again for high values. (See figure 11.2.) This pattern is not an anomaly: It implies that in all likelihood several experimental efforts will be needed to achieve positive net returns.

When there are significant costs of visibility to overcome, one or even two trials may not lead to a design good enough to justify the costs of redesigning other parts of the system. An example of this phenomenon can be found in the early history of Microsoft Windows. Operating systems are a quintessential architectural module, for they embody a host of commands (application programmer instructions, or APIs) that are visible to applications programmers. Thus a major change in an operating system may require changing as many as 50% of the hidden modules of the system.

When Microsoft sought to incorporate a graphical user interface (GUI) in its operating system, it did impose legacy design rules. It wanted Windows, the graphical system, to be compatible with its earlier, character-based operating system, MS-DOS, which had a large installed base of users. Therefore Windows was initially designed as a shell on top of MS-DOS. (Refer back to our discussion of the *porting* operator in chapter 5.)

Partly as a result of the MS-DOS legacy, the earliest versions of Windows were not very good, and did not gain wide market acceptance. These early experiments did not overcome the costs of visibility—that is to say, applications programmers did not change their coding practices to make use of the new Windows APIs.

The third version of the operating system, Windows 3.1, turned out to be good enough to win market acceptance.[5] It was still based on MS-DOS, and suffered

[5] For simplicity, figure 11.2 is drawn as if no learning would take place between successive trials. This is a fair assumption if experiments proceed in parallel, but is less appealing if they take place in series. Learning can be incorporated into the model by assuming that distribution of outcomes shifts from one trial to the next and that the "exercise price" for the next trial is

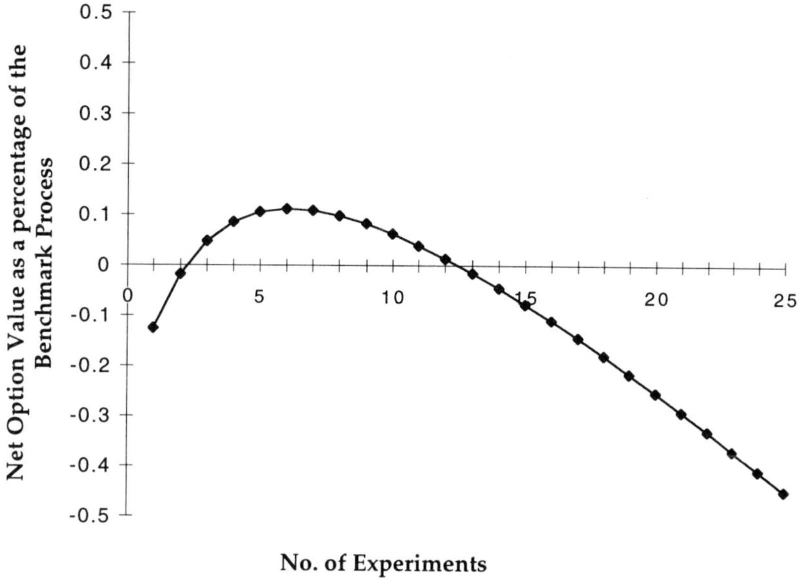

Figure 11.2 A module that requires more than one experiment.

on that account, but its attractiveness to consumers caused it to clear the "visibility hurdle." At that point, thousands of programmers adopted a new set of visible design rules (the Windows 3.1 APIs) and began writing applications that made use of the Windows GUI.

From the standpoint of design evolution, four things are worth noting about this episode. First, by making Windows compatible with MS-DOS, Microsoft effectively reduced the costs of visibility, but did not eliminate them. It could not eliminate all those costs because, to take advantage of Windows' GUI capabilities, programmers had to learn new instructions and coding procedures.

Second, the legacy design rules imposed to achieve backward compatibility were a burden that detracted from the performance of Windows. Windows 3.1 users, for example, had to live with short, cryptic file names because of the underlying structure

---

the best of the previous outcomes. The option values that result are analytically messy, but susceptible to Monte Carlo simulation. In terms of figure 11.2, learning between trials would cause the net option value functions to rise more initially, and fall off less steeply for high values of k. The optimal number of trials would rise, but other characteristics of the analysis would remain the same.

of MS-DOS, which still governed the heart of the operating system. But abandoning the legacy, and thereby incurring the full cost of visibility, might have been worse.

Third, in IBM-compatible personal computer systems, the operating system was an independent, upgradable module. Therefore the substitution of a new operating system for an old one was an option. Whether Microsoft liked it or not, that option was in the hands of users and software developers. In effect, software developers and consumers could "wait and see" how good each version of Windows was, and then decide whether or not to embrace it. Their fallback was to continue using MS-DOS, and so to be accepted, Windows had to be better than its predecessor.[6]

Fourth, because of the module's visibility, the implied "exercise price" of the "Windows option" was relatively high. This meant Windows had to be quite a bit better than MS-DOS before it could gain widespread market acceptance. In fact, it took three trials (with an unknown number of internal iterations) to get a design that was "good enough" to overcome the hurdle of visibility and justify the costs of redesigning major chunks of software.

When there is a high price to exercising a design option, we would expect that design to change very slowly. Moreover, if the high exercise price is due to costs of visibility, we expect designers to try to hold these costs down by imposing legacy design rules. For this reason, *architectural modules*—those that contain visible information—have a tendency to display more continuity in their designs across generations than hidden modules. The "best successor" to a particular architectural module is likely to be a close descendant of the previous design and not a radical new departure. As an example of this principle, the most valuable implementation of a GUI interface for IBM-compatible personal computers was Windows, a descendant of MS-DOS, and not a totally new operating system.

The visible information embodied in architectural modules will persist even longer than the modules themselves. Such information will have a tendency to be passed from one generation to the next in the form of legacy design rules.[7] This in

---

[6] Formally, the costs of visibility would be incurred only after the quality of the implementation was known.

[7] This is a form of memetic evolution: the visible information tends to be replicated many times over and to persist over time, despite potentially problematic performance. The survival of inferior, but widely adopted standards is a well-known phenomenon in the economics of networks. See, for example, Farrell and Saloner (1986 a, b); Katz and Shapiro (1986, 1992); Economides (1994).

turn suggests that contractual claims and property rights to visible design rules can be extremely valuable assets.[8] We return to this issue in volume 2.

A System Made Up of Heterogeneous Modules

We now look at the net option value of redesigning a whole system made up of large and small, hidden and visible modules. Our starting point is a breakdown of the hardware components for a computer workstation in 1990.[9] This list was compiled by Andreas Bechtolscheim of Sun Microsystems, and appears in John Hennessy and David Patterson's widely used text on computer architecture. The list does not include system software (except for the I/O interfaces) or any applications. To be fully functional the system would need an operating system, and various applications. These in turn might be small, medium, or large modules in their own right.

Table 11.1 shows our assumptions. It lists the modules of the system, their costs, and our estimates of the complexity, visibility, and technical potential for each one. Somewhat arbitrarily we defined a basic system to be a workstation with sixteen megabytes of random access memory (RAM), a color monitor, a keyboard, mouse, and a hard disk or a tape drive. Using Bechtolsheim's costs, this leads to a system with hardware component costs of approximately $5500 (in 1990 dollars). Direct costs of building the units would bring hardware costs up by 20% to around $6500.[10]

Also somewhat arbitrarily, we assumed that software costs would be equal to hardware costs. Given 50% gross margins for hardware and software, the fully-loaded base system would retail for around $26,000—this was in the lower end of the range for high-performance workstations at the time. (Personal computer systems of that era used cheaper versions of the same components: a typical fully-loaded IBM-compatible system at the time might retail for $3000 to $5000, and a Macintosh for $5000 to $7000.)

Plate 11.1 is a three-dimensional representation of the "value landscape" of net option values associated with experiments on the modules of this hypothetical workstation system. Each "hill" in the figure is a slice of the value landscape along the

[8] On the value of controlling visible information in a complex design, see Ferguson and Morris (1993); Morris and Ferguson (1993); Baldwin and Clark (1997a).

[9] Hennessy and Patterson (1990, p. 63).

[10] Ibid., p. 65.

Table 11.1 Modules of a computer workstation, circa 1990

	Cost	Relative complexity	Visibility	Technical potential
Hardware				
CPU Board				
Integer unit	$300	1.8%	100%	1
Floating point unit	$200	1.2%	100%	1
Memory management unit	$150	0.9%	0%	1
Cache memory	$150	0.9%	0%	1
DRAM/MB	$150	0.9%	0%	1
Video logic (color)	$500	3.0%	0%	1
I/O interfaces SCSI, Ethernet, floppy, PROM, time-of-day-clock	$100	0.6%	100%	1
Printed circuit board	$50	0.3%	0%	0.1
I/O devices				
Keyboard, mouse	$50	0.3%	0%	0.1
Display monitor (color)	$1000	6.1%	0%	0.5
Hard disk	$400	2.4%	0%	1
Tape drive	$400	2.4%	0%	1
CPU cabinet				
Sheet metal, plastic	$50	0.3%	0%	0.1
Power supply and fans	$55	0.3%	0%	0.1
Cables, nuts, and bolts	$30	0.2%	0%	0.1
Shipping box, manuals	$10	0.1%	0%	0.1
Manufacturing				
Assembly and test	$1200	7.3%	0%	1
Software				
System software				
Operating system	$400	2.4%	50%	1
Applications software				
Software development tools	$5000	30.3%	0%	1
Large engineering (per application)	$5000	30.3%	0%	1
Large commercial (per application)	$1000	6.1%	0%	1
Medium (per application)	$250	1.5%	0%	1
Small (per application)	$50	0.3%	0%	1
Total	$16,495	100.0%		
Approximate System Cost				
16-MB RAM, color monitor Disk or tape drive, CAE or CASE Other software	$13,000		Retail: $26,000	

dimension k_i, the number of experiments performed on the ith module. Table 11.2 shows the numbers behind the figure.

In constructing the picture shown in plate 11.1, we used the cost of each component divided by the total cost of all components as a proxy for the complexity of a module. We judged that the instruction set of the integer and floating point units, as well as the I/O interfaces, were visible to *all* of the system. The operating system, we assumed, was visible to the applications software, or about 50% of the system. The rest of the modules we assumed were hidden behind their respective interfaces.

Finally, we did not attempt to construct a finely-tuned estimate of the technical potential of each module, but we did distinguish between the electronic and software components, the monitor, and the nonelectronic components, such as the keyboard and mouse and the power supply. We assumed that the electronic and software components had twice the technical potential of the monitor and ten times the potential of the nonelectronic components.

The scale we used was percent of total system revenue. For example, Bechtolscheim's costs were based on a total market size of 100,000 workstations. Assuming an average selling price of $26,000, total system revenue would be $2.6 billion. According to our calculations, the first redesign of the memory management unit, for example, would enhance system revenue by about 3.8% (around $100 million) and cost about one-tenth of that ($10 million), yielding a net option value of 3.4% of total system revenue, or $88 million.

The figure shows that, even taking this broad-brush view of module differences, there is great heterogeneity in module net option values. Some modules (the visible ones and those with low technical potential) warrant almost no investment. Others (the large, hidden modules) offer significant gains to a small number of trials, but their economic potential is quickly exhausted. Still others (the medium and small hidden modules) are not tapped out so quickly—they show positive net option values for twenty-five or more trials.

This heterogeneity poses a significant challenge to designers and to managers for it means that there are no simple rules that can effectively guide investment decision making across all modules. Different strategies are needed for hidden and visible modules, for large and small modules, and for those with high and low technical potential. These strategies are made even more complicated by the opportunity to split large modules, and to apply other operators besides splitting and substitution at various points in the design. (We discuss the use of other operators in the next two chapters.)

Table 11.2 Value added by different modules for different levels of experimentation (% of system value to consumers)

	Number of Trials																				
	1	2	3	4	5	6	7	8	9	10	11	12	13	14 ...	20	21	22	23	24	25	Max
Software																					
Visible																					
Operating system	0	0	0	0	0	0	0	0	0	0	0	0	0	0	0	0	0	0	0	0	0
Hidden																					
Software development tools	10	13	13	9	4	0	0	0	0	0	0	0	0	0	0	0	0	0	0	0	13
Large engineering (per application)	10	13	13	9	4	0	0	0	0	0	0	0	0	0	0	0	0	0	0	0	13
Large commercial (per application)	7	12	15	16	17	17	16	16	15	14	12	11	10	8	0	0	0	0	0	0	17
Small (per application)	2	4	5	5	6	6	7	7	7	7	7	8	8	8	8	8	8	8	8	8	8
Medium (per application)	4	7	9	10	11	12	12	13	13	13	13	13	13	13	11	11	10	10	9	9	13
Hardware																					
Visible																					
Integer unit	0	0	0	0	0	0	0	0	0	0	0	0	0	0	0	0	0	0	0	0	0
Floating point unit	0	0	0	0	0	0	0	0	0	0	0	0	0	0	0	0	0	0	0	0	0
I/O interfaces	0	0	0	0	0	0	0	0	0	0	0	0	0	0	0	0	0	0	0	0	0
Hidden, electronic																					
Display monitor (color)	2	4	4	3	2	1	0	0	0	0	0	0	0	0	0	0	0	0	0	0	4
Assembly and test	8	13	15	17	17	17	16	15	14	12	11	9	7	5	0	0	0	0	0	0	17
Video logic (color)	6	9	12	13	14	15	15	15	15	15	14	14	13	13	8	7	7	6	5	4	15
Hard disk	5	9	11	12	13	14	14	14	14	14	14	14	13	13	10	9	8	8	7	6	14
Tape drive	5	9	11	12	13	14	14	14	14	14	14	14	13	13	10	9	8	8	7	6	14
Memory management unit	3	6	7	9	9	10	10	11	11	11	11	11	11	11	11	10	10	10	10	10	11
Cache memory	3	6	7	9	9	10	10	11	11	11	11	11	11	11	11	10	10	10	10	10	11
DRAM/MB	3	6	7	9	9	10	10	11	11	11	11	11	11	11	11	10	10	10	10	10	11
																					162

Note: Not shown are the option values of the nonelectronic modules (keyboard, printed circuit boards, etc.). In table 11.1, we assumed these modules had very low technological potential, which translates into a low option value. As a result, the value of experimentation in these modules is zero across the board.

Varying Rates of Change by Modules

The level of investment in a given module in turn will affect the rate of change (and improvement) in its design. Modules that elicit a lot of investment should change and improve rapidly; those that warrant less investment should change more slowly.

Hence our analysis of a system of heterogeneous modules suggests that rates of design turnover and improvement will vary significantly across modules and across levels of a design hierarchy. To illustrate, suppose for the sake of argument that the levels of investment in each module corresponded to the optimum. (As we show in chapter 15, this would be the predicted outcome if the surrounding markets were either monopolistic or highly competitive.) How many new designs would then be created for each module relative to the base rate of investment in an unmodularized system? In other words, if one new design per generation were warranted for an interdependent design, how many designs would be created in that same time interval for each module in a modular system?

The results are shown in table 11.3. Each number can be interpreted as a rate of change or turnover in the design of the individual modules.[11] For hidden modules, these endogenous rates of innovation range from two times the base rate for software development tools or large engineering applications, to twenty times the base rate for small application programs. Hardware modules and assembly and test processes fall in the intermediate range: they are predicted to change at five to thirteen times the base rate. For modules containing visible information, the predicted endogenous rates of change are zero. However, our assumptions do not take account of the possibility that the hidden parts of an architectural module might be redesigned, while leaving the visible information intact.

These results are only suggestive, however. In particular, the value that can be captured by a firm or firms, hence the levels of investment to be expected from private companies, will depend on the competitive structure of the submarkets that surround each module.[12] For this reason, it is difficult to predict the rate of change in designs from technological data alone. Nevertheless, as we try to understand the dynamics of innovation in the submarkets defined by individual modules, the module's total net option value gives us a place to begin. We treat questions of market

---

[11] The rate of change or turnover in designs is an operational measure of "turbulence" in a marketplace. At this point most studies of "turbulent" environments do not define the term, nor do they attempt to compare levels of turbulence across markets.

[12] Sutton (1991, 1998).

Table 11.3 Endogenous rates of innovation for the modules of a computer system: An illustrative example

	Endogenous rate of innovation
Software	
Visible	
Operating system	0
Hidden	
Software development tools	2
Large engineering (per application)	2
Large commercial (per application)	6
Small (per application)	20
Medium (per application)	10
Hardware	
Visible	0
Integer unit	0
Floating point unit	0
I/O interfaces	0
Hidden, electronic	
Assembly and test	5
Display monitor (color)	6
Video logic (color)	8
Hard disk	8
Tape drive	8
Memory management unit	13
Cache memory	13
DRAM	13

Note: The base rate of innovation for an interdependent system = 1.
Explanation: We assumed that one experiment on an interdependent system would just break even, hence the "base rate of innovation" is one new system per (unspecified) unit of time. We then asked: Relative to this base rate, how rapidly would new *modules* come to market? The results are shown in the table.

structure and investor expectations more formally in chapter 15. There we will argue that the modules with high total net option values are precisely those that are most likely to attract entry, and may as a result turn out to have fiercely competitive submarkets.

In the next two chapters, we see how the other four operators—augmenting, excluding, inverting, and porting—can be used to create value in an already modularized system, and how these operators give rise to even more complex and heterogeneous systems.

12 The Value of Augmenting and Excluding

If splitting and substituting were all that designers could do with a modular design and task structure, the resulting systems would be relatively simple. Systems with different initial design rules would be walled off from one another, and module roles would be limited to those conceived of in the initial system. However, the four additional operators, used in conjunction with splitting and substitution, can give rise to

- modules with new functions;
- new systems;
- new sources of visible information;
- architectural modules visible to some, but not all parts of a system;
- initially incompatible systems that can communicate (and may even run the same programs) via hidden "translator" modules.

Over time, applying the six modular operators at different points and in different combinations can create a rich and variegated set of modules bound together by overlapping sets of design rules. Then it becomes almost impossible to define sharply the boundary of any one system, for workable combinations can be formed out of the elements of many systems.

These conditions may in turn lead to the emergence of a large, dispersed "modular cluster" of firms and markets, such as we see in the computer industry today. We will look at the factors that contributed to this dispersal of value across many firms and submarkets in chapters 14 and 15, and in volume 2. In this chapter and the next, however, we remain focused on the evolution of the designs themselves.

The Augmenting Operator

In a modular system, designers are not limited to splitting modules and improving those modules through substitution. They can also perceive new functions and design new modules to fulfill them. Thus modular systems are open-ended in their potential to create value.

If we look down the list of modules in a workstation of 1990 (see table 11.1), and compare it with the modules of System/360 circa 1970 (figure 8.4 or figure 9.3), we can see that some of the modules in the 1990 system, like disk and tape drives, were present in the earlier system. Other modules in the 1990 system brought new technology into a preexisting role. For example, DRAMs, first introduced by Intel Corporation in 1971, performed the same function as core memory but were based on different technical principles.

However, the list of 1990 modules also includes elements that had no counterparts in System/360 or in any other computer of that era. Many of these new modules were software applications. In the 1960s, almost all software was custom-built. In the language of our theory, at the time of the System/360 modularization and for some years afterward, the tasks of software design and production were highly interconnected with the tasks of use. As a result, software applications were conceived and coded by programmers in management information systems (MIS) departments for use within one company.

The first split to occur in software was between the tasks of software design and production on the one hand, and the tasks of use on the other.[1] This split began to take place in the 1970s, as value-added resellers (VARs) emerged for minicomputers. (See the discussion of Digital Equipment Corporation, later in this chapter.) The splitting of tasks related to software became more pronounced in the 1980s as personal computers opened up huge new markets for "packaged" spreadsheets and word-processing programs.

---

1 Software "production" involves copying the code onto some medium and distributing it. To the extent that coding is automatic (e.g., performed by a compiler), it, too, is a production task by our definition. However, nonautomatic coding is a form of design. Today, optimizing compilers and other automatic code-generating programs (IDEs) are making it increasingly easy to split the conceptualization (design) and the coding (production) of software applications. Modularity has been recognized as an important design goal for software from the early 1970s, but the goal has proved elusive (Parnas 1972a, b; Constantine, Myers, and Stevens 1974; Myers 1975, 1979; Schach 1993; Salus 1994.) In 1998, most large software programs do not have modular designs or task structures. We will come back to this issue in volume 2.

The emergence of never-before-imagined software applications is illustrative of the modular operator "augmenting," which we introduced in chapter 5. To see how augmenting "works" in the context of a modular system, let us look at the example of packaged spreadsheet applications. The origin of these programs is well known: in 1978, Dan Bricklin and Bob Frankson constructed the electronic analog of an accountant's columnar pad and endowed it with automatic recalculation capabilities. The marriage of a familiar interface (the pad) with computational power (the recalculation) created an enormously appealing product. The program, VisiCalc, which ran on Apple II computers, was the first of the "killer apps" which fueled the demand for personal computers. A host of novice users bought Apple IIs because they "needed" a personal computer to run VisiCalc. Many turned on their machines for no other purpose.

VisiCalc had no predecessor. There were accounting and economic modeling programs that ran on IBM mainframe computers, but they lacked VisiCalc's intuitive user interface and its low price. (To use VisiCalc, a person had to buy an Apple II, which cost about $1500. Buying a mainframe to do personal spreadsheet calculations was unthinkable. Many users did have access to time-sharing systems, but such systems generally used teletype or line printers for tabular output, which made recalculations extremely tedious.)

VisiCalc itself was a hidden module in a larger system. Bricklin and Frankson did not set out to design a VisiCalc computer from scratch. They took the concept of a personal computer, and specifically, the Apple II, with its cathode ray tube (CRT) screen (which was capable of graphical display), its built-in high-level language (BASIC), its well-integrated floppy disk storage module, and its low price, and went on from there.

The Apple II itself was a system built out of preexisting hardware and software modules—a microprocessor from Motorola, the BASIC computer language, a floppy disk drive, and a CRT screen. These modules were bridged by a visible instruction set—BASIC and a simple I/O command structure. Significantly, most of the information visible to software developers was already known to programmers like Bricklin. In essence, then, the Apple II was a small, hardwired BASIC compiler, plus a primitive operating system, built around a Motorola microprocessor and some I/O devices.

The Apple II became the architectural context for Bricklin and Frankson's spreadsheet design. Because the system was already there, the inventors of VisiCalc did not have to be hardware designers. They also did not have to communicate with the designers of the Apple II, Steven Jobs and Stephen Wozniak, to implement their idea. They could simply treat the Apple II's visible information as a set of design rules, and proceed independently. VisiCalc thus emerged as a new hidden module for the Apple II. Having no predecessor, VisiCalc *augmented* the Apple II system.

The Value of Augmentation

We can understand both the value of VisiCalc and its evolutionary path using the valuation tools developed in the previous two chapters. The Apple II and its software constituted a modular system. The modules created a "value landscape" similar to that graphed in plate 11.1. Every module had its own value function, reflecting the interplay of complexity, visibility, and technical potential. Each module therefore had an endogenous "peak" reflecting the optimal number of experiments worth mounting on that module. Within the system, hidden modules were easy to change, whereas visible design rules were more difficult and expensive to change.

Augmenting a system adds a new module, hence a new source of value to the system. Thus VisiCalc increased the number of Apple II modules by one: in the "value landscape" of the Apple II, a new "peak" arose. The net option value function of the new module was of the same general form as other modules [see equation (11.2)]:

$$\mathrm{NOV}_{\mathrm{VisiCalc}} = \max_{k_V} \underbrace{\sigma(n_V)^{1/2}Q(k_V)}_{\text{Option value}} - \underbrace{cnk_V}_{\substack{\text{Costs of}\\ \text{experiments}}} - \underbrace{Z_V}_{\substack{\text{Costs of}\\ \text{visibility}}}. \qquad (12.1)$$

Equation (12.1) indicates that new modules have the same ability to evolve through repeated experimentation and substitution as do existing modules. As before, the optimal number of trials, k, for the module in question is endogenous, and depends on the module's technical potential (measured by σ), its complexity (n), the cost of experimentation (c), and the cost of visibility (Z).

We know from our previous analysis that small hidden modules with low costs of design and high technical potential warrant a large number of experiments. VisiCalc, the exemplar of spreadsheets, was a small hidden module with a low cost of design and high technical potential, hence it invited substitution by promising high rewards to potential experimenters. Aware of these incentives, a number of other companies quickly jumped into the marketplace with their own spreadsheet programs.

But—and this is the power of augmentation—when a concept is truly new, it adds value to *all* computer systems, not just to the system that plays host to the initial implementation. Thus with VisiCalc before them as an example, designers could begin to imagine spreadsheet programs on everything from an Apple II to a Cray supercomputer. It was even possible to imagine spreadsheets on systems yet to be designed. Thus in the "value landscape" of every computer system, a new "peak" emerged.

This immediate dispersal of the concept across the full range of computer systems is an inherent feature of the augmentation operator. For VisiCalc's designers, the power and novelty of the concept meant that their product was vulnerable not only to spreadsheet experiments implemented on the Apple II but to experiments implemented within many other computer systems as well.

In modular systems, selection most often works at the level of hidden modules, but it does not work only at that level. As we saw in chapter 9, the unit of selection in a modular system may be a module, a group of modules (a subsystem), or a whole system. Which level bears the brunt of selective pressure at any point in time depends on the net option value tradeoffs we have been discussing all along. Specifically,

Is the overall system flexible enough to accommodate improvements in its modules? If so, change will occur in modules. The system's design rules may survive a very long time.

If the system is not flexible, what is the cost of redesigning the whole system from scratch? If the cost is low, then the system itself will become the focus of experimentation. Initial versions of the system may then be replaced by later, better versions.

The Apple II was both inflexible and relatively cheap to design, which did not bode well for its long-term survival. Although it was built out of off-the-shelf components, its architects, working under the joint constraints of a limited budget and ignorance, did not develop a flexible, comprehensive set of design rules, nor did they maintain strict partitions and information hiding with respect to their own design decisions. Thus the Apple II was only partially modular. In particular, the hardware partitions were not robust and the interfaces were not clean. Because of these inherent limitations, the Apple II was not capable of becoming the basis of a flexible, long-lived computer family.

Given the shortcomings of its design structure, the fact that the Apple II had been assembled quickly out of cheap, off-the-shelf components made it prey to competition. Anybody could buy the same (or better) off-the-shelf parts, and there were many computer designers who had the knowledge needed to wire those parts together into a small, cheap, but functional system. For their part, Apple II purchasers had few ways to upgrade and reconfigure their systems. Upgrading or expanding meant buying a new system, which, because of Apple II's architectural limitations, would not be an Apple II.

Given the large market demand for spreadsheets revealed by VisiCalc's success and the low cost of redesigning both the module and the system, it was highly improbable that the first implementation of a spreadsheet program would turn out to be the best. Indeed, people quickly figured out how to make better spreadsheets, as well as how to make personal computers that were better at running spreadsheets. Thus in rather short order, the VisiCalc-Apple II combination was unseated by the Lotus 1-2-3-IBM PC combination.

Augmenting—The Wild Card Operator

To summarize, *augmenting* is the act of adding a new module to a preexisting modular system. A modular system design makes augmentation relatively easy for designers, and, potentially, users as well.[2] Users can address their own specific needs by adding modules to a basic system. For their part, designers can apply their specific knowledge to a particular module, taking the rest of the system as given. The purchasers and designers of new hidden modules do not have to coordinate their decisions or clear their designs with the system's architects. They only have to understand and conform to the system's visible design rules, for example, the nature of the ports and way in which the processor sends and receives instructions. The system's architects in turn do not need to be aware of augmentation possibilities when they design the initial system.

In advanced market economies there are many modular systems that give rise to augmentation opportunities. These include transportation systems like airlines, communication systems like telephone networks, and power systems like electricity grids. In each, there is a high degree of modularity between the tasks of design, the tasks of production, and the tasks of use. An electrical company does not need to know what customers use electricity for, as long as their appliances are "plug-compatible" with its system. And inventors of new electrical appliances (like computers) do not have to design a power supply network from scratch to support each device.

Modules created via augmentation may be valued in the same way as other hidden modules. However, because they embody new concepts, such modules change the "value landscape" not only in their original system but in all systems. Therefore, unless the original system is very flexible, it is not likely to be the system of choice

---

[2] There are some artifacts, which, though modular in design, have their parts fused together in production and hence are not modular in use. An example of this is a book.

for the new concept in the long run. Major augmentations are thus a stimulus to both new system development and to intersystem competition.[3]

The opportunity to augment modular systems presents several problems for the construction of a theory of design evolution. Fundamentally, augmenting is a "wild card" operator: it is difficult to place a value on things yet to be invented, or to predict when and where those inventions will occur. In effect, then, the opportunity for augmentation adds an *X of unknown magnitude* to the value of each modular system.

From an analytic perspective, it is tempting to ignore those *X*s since we do not understand them very well. Yet much of the value in computers today arises from augmentations that took place in the past. Moreover, a major augmentation may change users' preferences with respect to their systems, as the case of VisiCalc shows. Thus the competitive consequences of a successful augmentation may be complicated, and not entirely benign for the inventors of the new concept.

The Exclusion Operator

Just as one can think of adding modules to a system, one can think of subtracting them. The exclusion operator is thus the logical opposite of augmentation and its economic complement. The purpose of excluding one or more modules may be to reduce cost or to protect a competitive position.[4] In this chapter, we are interested in the cost-reducing aspect of exclusion.

The exclusion operator works to reduce cost in two ways. First, *users,* when selecting from a broad, compatible family, may elect to exclude some modules from their individual configurations. When they do so, they may retain the option of adding

[3] This is the implicit logic behind the popular theory of "killer apps." Briefly, this informal theory, which is associated with Bill Gates, the founder and CEO of Microsoft Corporation, holds that every ten years or so a major new application arrives on the scene. This "killer app" changes the terms of competition in ways that can lead to the downfall of formerly dominant firms. In our language, "killer apps" are augmentations that change the value landscape for all computer systems. Given the widespread effects of augmentations on value it is not surprising that they would tend to shake up the existing order (Gates 1996).

[4] An example of a competitive use of the exclusion operator is when the architect of a system excludes interfaces that would allow other vendors' hidden modules to function in that system. For example, when Intel Corporation introduced MMX, Microsoft did not incorporate those instructions in the next release of the Windows operating system. (We are grateful to Nitin Joglekar for pointing this out to us.)

the excluded modules later (thereby augmenting those systems). *Designers* may also initially choose to exclude some modules from a particular system. However, if they maintain a strict partition of information in visible design rules and hidden design parameters, they will be able to add implementations of the excluded modules to the system at a later date.

In the computer industry, the quintessential example of the use of the exclusion operator is the case of minicomputers. System/360 was a broad, compatible line of computers, which, as we have discussed, made extensive use of the splitting and substitution operators. System/360's breadth and modular design created a set of options to customize and upgrade computer systems that were extremely valuable to a large set of users. (See box 12.1.) As it turned out, however, there was another set of users who did not value these configuration options highly enough to pay a premium for a flexible system. This other set of users became the potential purchasers of stripped-down systems that excluded most of the modules (and therefore the options) embodied in the large family. They created a niche in the computer marketplace that was filled by *minicomputers.*

Minicomputer manufacturers in turn showed that there was a different path of design evolution for a large compatible family. Rather than being created by splitting and substitution like System/360, a family could be developed by first designing a narrow system (but one with a spacious modular architecture), and then adding compatible modules over time. Thus the process for arriving at the family was an initial *splitting* of the design, followed by *exclusion,* followed by systematic *augmentation* within the initial architecture. This was an attractive design strategy for companies with limited resources, which in 1970 meant everyone in the computer industry except IBM.

Nowhere was this strategy pursued more vigorously than at Digital Equipment Corporation. The history of DEC in the 1960s and 1970s demonstrates that exclusion was an effective way of bringing a low-cost system to the marketplace. Essentially, exclusion made experimentation on whole systems feasible. However, just as the SPREAD designers found that conceiving of a modular architecture was easier than constructing one, so, too, did DEC's designers discover that it was difficult to build all the necessary design rules into a first system, especially if that system had to be offered at very low cost. In fact, it took two attempts—the PDP-11 and the VAX—before DEC was successful in this endeavor.

DEC's experience with the PDP-11 also shows how modular designs can elicit coinvestment by third parties, thus spurring design evolution. But once coinvestors entered the picture, DEC found it difficult to maintain control over the system as a

Box 12.1 The Problem of Pricing Minimal Systems in a Broad, Modular Family

A broad, modular family of compatible computers, like System/360, provides users with a set of opportunities to upgrade and otherwise reconfigure their systems at low cost as the state of technology or their own needs change.[1] These "options to configure and reconfigure" are one of the mechanisms by which the option values of a modular design are realized by users.

The high demand for System/360 showed that, for the great majority of computer users, reconfiguration and upgrade options were very valuable. But for some, the ability to swap modules in and out of a system was of little or no value. Process control, for example, involved inserting a computer at a control point of a complex production system. There was no thought (at that time) of changing the controller: the goal was simply to replace mechanical, electrical, and human controls with cheaper, faster, and more reliable electronic controls. Hence, future upgrades were not very important to purchasers of computers used for process control.[2] Price, however, was critical.

The mere fact that users place different values on reconfiguration options presents an interesting problem for pricing the units of a modular product family. If the modular units are sold "unbundled," and the seller cannot charge different prices for the same good, options to reconfigure "come with" the initial purchase of a small (minimal) system. The question then becomes whether to set a high price on a minimal system or a low one. For a monopolist, the value-maximizing solution to this problem is to set a relatively high price on minimal systems in order to capture a large fraction of their option value to users.[3]

Naturally, users who do not place a high value on the embedded options will not find the high-priced systems attractive.[4] Those users in turn may be a target market for alternative systems that do not provide the reconfiguration and upgrade options of a large modular family. The alternative systems, which exclude many modules, will initially cost less to design than a full-blown family. Still—and this is key to understanding DEC's strategy in the 1970s—options to add modules later may remain with the designers, as long as the initial system architecture is modular and can accommodate subsequent augmentation.

1 If a system design is interdependent, its components will be irresponsibly linked together. In that case, users can only purchase multiples of the whole system.

2 Many of the production systems—assembly lines, chemical plants, and others—were in effect sequential task structures optimized for throughput. Japanese manufacturing principles were not widely appreciated at this time. Thus, the goal of a flexible, responsive production facility, which might have led to demand for upgradable process controllers, lay far in the future.

3 This is the first time in our analysis that the social optimum and the private optimum differ, so that it potentially matters who owns what, and how much value is captured by different parties. We take up this issue in chapter 15 and in volume 2.

4 We are assuming that the seller cannot charge different prices to different users for the same physical system. This means that the seller cannot sell the reconfiguration options separately. In fact, there are clever ways of bundling the physical products, which lead to the unbundling of reconfiguration options. However, such pricing strategies are difficult to implement, and subject to legal challenge.

whole. Investors in complementary modules could and did use the operators *inversion* and *porting,* which are discussed in Chapter 13, to set up new "apexes" of visible information and to insulate themselves from DEC's control of their systems' design rules.

The Value of Exclusion

The opportunity to design a minimal system can be valued using the options valuation methods explained above. However, although the basic elements of the analysis are the same, the interplay of the modular operators leads to a more complex option structure, because exclusion opens up the possibility of experimentation on whole systems. Rather than designing all parts of the system at once, as IBM did with System/360, the exclusion operator invites designers to introduce a system, and "see if it works." First, with a target market and price point in mind, the designers must decide what functions to include in the initial system and whether to establish design rules that will allow later substitutions and augmentations. The designers must then implement a design task structure that corresponds to their design strategy.

If the resulting system's combination of functionality and price is appealing to users, the system will succeed in the marketplace, and orders will be forthcoming. Eventually customers will request additions and upgrades, in which case the substitution and augmentation plans for the system can be implemented. If high levels of demand do not materialize, however, the system design can be abandoned, and a considerable portion of the total design cost will have been saved.[5]

If system-level experimentation takes place and is followed by module-level experimentation, then we have a new, more complex option value structure. In terms of our initial assumptions (refer to the first sections of chapter 10), we now explicitly allow the "wheel of random outcomes" to spin twice: once at the system level and again at the module level. Figure 12.1 provides a schematic diagram of this process.

---

[5] We are oversimplifying. Obviously, the second-stage outcomes will not be simply black and white. In reality, there will be a whole set of contingent plans that depend on the particular level of demand revealed in the first stage. However, these plans do not have to be spelled out in great detail—indeed it is economical to defer much of the detailed planning until the firm has some data about the state of demand. For first-stage valuation purposes, these complex second-stage events can be approximated by a two-pronged option value—a valuation conditional on proceeding and a valuation conditional on abandoning the system. A simple expected value is not sufficient because of the nonlinear structure of option payoffs.

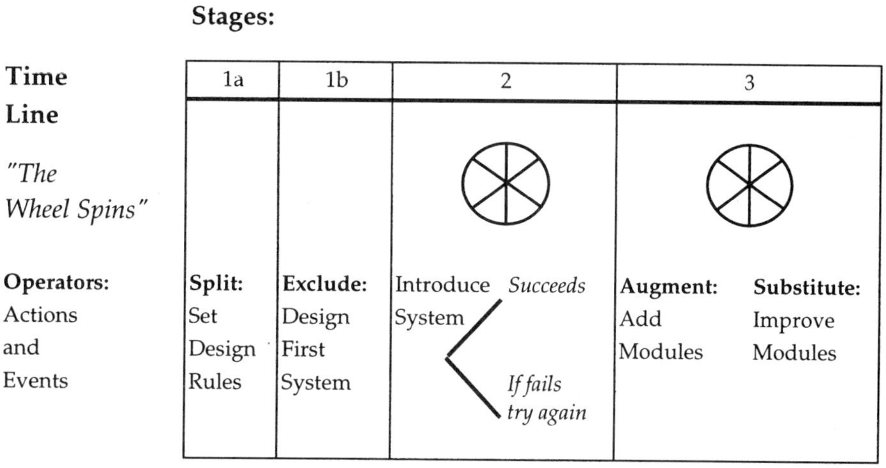

Figure 12.1 System design: a multistage design process that uses four operators and two levels of selection.

To understand the value of the options embedded in this design process, we must work by backward induction from the later stages to earlier stages. We have already seen that a successful modular system constitutes a portfolio of options that allows designers (or users) to select the best module designs. The system value was captured in equation (11.1), which we reproduce here:

$$\text{System value} = S_0 + \text{NOV}_1 + \text{NOV}_2 + \ldots + \text{NOV}_j . \tag{11.1}$$

The individual module net option values in equation (11.1) each have the form of the NOV_i in equation (11.2). In other words, each is an optimized algebraic sum of an option value minus costs of experimentation and visibility.

However, as figure 12.1 indicates, systems themselves must pass through an experimental stage before they can be deemed successful. At this earlier stage, each experimental system is an option in its own right. It may be developed or not depending on whether there is sufficient demand for it in the marketplace. Mathematically, then, the net option value of one system can be written as follows:

$$\begin{aligned} \text{NOV}_S &= \underset{\substack{\text{Proceed/}\\ \text{don't proceed}}}{\max}[\text{System Value} - \text{Design Cost}, 0] \\ &= \underset{\substack{\text{Proceed/}\\ \text{don't proceed}}}{\max}[S_0 + \Sigma\,\text{NOV}_{\text{Substitutions}} + \Sigma\,\text{NOV}_{\text{Augmentations}} - \text{Design Cost}, 0] \end{aligned} \tag{12.2}$$

This expression indicates that, after introducing a minimal system and gauging user demand for it, designers (or their firms) may estimate the market value of the options generated by the system.[6] They then have an option to proceed with further system design or not. The value of proceeding is captured by the first term in equation (12.2); the value of not proceeding is zero. The option value of the system lies in the fact that the designers may pick the maximum of those two values.

The value of a collection of experimental systems can then be expressed as

$$\text{Value of } m \text{ system experiments} \;=\; \sum_{m} \text{NOV}_S \,. \qquad (12.3)$$

Equation (12.3) is recursive in the sense that there are two levels of options embedded in this expression: options to develop systems and options to develop the modules of systems. The recursive nature of the valuation makes it hard to describe in natural language. In the technical language of option theory, however, each experimental system is an "option on a portfolio of options," and the aggregate encompassed by equation (12.3) is a "portfolio of options on a portfolio of options."

The same property of recursiveness makes the total value of a particular set of system design opportunities very hard to compute. Finance theory tells us that this value exists, and gives us algorithms [summarized by equations (11.1), (11.2), (12.2), and (12.3)] for calculating it. But the computations themselves are hard to implement for they depend on data that are not readily available. Thus any calculation of the "value of *m* system experiments" will be at best a very imprecise approximation of the "true" value of those options.

However, just as modularity creates order in complex designs, so too does modularity create order in intertemporal valuation and decision-making processes. What is difficult from a centralized, computational vantage point becomes quite manageable for future decision-makers operating in a decentralized fashion. It is manageable, that is, as long as competent decision-makers are distributed at the key decision-making nodes of the modules, and operate under a sufficiently complete set of design rules. (See box 12.2.)

In summary, by utilizing the exclusion operator, designers can conserve design resources by deferring investments until they have some indication of the market's acceptance of the overall system. If the system design is successful, resources can

---

6 The market value of the system might be observed if the company that owned it was a "pure play" in the design and had issued publicly traded stock. However, even assuming that the market has perfect information about the value of the system, the inference problems with respect to the time series of stock prices are tricky. (See Bodie and Merton 1995; Merton 1997.)

Box 12.2 Valuing Systems as Options on Options

In principle, valuing a set of complex options on options is quite feasible; the algorithms for valuing endpoints and then collapsing endpoints into values are very clear and robust. Under a set of well-known assumptions there is no problem with recursive substitution, that is, with using values instead of raw payoffs as the endpoints of probability outcomes and then defining contingent claims and deriving optimal values from those endpoints.

As one tries to operationalize this valuation procedure, however, one quickly reaches technical limits. In particular, our ability to turn *specific information* that will be available to decision-makers when and if they need it into *general information* that can be used to value a complex contingent strategy today is limited.[1] In the case of modular design processes, the relevant specific knowledge will not exist until events call it into being. It is scenario-specific knowledge, which will be available if and only if a particular set of events transpires. It is impractical to bring that specific knowledge back to a central point in the present, analyze all the decisions that will be made, and thereby calculate precisely the market value of the whole. For this reason, it is difficult to compute aggregate values of complex portfolios of design options, even though in principle we know those values exist.

As computation and valuation technologies get better (and what we hope we are presenting in these chapters is better valuation technology for design processes), designers can look deeper into the intertemporal structure of their designs, and use this insight to improve their initial design decisions. But an evolving, modular system of designs is an open-ended system, and thus the designers can never look all the way to the end of the process.

The computational problem mirrors the management problem. If it were easy to compute the value of the whole, it would be correspondingly easy to manage the evolving system from a central vantage point. Hence, the complexity of the option valuation problem is telling us something about the difficulty of managing a set of evolving modular design processes within the confines of a single, centralized decision-making organization.

In contrast, advanced capital markets allocate resources in a decentralized manner, based on limited, local computations of value. Hence a capital market may be a more efficient mechanism for selecting good designs and allocating scarce resources than a centralized organization.[2] We return to these issues when we discuss the emergence of the modular cluster in chapters 14 and 15, and in volume 2.

---

[1] Hayek (1945).

[2] In the language of John Holland's theory of complex adaptive systems (Holland 1976), a decentralized capital market and a centralized organization represent two different types of "adaptive plan." We are suggesting that the capital market's adaptive plan is more efficient than some alternative, centralized organization's adaptive plan. This suggestion carries echoes of Jensen and Meckling (1992); Radner (1992); Sah and Stiglitz (1986); Williamson (1985); Merton and Bodie (1995); as well as, of course, Hayek (1945, 1988); and Alchian (1950).

then be allocated to exploit the options embedded in the system. Conversely, resources will not be consumed in building a full set of modules for a system that is doomed to fail. The value of the exclusion operator rests on this principle of resource conservation and efficiency.

Digital Equipment Corporation

The history of DEC, as we have said, exemplifies the use of the exclusion operator to experiment with new system designs. DEC's technological trajectory differed from IBM's in that

- it discovered new markets by offering minimal systems;
- its architects consciously used exclusion to conserve design resources;
- the architects then planned to exploit successful systems by augmenting them, that is, by converting bare modular architectures into full-blown product families.

The last point of this strategy turned out to be harder to execute than DEC's architects thought it would be. As a result, their first attempt—the PDP-11 product line—did not fulfill its design goals. In purely economic terms, however, the PDP-11 product line was very lucrative, and it solidified DEC's position as the premier minicomputer company. Thus in the 1970s, DEC became the second most successful computer systems manufacturer after IBM.

DEC's experience with exclusion can be traced to its beginnings in the late 1950s. As a condition of its initial equity financing, DEC did not at first even sell whole computers; it sold "logic modules," which knowledgeable users could hook up with I/O devices, according to their own highly specific needs.[7] DEC designers also used the logic modules internally as building blocks for larger computer systems. In the first eight years of the company's existence, they experimented with a number of system designs.[8] Then in 1965, one of the designs found a profitable market niche.

---

7 According to DEC insiders,

The principal backer of DEC, American Research and Development . . . was somewhat skeptical that a computer company could be successful [against IBM and other competition]. They were enthusiastic, however, about the business possibilities in logic modules for laboratory and system use. (Bell, Butler, Gray, McNamara, Vonada, and Wilson 1978, chapter 6 in Bell, Mudge, and McNamara, p. 123.)

8 DEC generated so many different systems designs that insiders felt it was in danger of becoming a specialized systems design boutique. Its designers viewed computers as being con-

Very quickly thereafter, the concept of "a minicomputer" crystallized around the DEC PDP-8.

In fact, minicomputers were only one step up in complexity from logic modules. Gordon Bell, DEC's chief designer and a pathbreaking computer architect, defined them in the following way:

> *MINICOMPUTER: A computer originating in the early 1960s and predicated on being the lowest (minimum) priced computer built with current technology. . . . Minicomputers are integrated . . . on a* dedicated *basis (versus being configured with a structure to solve a wide set of problems on a highly general basis).*[9]

This definition pointed to the two complementary elements that make *exclusion* a valuable operator: a low price and a low value placed by users on the options embedded in a large, compatible computer family.

The PDP-8

The PDP-8, the original minicomputer, was a stripped-down machine in every sense of the word. It processed only twelve-bit words, thus the highest number it could compute in a single arithmetic operation was 2^{12} or 4096. Bell and Newell characterized its instruction set as "about the most trivial" of any general-purpose computer.[10] It had a small (but fast) memory, a limited number of peripheral devices, and no operating system. It had no software, other than the code the user programmed. But it fit in half a cabinet, and could sit on or under a desk.[11]

In 1965, a stand-alone PDP-8 without peripherals cost $18,000: ten years later, in 1975, the same machine could be purchased for $3000. A small IBM computer in 1970 might cost $100,000 to purchase, and would need to be combined with equally costly peripheral devices.[12]

---

structed out of basic building blocks, and "DEC could afford to stock an ample supply of basic modules" (Bell et al. 1978, p. 140). This "off-the-shelf" approach to systems design is eerily reminiscent of the approach used to design the Apple II (but not the Macintosh), the first IBM PC, and Sun Microsystems's first workstations.

9 Bell, Mudge, and McNamara (1978, chapter 1, p. 14).

10 Bell and Newell (1971, chapter 5, p. 120).

11 Bell and McNamara (chapter 7 in Bell, Mudge, and McNamara 1978, pp. 175–208) and Bell and Newell (1971, chapter 5, pp. 120–133).

12 Comparative costs of IBM System/360–30 are taken from Bell and Newell (1971, pp. 562 and 580).

By excluding many parts of the system, the PDP-8's designers addressed the needs of price-sensitive, but power-hungry users. As a result, this machine became the first non-IBM computer to ship in large volumes. Over its fifteen-year life span, its basic instruction set (the core of its visible information) was implemented in ten different physical machines, and over 50,000 units were eventually produced and shipped to users.

The initial demand for the PDP-8 created high profits for DEC, but even higher growth, leading to growing cash shortfalls. The cash shortage in turn led to DEC's legendary IPO (initial public offering) in 1967. DEC's IPO was one of the seminal events in the creation of the modern venture capital industry because it revealed American Research and Development's high rate of return on its initial seed capital. After the public offering, the rates of formation of venture capital funds went up dramatically.[13]

Nevertheless, as a design, the PDP-8 instruction set, which was the only thing common to all implementations, left a lot to be desired. Its principal shortcoming was that it was not what is now called a "scalable" architecture. The instruction set did not allow designers to take advantage of either the better hardware (medium- and large-scale integrated circuits) or the knowledge about systems design and applications programming that was accumulating in the late 1960s and early 1970s. Thus, as the size of their installations increased, users suffered from "diminishing returns with respect to memory, speed, efficiency, and program development time." [14]

The PDP-11

Conscious of the shortcomings of the PDP-8, and intrigued by the possibilities revealed by the modular design of System/360, Gordon Bell and his colleagues proposed to create a new set of design rules for the PDP-11 product family. Their experience with the PDP-11 illustrates both the value and the limitations of an exclude-then-augment design strategy for modular computer systems.

Like System/360, the PDP-11 was conceived of from the beginning as a modular, compatible family of computers.[15] Designers' and users' options were explicitly built

---

[13] Bygrave and Timmons (1992).

[14] Bell, Cady, McFarland, Delagi, O'Loughlin, Noonan, and Wulf (1970, reprinted in Bell, Mudge, and McNamara 1978, chapter 9, pp. 241–262).

[15] The design goals for the PDP-11 family were remarkably similar to those of System/360. This is not surprising, since Gordon Bell had studied System/360 carefully and concluded that

into the original design. The family was to include processors that spanned a wide range of performance and price, that operated with a common set of peripheral devices,[16] and that ran compatible software. The system's architects envisioned that as technology (primarily integrated circuit technology) improved, new models with significantly better performance would be added to the family.

But, unlike IBM, which designed all parts of the System/360 family in parallel, DEC approached the market more cautiously. The first PDP-11 (PDP-11/20) was carefully inserted into DEC's existing product line. Its price was constrained to lie between $5000 and $10,000, and its performance had to be "the same as" a PDP-8's.[17] DEC's top managers and the salesforce felt that a new computer with these levels of price and performance would find a ready market.

Consistent with the strategy of exclusion, to achieve the low target price DEC's designers created what they termed a "bare machine architecture." The PDP-11/20 had no floating point hardware, no integer multiply-and-divide capability (except as an option), and no ability to carry out multiple bit shifts. The absence of these hardware elements meant that complex arithmetic operations had to be constructed out of simpler ones, which slowed down the performance of many tasks. Initially, moreover, the PDP-11/20 was shipped with no systems software, although there were plans to add a disk operating system and a FORTRAN compiler at a later date.

Of course it was the exclusion of all these "nonessential" elements that made it possible for DEC to offer a fast and versatile machine at a very low cost. The PDP-11/20 was a stripped-down core of a computer system, but it was a powerful core with very well-designed interfaces. For example, it had a very complete and flexible instruction set (this was important, because when the PDP-11 was introduced, most programmers still used the machine instructions directly). Furthermore, hardware interconnection was easy because all I/O and memory devices were connected via a single chained switch—the Unibus (see box 12.3).

The design hierarchy diagram for the PDP-11/20 is shown in figure 12.2. At the top level were the basic machine specifications: the PDP-11/20 was a sixteen-bit computer with eight internal registers that operated on integer and Boolean data

---

there was nothing particularly distinguished in the design of the individual components (Bell and Newell 1971, pp. 561–587).

16 At this time, DEC was gearing up its peripherals business. Standardized peripheral interfaces were "a deliberate attempt to prolong the service lives of Digital's peripheral equipment" (Levy 1978, p. 277).

17 Bell et al. (1978, p. 243).

Box 12.3 The Unibus—Part of the Visible Information of the PDP-11

A major hardware innovation in the PDP-11 was its use of a chained switch or "bus" to connect different parts of the system. The PDP-11's Unibus was a novel structure. It allowed any component to communicate with any other component, and made "transactions" (data transfers and interrupts) independent of both the response times of the devices and the physical distance between them.

The Unibus and later bus designs were in effect an application of network technology deep within the computer. Their impact on the overall design was to support modularity. In the first place, buses standardized and simplified peripheral and memory interfaces. As a result, "any company, not just DEC, [could] easily build components that interface[d] to the bus."[1] DEC in fact sold Unibus interfacing modules, and an active industry of suppliers of memory and peripherals soon grew up.

Second, the asynchronous data transfer of the Unibus made it possible (within limits) to introduce memory subsystems with progressively higher speeds, without changing the characteristics of the bus itself. This meant that different memory technologies could be intermixed in the same system. The early 1970s were a time when memory types were changing and memory prices dropping very rapidly. Although the PDP-11's architecture predated the industry shift from core to semiconductor memory,[2] for a time the Unibus made it possible to incorporate the new DRAMs, EEPROMs, and bipolar cache memories into PDP-11 systems without drastically changing the basic layout of the computer.

---

[1] Bell and Mudge (1978, p. 388).

[2] The first PDP-11/20s were shipped in mid-1970. Intel announced its first DRAM and EEPROM products in November 1971. Cache memory, initially reserved for very large computers, was incorporated in the PDP-11/70, introduced in 1975.

types.[18] Below the top level, the visible information was partitioned into two major blocks: (1) the instructions, which were visible to programmers; and (2) the Unibus specifications, which were visible to hardware designers.

At the third level, the hidden modules offered by DEC were the processor and memory, console, teletype, paper tape reader, paper tape punch, disk drive, and a clock. That was all. There were no other processors or peripherals. There was no software of any kind—just a list of machine instructions.

---

[18] Boolean data are vectors of ones and zeros. The on/off states of a process can be represented as a Boolean vector; hence this data type was very important for process control applications.

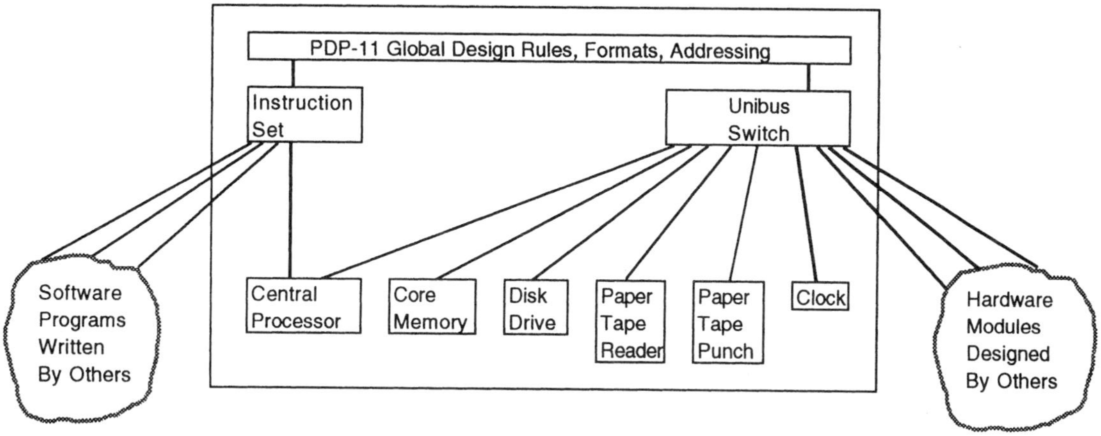

Figure 12.2 Design hierarchy of the DEC PDP-11/20 minimal system introduction in 1970 (compiled from Bell et al. 1970).

DEC's Theory of Design Evolution

Although the first system brought to market was minimal, DEC's designers had an explicit theory about how the PDP-11 family would evolve under their oversight. This theory of design evolution seems to have crystallized in the DEC engineering establishment around 1974. As explained by Craig Mudge, one of Gordon Bell's colleagues: "[A]dvances in technology can be translated into either of two fundamentally different design styles. One provides essentially constant functionality at minimal price (which decreases over time); the second keeps cost constant and increases functionality."[19]

This theory is summarized in figure 12.3, which is adapted from several of DEC's internal engineering appraisals of the PDP-11 family. The figure shows that with improving technology, every two years, designers could create *two* new family members for each old one: one equivalent in price; the other equivalent in functionality. Thus the initial PDP-11/20 was replaced by the 11/40 (same price, higher function) and the 11/05 (lower price, same function).

DEC's plan for the development of the PDP-11 family was based on both the exclusion and augmentation operators. One path of development was to design a series of minimal machines (repeated exclusion). The other paths involved adding

[19] Mudge (1977).

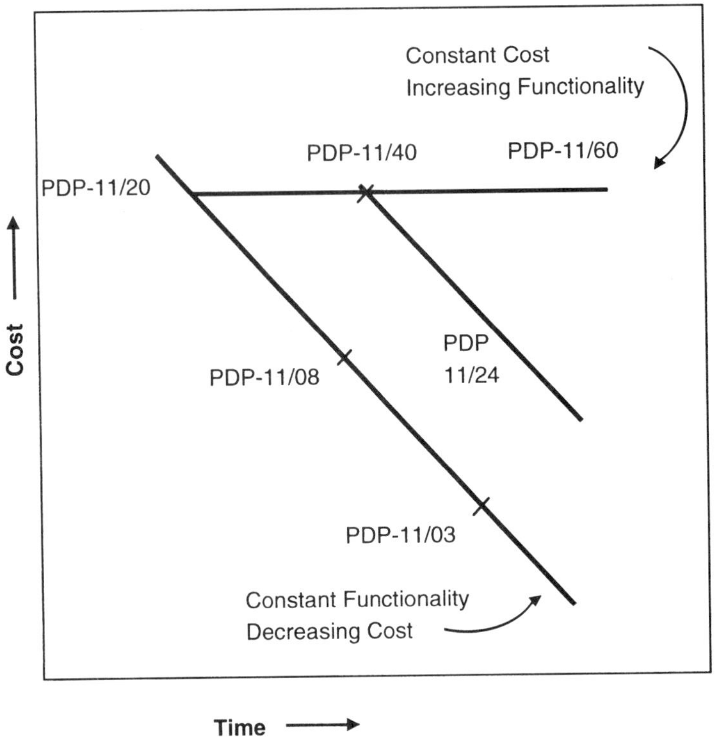

Figure 12.3 DEC's theory of minicomputer design evolution: PDP-11 family (adapted from Mudge 1977).

functionality to the designs (augmentation) while staying within a price range for which there was known to be market demand.[20]

Over multiple rounds, this process would lead to a product family spanning an ever-wider range of price and performance. Modularity meant that different parts of the family could evolve independently: peripherals, for example, would not have to evolve in lockstep with the processors. However, robust and complete initial design rules were needed to ensure that the visible information stayed relatively constant, so that hardware and software would be compatible across the family and through time.

An implication of this theory of design evolution was that DEC would be able to "bootstrap" the design of a large computer family. It was not necessary to follow

---

20 Another possibility was to offer higher functionality at "somewhat" higher prices. However, DEC designers believed this trajectory held less economic potential than the first two.

IBM's route in designing System/360, with its massive upfront costs[21] and accompanying technical and financial risks. Instead, according to this theory, DEC could build up a modular family of computers over multiple generations according to the plan laid out in figure 12.3. Each "implementation generation,"[22] as it was called, would pay for the development of the one that followed it.

For a young, high-growth company that was chronically short of cash,[23] this was a very appealing possibility. All that was necessary to make this vision come true was to select the right design rules—architecture, interfaces, and tests—for the first implementation. That is what DEC's designers tried to do with the initial PDP-11 architecture.

Coinvestment in the Design Architecture by Other Firms

The PDP-11's architecture and its clean interfaces—its modularity, in other words—meant that the design held within it the promise of a very large "shadow set" of potential augmentations. These augmentations included additional processors, components, peripheral devices, and software for specific applications. DEC itself planned to offer the processors and memory, but in 1970 it was not in the business of making components or developing software. Thus, at least initially, most of the augmentations would have to come from outside of DEC. For the family to achieve its full potential, therefore, other companies would have to design and offer hidden modules. In other words, users and suppliers of equipment would have to *coinvest* in the DEC PDP-11 architecture.

Coinvestment in an architecture generalizes the classic notion of a supply chain, which is the basis of theories of vertical integration in both economics and strategy.[24] Coinvestors can be suppliers, customers, or providers of complementary products

---

[21] The cost of System/360 was estimated at $.5 billion for R&D and $4 to $5 billion for new capital equipment (Wise 1966a; Evans, in Aron et. al. 1983).

[22] Bell and Mudge (1978, p. 381).

[23] DEC Annual Reports.

[24] Industry supply chains map the flow of product from raw material, through manufacture, distribution, and sale. They are also known as value-added chains (Porter 1985). How much of the supply or value-added chain falls under the control of one firm is the classic question of vertical integration. (See Coase 1937; Stigler 1951.) Modern formulations of this theory, based on information and contract enforcement costs, have been put forward by Williamson (1976); Grossman and Hart (1986); Baker, Jensen, and Murphy (1988); and others.

and services. Their essential characteristic is that all look to the same set of visible information in formulating their designs.

Systems that are based on the exclusion operator both invite and require coinvestment by third parties. Because they are minimal, such systems do not satisfy all (or even most) user needs; because they are modular, they can be augmented with a large variety of hardware devices and software applications. Thus one measure of the architecture's success is the extent to which it attracts coinvestment.

Although it is hard to get precise data, the PDP-11 was extremely successful in this respect. Large numbers of original equipment manufacturers (OEMs) developed specialized suites of hardware and software based on a PDP-11 core. Moreover, with respect to hardware, Gordon Bell reported in 1978: "[T]he Unibus has created a secondary industry providing alternative sources of supply for memories and peripherals. With the exception of the IBM 360 Multiplexer/Selector Bus, the Unibus is the most widely used computer interconnection standard."[25]

High levels of coinvestment by third parties are indicative of an architecture's success. Nevertheless the modular cluster type of industry structure that results creates a predictable strategic dilemma for all parties down the road. The crux of the issue is the fact that the most profitable hidden modules will later be the most attractive targets of investment for both the architect and the coinvestors. This in turn leads to the following intertemporal paradox.[26]

When an architect firm first introduces a modular system that makes use of exclusion, that firm by definition needs to attract coinvestment to exploit fully the design's potential. However, if and when the design becomes established, the architect firm will want to increase its returns by adding new modules. Therefore, its managers must worry about the possibility that coinvestors will have preempted the most profitable hidden modules of the system. Symmetrically, coinvestors must be concerned that the architect firm will home in on profitable niches that they uncover.

Because of these embedded conflicts of interest, the cluster of coinvestors that surrounds any particular modular platform is a very delicate structure. We will look in greater detail at these conflicts and the long-run dynamics of such clusters in volume 2. At this point, however, we simply want to mark the fact that a cluster of firms is a natural (indeed a necessary) short-term consequence of an architect firm's use of the exclusion operator to experiment with new modular system designs.

---

[25] Bell and Mudge (1978, p. 388).

[26] In the language of game theory and competitive strategy, this is known as a "time-inconsistency."

13 The Value of Inverting and Porting

Except for splitting, the operators we have looked at thus far treat modules as "black boxes." Indeed one of the objectives of splitting is to create black boxes by "hiding" as many parameters as possible behind a module's interface. However, independent hidden modules may still have many similarities embedded in their microstructures. The inversion and porting operators enable designers to seek value by building on those similarities.

We first saw the inversion operator at work in chapter 6 in the case of IBM's standardized circuits. Soon after IBM split computer design and computer production, it faced an overwhelming expansion in the variety of circuits it had to produce. The Standard Modular System (SMS) restricted computer designers to a very small number of standard circuits, which in turn could be produced in high volumes at low cost. By mixing and matching standard circuits, the company was able to achieve significant manufacturing economies across its very diverse product line.

The same sequence of events often occurs in the wake of a modularization. Immediately after the split, there will be a rash of experimentation, but sooner or later such experiments display diminishing returns, and give rise to unmanageable amounts of variety. When the benefits of further experimentation no longer justify the cost, additional design rules are called for. These new design rules are created via the inversion operator.

Useful collections of new design rules may become modules in their own right. However, critical parts of these modules must be visible to other modules. We call modules that embody visible information *architectural modules,* to distinguish them from the hidden modules we have focused on to this point.

Operating systems are an example of an architectural module that emerged through inversion. As with other modules, once the operating system became a well-defined component in modular computers, it became the focus of investment and experimentation. However, the visibility of the operating system made it hard to separate from the system's global instruction set, which was often hardwired into the circuits of the CPU.

Unix was an operating system developed in the early 1970s, which was initially designed to run on Digital Equipment Corporation PDP-7 and then PDP-11 computers. Interestingly, Unix was created not by employees of DEC, but by software designers working at AT&T's Bell Laboratories.

One critical fact about Unix was that its own internal structure was highly modular. Unix's designers created a strict partition between design rules and its hidden (software) modules, and adhered to the principles of information hiding throughout their design. As a result of their discipline, the core of Unix was very small, and the interfaces between independent software modules (programs) were very simple and clean. In addition, Unix was written in C, a high-level language that was independent of any processor's instruction set. The fact that Unix's source code was processor-independent meant that it was portable to other computer systems, and not constrained to run only on DEC equipment.

Within the "design space" of operating systems, Unix was a landmark. Its evolution provided designers with an object lesson on the value of modularity and portability, as well as an example of how to achieve modularity and portability in software. In this chapter, we use Unix to illustrate the mechanics and the option values of the inversion and porting operators.

Why Invert?

For inversion to be possible, there must be modules within modules. Thus the starting point is a system that has been split and (at least conceptually) split again. To fix ideas, let us focus on the PDP-11/20, as it was initially shipped out of DEC in the early 1970s. In keeping with the principle of exclusion, the first PDP-11s were bare machines with no systems software. (A paper-tape assembler was shipped with the hardware.)

Initially DEC did not even supply "boot ROM" (read-only memory) to get the machines started. According to Neil Groundwater, who was placed in charge of a new PDP-11/20 at New York Telephone in the summer of 1972,

Booting a PDP-11/20 consisted of loading a starting address [a location in read-only memory] . . . into the console switches and hitting the Execute switch. The boot ROM was literally a

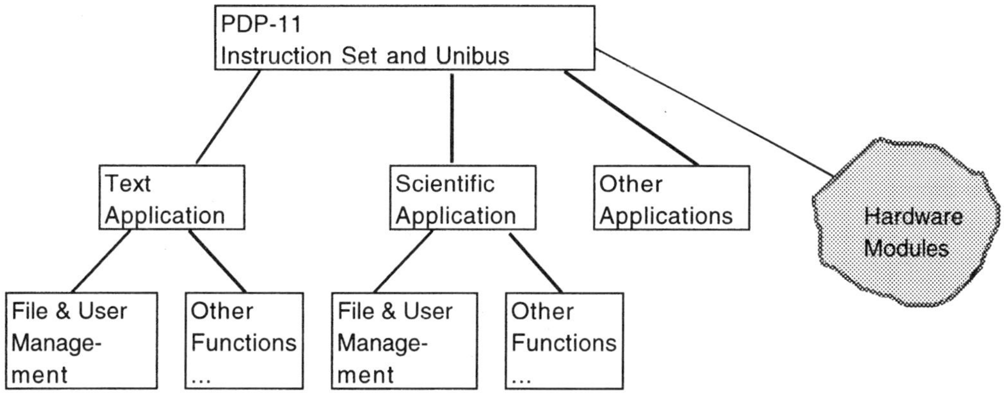

Figure 13.1 Design hierarchy for PDP-11 software modules before inversion.

board full of diodes that came from DEC with all the bits full and they were snipped out to create the instructions for booting.[1]

Figure 13.1 presents a design hierarchy diagram for the software of a PDP-11/20 at this time. At the beginning of the PDP-11 family's life, all software applications were designed by users and were coded in machine language. Thus, the figure depicts a series of software modules "looking up" to the PDP-11 machine instruction set and Unibus specifications.

As the PDP-11 was coming to market, programmers were divided as to whether high-level languages enhanced their productivity enough to justify the "runtime" inefficiency of compiled code.[2] DEC's designers clearly envisioned that some code for PDP-11 systems would be generated by human programmers, while other code would be generated by compilers or other automatic translators. As a result they put a lot of effort into making the machine-language instructions work for a variety of programming styles.[3]

---

[1] Neil Groundwater, quoted in Salus (1994, p. 47). One bit equaled one diode: Groundwater was essentially programming with scissors, one bit at a time.

[2] Brooks (1995). Around 1975, the debate was resolved in favor of compilers. By the late 1980s, compilers had improved to the point where they could outperform most human programmers in writing efficient code. It was then feasible to tailor machine instruction sets to compilers. This transition led to the so-called reduced instruction set computer or RISC approach to computer architecture. See Hennessy and Patterson (1990, chapter 3, pp. 89–132.)

[3] Bell et al. (1978, pp. 243, 322, 365–378, 383).

However, every software application made use of some of the same basic functional elements. Some code was needed to manage the hardware resources of the machine, to shift blocks or lines of instructions between the processor and memory, to keep the processor busy, to keep track of addresses, to create and retrieve files, and to perform a host of other "housekeeping" chores. Other functions common to all applications had to do with the act of writing code. Programmers needed to be able to enter code in machine-readable formats, edit it, run small sections, trace the execution of a program, interrupt execution, hook components together, and generate error messages.

Both housekeeping and programming tasks were extremely repetitive, time-consuming, and tedious, but they had to be done if the machines were to function. In figure 13.1, we have lumped them all together in "submodules" within each application labeled "File & User Management." User management was especially important on machines that gave more than one user access to the central processor. (Such "time-sharing" systems were a hot area of computer research in the late 1960s and also represented a competitive opportunity because they were an area in which IBM's System/360 did not shine.)

The opportunity implicit in this design hierarchy was to develop a common solution to the repetitive tasks of housekeeping and programming and to utilize that solution across a wide range of software application modules. This change, in turn, required two steps: (1) *splitting* the original applications to isolate the common functions; and (2) developing a common solution and imposing it on the applications. The second step is a classic application of the *inversion* operator, which we described in Chapter 5.

These steps—splitting and inverting—could be taken at the same time. Note, however, that the first step involved an application of the *splitting* operator, discussed in Chapter 10. In figure 13.1, we assumed that such a split had already taken place. Hidden-module experiments would then include a set of experiments aimed at the functions common to all the hidden modules.

Imposing a common solution on all modules effectively terminates such "natural" experiments. It follows that one cost of inversion is the loss of better solutions that might be discovered in the experimental process. Ending this experimentation increases the economic value of the whole only if the experiments themselves have reached the point of diminishing returns.

The Value of Inversion

The history of computers provides many examples of design processes that begin with a period of active, fruitful experimentation, go through a phase of diminishing returns, and eventually reach the point of tedious repetition. The best time to apply the inversion operator is during the second phase, when the costs of redundancy first begin to outweigh the benefits of experimentation.

Such inversions are often highly controversial. They take away designers' prerogatives, because some parts of the design must be "frozen" for the inversion to work. Moreover, inversions require designers to learn new rules and perhaps new skills. Finally, at least in the short run, inversions reduce the total amount of design work to be done, hence they may directly threaten the designers' livelihoods.[4]

As with the other operators, the value of inversion can be expressed as an algebraic sum of costs and benefits. For example, suppose there was a proposal to invert the File & User Management functions of the DEC PDP-11 family. Let us assume that a File & User Management module contained *n* design parameters (this might be the number of lines of code devoted to these functions). Initially the functions would be performed by components (knots of structure) within *m* different hidden modules. Once inverted, however, the same functions would be performed by *one* architectural module, which, by definition, would be visible to the next generation of the same *m* hidden modules.

The net option value of inverting the File & User Management functions and placing them in an architectural module is

$$\mathrm{NOV}_{\mathrm{Inv}} = \underbrace{V_{\mathrm{Inv}}(j,k)}_{\substack{\text{Option}\\ \text{value of}\\ \text{architectural}\\ \text{module}}} - \underbrace{C_{\mathrm{Inv}}(j,k)}_{\substack{\text{Cost of}\\ \text{designing}\\ \text{architectural}\\ \text{module}}} - \underbrace{\sigma_{\mathrm{Inv}} n^{1/2} Q(m)}_{\substack{\text{Option value}\\ \text{lost in hidden}\\ \text{modules'}\\ \text{experiments}}} + \underbrace{cnl}_{\substack{\text{Cost}\\ \text{savings in}\\ \text{hidden}\\ \text{modules'}\\ \text{experiments}}} - \underbrace{Z_{\mathrm{Inv}}(m)}_{\substack{\text{Costs of}\\ \text{visibility}}}. \qquad (13.1)$$

The first term in equation (13.1) denotes the option value of the new architectural module that will be created through inversion—we discuss what makes up this value below. The second term represents the cost of designing the new module, which would have its own (internal) modular and experimental structure. Thus both the option value and the cost of inversion will depend on the number of modules within the architectural module (j), and the number of experiments on each of those modules (k).[5]

[4] Inversion increases the efficiency of the design process, which may in turn make new applications and modules feasible. Thus over the longer term, inversions may increase the total demand for designers' efforts. But the increase in demand may pop up in a different part of the modular design structure, where the former designers' specific knowledge is not relevant.

[5] More generally, the new module may itself be made up of asymmetric modules, in which case the value and cost functions would be replaced by a sum (over j) of module experiments, as in equation (11.1). This is another example of recursion in value functions.

The third term in the equation represents the option value that would be realized if the File & User Management function were to continue to be designed in each of the *m* existing hidden modules. The fourth term is the cost of those design efforts. These costs will not be incurred if the inversion succeeds.[6]

The fifth term reflects the cost of redesigning the hidden modules that "look to" the architectural module. If the hidden modules have not already been split to isolate the common functions, then they may need to be redesigned from scratch. But if they have previously been split (as we assumed), the costs of visibility will be limited to the cost of redesigning the interfaces—the bulk of these hidden-modules designs will be untouched.

The expression shows just how complex are the design tradeoffs that attend an inversion. Note that the value of further experimentation (term 3) exerts a pull in one direction, while potential economies of scope (term 4) pull in the other direction. However, in contrast to what we saw in the previous net option value equations, (11.2) and (12.1), the number of experiments being run (*m* in this expression) is *not* the result of an optimization. Instead, the number of experiments is determined by the number of modules that use the File & User Management function. This number may be much higher than that needed to explore alternative designs, in which case the cost savings of term 4 will be higher than the lost option value of term 3.[7]

The most interesting terms in the expression are the first two—the option value of the architectural module and its costs of design. The architectural module itself may contain one or more modules and is subject to experiments within its modular structure. The number of submodules in the architectural module is a choice its designers will make, subject to constraints imposed by the state of knowledge at the time. And the optimal amount of experimentation on the new module will in turn be a function of how it is split.

Hence the new architectural module will have its own evolutionary trajectory, determined by its own internal modular structure. For example, in the case of the inverted File & User Management subsystem, at one extreme, this architectural module might be designed in a highly interdependent fashion. In that case, experiments would have to be conducted at the level of the module itself. This could be expensive,

---

[6] Using the same item in many places is a classic economy of scope.

[7] In the real world, we have further complexities of communication. How do designers, dispersed over hidden modules, learn of good solutions to common functions? This problem is often solved via forums for information exchange and solution sharing.

especially if the interfaces with the rest of the system [Z_{Inv} in equation (13.1)] had to be redesigned each time.

At the other extreme, the File & User Management architectural module might be split into a number of submodules, which themselves adhered strictly to the principles of modular independence and information hiding. In that case, the substitution, augmentation, and exclusion operators (discussed above) would be applicable to the design of the inverted File & User Management module itself. Design evolution could then occur within the framework of the architectural module's own design rules.

Unix

Unlike its predecessors, Unix was a flexible, modular operating system, which was itself capable of significant design evolution. Its internal modularity allowed it to evolve and develop new functions. As a result, Unix continues to be an important operating system today, long after the PDP-11 family for which it was initially developed has passed out of general use.

Readers may recall that the initial PDP-11s were shipped as "bare machines," lacking even the most basic systems software. DEC's long-term game plan was to add system software later as the new family became established.[8] This plan was consistent with DEC's basic "exclude-and-augment" approach to computer systems development, which we discussed in chapter 12.

Nevertheless, virtually every PDP-11 shipped needed some form of file and user management software. For original equipment manufacturers (OEMs), who accounted for much of the demand, such software was an important part of their proprietary value added. Although writing operating system code (in machine language!) increased their initial costs, it also served to lock in their customers; the more opaque their operating system, the more difficult it would be to reverse-engineer. Hence, for many OEMs a densely interdependent, difficult-to-imitate operating system was an important barrier to entry, protecting their market positions.[9]

[8] Bell et al. (1978, chapter 9).

[9] In general, it is difficult to protect intellectual property rights vested in software. (Davis, Samuelson, Kapor, and Reichman 1996.) This creates economic incentives to make code incomprehensible. However, opaqueness is a two-edged sword, for "spaghetti code" is as hard for the owner to maintain and upgrade as it is for third parties to imitate.

However, an equal number of PDP-11 purchasers were end users or managers of shared computer facilities. These purchasers were not interested in building up an OEM franchise. Many were in academic or scientific fields and operated under fairly tight budgets. For the most part, these individuals were sophisticated users of computers, hence they knew how difficult and time-consuming it was to write a good operating system from scratch. To avoid the cost of writing an operating system, and, equally important, to avoid the risk of having it turn out to be a dud, these users were eager to license an operating system as long as it did not cost too much.

With respect to the net option value of a proposed inversion [equation (13.1)], the number of potential independent experiments (m) was quite large. In the absence of an inversion, each PDP-11 purchaser would have to write and implement its own operating system. Thus there was a high ex ante probability that the cost of redundant effort would outweigh the value of additional experimentation by the purchasers.

Moreover, at the outset of the PDP-11's life cycle, the costs of visibility were at an all-time low: the modules that would "look to" the operating system were mostly to-be-written software applications. Hence, in the early 1970s, the last three terms in equation (13.1) weighed heavily in favor of inversion. Nevertheless, the net option value of any specific inversion would depend on the quality and cost of the candidate architectural module, represented by the first two terms in the expression.

The challenge of inversion was taken up by two computer scientists at AT&T's Bell Laboratories in 1969 and 1970. Ken Thompson and Dennis Ritchie had worked on a project to design a large-scale operating system (MULTICS), that would support interactive computing for up to a thousand users. The project, a joint effort of MIT, GE, and Bell Laboratories, had foundered badly. While some of the underlying concepts would prove to be excellent (e.g., a tree-structured file system, a "shell" to do command interpretation), MULTICS itself was a failure. When Bell Laboratories withdrew from the project, the prototype time-sharing system could only support three users at one time without crashing.

Ritchie and Thompson set out to create an operating system that would be easier to work with, and much cheaper to run. They started with a file system that Thompson had built for a PDP-8. Without a grand plan, but with a clear sense of what they wanted, Ritchie and Thompson and their collaborators at Bell Laboratories began to add commands (e.g., *mv:* move or rename file; *chdir:* change working directory) and utility programs (e.g., an editor that would work with the file system). By the fall of 1970 the new operating system—dubbed Unix—was up and running on a brand-new PDP-11/20 (which, as we know, arrived as a bare machine).[10]

---

[10] Salus (1994, p. 36–37).

Unix was a breakthrough design. It was at once powerful, flexible, small, simple, elegant, and easy to use.[11] Moreover, because of antitrust constraints, AT&T's policy was to license the operating system for a nominal fee, hence Unix was cheap.[12] Thus for many users, the net option value of Unix [the sum of all five terms in equation (13.1)] was resoundingly positive. Many inversions must be imposed on users and designers by a central authority. Not so with Unix: it was voluntarily adopted by hundreds of users at scores of computer facilities around the world. And because it was fundamentally a modular design, users could make it better by adding modules (augmenting) and by improving those in existence (substitution).

The Design of Unix

The genius of Unix was that it was modular both in its own design structure (static design modularity) and in the way it managed computer resources (dynamic design modularity). Static design modularity is the type of modularity we have focused on thus far. The individual components of Unix were carefully partitioned so that design tasks on different modules could be carried out independently. Thus Unix was easy to upgrade and augment; its design invited tinkering, and its users soon developed mechanisms for sharing good solutions to commonly encountered problems. Static design modularity also made porting the operating system easier; the essential parts of the system were small (measured in terms of lines of code), and the additional utilities could be ported piecemeal.

However, what set Unix apart was the way it managed computer resources, in particular, the modularity of its dynamic memory-to-process or management function. Memory-to-processor management, or memory management for short, governs the "chunking" of code as it is executed. It is an essential role in the internal operation of a computer. The efficiency of dynamic memory management greatly affects the speed of a computer system.[13]

At the time of Unix's development, dynamic memory management was often embedded in the code of application programs.[14] However, it was a hardware-dependent function, which had to be performed by all programs that ran on a given type of

---

[11] Ritchie and Thompson (1974).

[12] License fees varied by user and the nature of use. In some cases the software was conveyed royalty-free under a simple letter agreement. A relatively unrestricted commercial license cost $20,000 in 1976 (Salus 1994, pp. 60, 180).

[13] Hennessy and Patterson (1990).

[14] Myers (1975). The function was then called "pagination."

hardware. For this reason alone, the function was a clear candidate for inversion. Moreover, what was optional in a single-user or a batch-processing system was essential in a time-sharing system. If multiple users have interactive access to a single processor, that system must have protocols for dynamic memory management that are common to all programs, whatever their source.

Unix created a modular memory-to-processor management system through its definition of *images* and *reentrant processes,* and by using a command line interpreter called the "Shell." In Unix, all processing took place through the execution of an "image," which was the current state of a virtual machine. (The term "virtual machine" is anachronistic; in their initial article, Thompson and Ritchie called it a "pseudo computer.") Images resided in primary memory during execution of a process, but could be swapped out to secondary memory if a higher-priority process needed the space. When the initial process was reactivated, the swapped-out image was returned to primary memory, and execution continued from where it left off.

In Unix, every image had a standard form: in addition to space indicating general register values, the current directory, open file addresses, and so on, each was divided into a write-protected program text segment (which could be shared among processes executing the same program), a writable data segment, and a stack. Thus in a system operating under Unix, chunks of executable code came in a standard "size" and "shape."[15]

Strict modular independence and information hiding across processes was further enforced by segregating the "kernel" of Unix from the Shell. The kernel was the subset of compiled Unix code that was loaded into the computer's memory at startup, and remained there until the system shut down. The kernel managed all hardware, memory, job execution, and time sharing. Although it was the heart of Unix, consistent with the principle of information hiding, its internal workings were invisible to users.[16]

A Shell isolated the kernel from the user and vice versa. Users could choose among different Shells according to their taste (e.g., the Bourne, Korn and C Shells are shipped with most versions of Unix today). In contrast to the kernel, a Shell was *not* loaded when Unix booted (i.e., when a computer managed by Unix was turned on). Instead, Shells were programs, and an instance of a specific Shell would begin execution whenever a user logged onto the system.[17]

---

[15] Ritchie and Thompson (1974, p. 370).

[16] Ibid., pp. 365–370. The name "kernel" is an anachronism that does not appear in the initial article.

[17] Ibid.

Shells were also highly recursive structures. A Shell (the parent) could create another Shell (the child) to perform a particular subset of tasks. The state of the parent would not be changed by the child, until the child's tasks were complete and the results recorded in the child's "standard output" location. At that point the parent Shell would resume execution: it might ask for a new command, or it might have instructions to read the child's output and manipulate it in various ways.

Pipes and Software Tools

One of the key features of the Shell program, and a breakthrough in the design of Unix, was the invention of "pipes." Pipes were possible because every instance of a Shell had its own standard places to receive input and write output. (In a time-sharing system these would generally be the terminals used to dial into the system.) Standard input and output locations could be overridden by a user's command. But the standard output of one process could also be converted into the standard input of another by means of an intermediate file known as a pipe. As described in the initial article on Unix:

> A read *using a pipe file descriptor waits until another process writes using the file descriptor for the same pipe. At this point, data are passed between the images of the two processes.* Neither process need know that a pipe, rather than an ordinary file, is involved. *[Emphasis added.]*[18]

The power of pipes can be seen most easily via an example. Suppose a programmer wanted to find all lines in "oldfile" containing the character string "Unix," to sort those lines alphabetically, to save the output to "newfile," and to display the output on the programmer's Teletype or screen. Without pipes, the programmer would have to create temporary files between each of the commands, making sure that the output arguments of the prior command were consistent with the input arguments of the next command. (The programmer would also have to remember to delete all temporary files when the job was done.)

With pipes in Unix, the programmer could type in the following line containing three commands and two pipes (denoted by vertical lines):

grep "Unix" oldfile | sort | tee newfile.

The computer would then execute the sequence of commands; it would not need to create the intermediate temporary files.

---

[18] Ibid.

Pipes simplified the interfaces between commands, making it easy to construct chains. In essence, then, pipes were a modularization of the microstructure of software code. Theorists in the new field of software engineering were already debating the pros and cons of modularity—it was a major issue in the "holy wars" of structured design that raged in the early 1970s.[19] But Unix's designers ignored the debate and instead achieved modularity in the microstructure of their program.

Pipes in turn created a new way of thinking about software, which became known as the "tools" approach. According to this new way of thinking, a command was itself a self-contained module, which was flanked by an input stream and an output stream. Hence the focus of programming for users of Unix shifted from programs and subroutines to single commands and strings of commands. Unix programmers sought regularity in their code structures down to the level of commands, that is, partial lines of code.[20]

Regular structures became a principle of design among the community of Unix programmers and users. A "Unix philosophy" was codified and spread in this community. It was

- write programs that do one thing and do it well;
- write programs that work together;
- write progams that handle text streams, because that is a universal interface.[21]

In effect, this philosophy was an affirmation of the principles of modularity. It instructed programmers to do the following: break code down into the smallest possible functional units and find the best implementation of that function ("programs that do one thing and do it well"); hide the contents of a module by making its input and output a text stream ("the universal interface"); and create large systems out of modular primitives ("write programs that work together").

The Modular Structure of Unix

The fact that Unix was (and is) modular in both its static and dynamic design structures makes it difficult to construct a task structure matrix (TSM) or a design hierar-

---

[19] See Constantine et al. (1974); Myers (1975); Kernighan and Plauger (1976).

[20] This search for regularity in the fine structure of designs is very reminiscent of Mead and Conway's search for regularity in the layout of transistors and gates on chips, as well as Maurice Wilkes's search for regularity in the microprograms of code execution. Refer back to Chapters 3 and 6.

[21] Salus (1994, pp. 52–53).

chy diagram that fully captures its design. The difficulty arises because developing Unix modules (commands and applications) and running computers under the Unix operating system involve two different task domains. Unix is modular in both domains: its static design modularity (program and file structure) contributes to its dynamic design modularity (runtime configuration), but the two are not identical.

When it is running, Unix consists of a memory-resident kernel and a number of active processes (a background process and at least one for each logged-in user). The top-level process for each user is usually a Shell. In general there will be processes within processes, and Shells within Shells. In this dynamic, interactive mode, users are shielded from the hardware and from other users by strict information-hiding protocols—standard images, standard I/O definitions, pipes, and so on. There is always a modular, hierarchical structure that describes the active processes, but this structure is a product of the moment. Processes come into existence as commands are entered, and wink out again when their tasks are complete.

The dynamic design modularity of Unix made it a big system in a small package. With Unix, it was possible for a minicomputer to operate as a multiuser, interactive, time-sharing system. In fact, in the eyes of its architects, one of the design's most important achievements was to show that "a powerful operating system for interactive use need not be expensive either in hardware or in human effort."[22]

However, it was the static design modularity of Unix—its program and file structure and its pipes—that allowed it to evolve. Not only was the Shell separated from the kernel; the kernel itself was compartmentalized and modular. In addition to its executive parts, the main pieces of the kernel were

- a tree-structured directory of files;
- an index list of files containing the physical addresses of the contents plus other information (the "i-node list");
- a list of devices present in the particular installation.

The inclusion of device driver files in the operating system was a deliberate inversion, as Ritchie and Thompson noted: "Given the requirement that all programs should be usable with any file or device as input or output, it is also desirable from a space-efficiency standpoint to push device-dependent considerations into the operating system itself."[23]

---

22 Salus (1994).

23 Ritchie and Thompson, (1974 p. 374).

The separation of the file directory, which was visible to users, and the i-node list, which contained the physical addresses of files, was an unusual feature of Unix. This design feature was an example of information hiding, and the creative application of the *splitting* operator. It meant that system-level file maintenance and diagnostic tasks could be implemented on a "flat file" database with one record per file. At the same time, users could organize their files in any kind of tree structure they chose, they could have "copies" of the same file in different directories, and different users could share the same file.

As far as Unix was concerned, a program was simply another file, with various citations in the directory and one index number pointing to its physical location in the i-node list. Because programs were simply files, it was very easy to add new utilities to the operating system in the form of short files. These in turn could be called up and executed by the Shell—they did not have to be loaded when the system was turned on.

In 1974, when the initial article on Unix was published, Bell Laboratories programmers had augmented the basic system with "over 100 subsystems, including a dozen languages."[24] In our language, these subsystems were hidden modules in the greater Unix system.

After the original article was published, a number of sites obtained Unix licenses, and the number of hidden modules (applications and utilities written for Unix) grew dramatically. Peter Salus describes Unix's subsequent evolution in this way:

> *Something was created at BTL [Bell Telephone Laboratories]. It was distributed in source form.*[25] *A user in the UK created something from it. Another user in California improved on both the original and the UK version. It was distributed to the community [of Unix users] at cost. The improved version was incorporated into the next BTL release. . . . There was no way that [AT&T's] Patent and Licensing could control this. And the system got better and more widely used all the time.*[26]

Salus is describing a process of decentralized design improvement, which lies outside the control of any single individual or firm. This is the process of "design evolu-

---

[24] Ibid.

[25] That is, in a format that allowed the licensees to read the source code and thereby understand the structure of the system. Source code distribution makes it easy to reverse-engineer, modify, or port a piece of software, but correspondingly makes it more difficult to enforce intellectual property rights. Had AT&T been intent on making money off of Unix, the company would probably not have distributed source code as freely as it did.

[26] Ritchie (1984); Salus (1994).

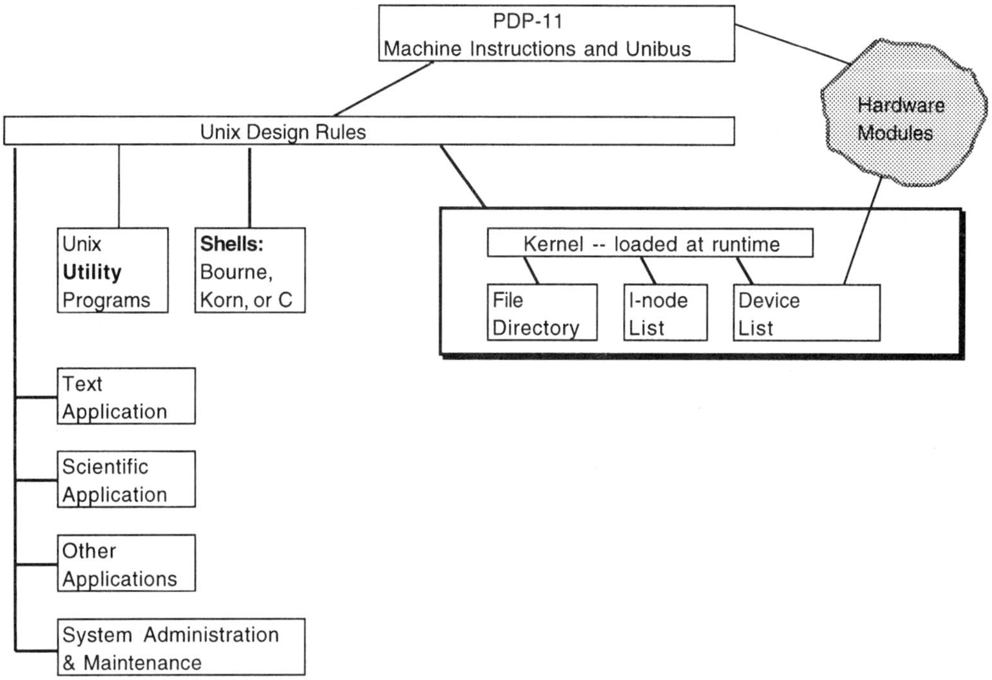

Figure 13.2 Design hierarchy of the PDP-11 after inversion of the Unix operating system. (From Ritchie and Thompson 1974, and various Unix operating manuals.)

tion," described in chapter 9, which can occur if (and only if) the underlying design structure is modular.

The PDP-11 Design Hierarchy after Unix

Figure 13.2 shows the design hierarchy of a PDP-11 computer system running under Unix. There are two things to notice. First, the Unix design rules "sat" between the PDP-11 instruction set and Unix and non-Unix software. The instruction set was still indirectly visible to programmers, because Unix was (initially) coded for a PDP-11, and because descriptions of hardware (at a particular installation) were part of the kernel that loaded at runtime. However, once Unix was installed, programmers did not need to have detailed knowledge about the hardware system in order to perform most of their programming tasks.

Programmers did not even have to know "everything there is to know" about Unix to perform their tasks. They had to know some things about its architecture (programs are files, the tree structure of the directory, etc.) and its conventions (standard input and standard output). Knowing the contents of the library of utilities could save them time and effort. But programmers and other users did not have to know how the file-sharing system actually worked (e.g., the mechanics of the i-node list) or what was loaded in the kernel vs. what was called up as needed.

Hence Unix itself had a highly modular internal design structure. The figure depicts this modularity by showing the modules of Unix "looking to" the Unix design rules, instead of to Unix as a whole or even the Unix kernel.

The internal modularity of Unix in turn was an object lesson for programmers. For example, in their book *Software Tools,* Brian Kernighan and P. J. Plauger acknowledge their debt, not only to the designers of Unix (who are cited separately) but to the design itself:

> *Finally it is a pleasure to acknowledge our debt to the UNIX operating system. . . . Many of the tools we describe are based on UNIX models. Most important, the ideas and philosophy are based on our experience as UNIX users. Of all the operating systems we have used, UNIX is the only one which has been a positive help in getting a job done instead of an obstacle to overcome. The widespread and growing use of UNIX indicates that we are not the only ones who feel this way.*[27]

Porting

An express goal of the creators of Unix was to give programmers the feeling of working with a generic machine, and to shield them from the vagaries of particular sets of hardware. However, as an operating system, Unix had to "know about" the hardware resources it was managing. Figure 13.2 shows that Unix initially displayed two types of hardware dependence: (1) on the PDP-11's instruction set because Unix was written in PDP-11 assembly language; and (2) on the device list, which served as a *translator module* between Unix and an individual PDP-11 facility.

Of the two types of dependence, the first was much more constraining. The device files were a group of hidden modules in Unix, grouped for convenience in a single directory. When Unix was installed at a new facility, or when hardware changed, the system administrator needed to change the device files (as well as some other user-related files), and reboot the system. While this was not a trivial task, it was not particularly daunting either. Depending on the system, it might take anywhere from a day to a week to get Unix up and running at a DEC installation.

---

[27] Kernighan and Plauger (1976).

In contrast, instruction set dependence permeated every line of code. This was not acceptable to Unix's creators, who were not DEC employees or wedded to DEC equipment.[28] If a cheaper, more powerful computer arrived on the scene, or if the management of Bell Laboratories purchased a system from another vendor, Ritchie and Thompson wanted to continue to enjoy their familiar Unix environment. Hence they wanted to make the operating system *portable.* To do this, they had to break the connection between Unix and the instruction set of a PDP-11.

The Mechanics of Porting

Porting takes a hidden module developed for one system and makes it work in other systems. It requires three steps:

1. *splitting* the module to be ported into system-dependent and system-independent functions;
2. *finding a representation* for the system-independent parts. This representation specifies the design rules and identifies the hidden modules of the portable system;
3. designing *translator modules* that make the portable system compatible with other systems.

As with inversion, the first step toward porting is to apply the splitting operator to the design of a module. However, inversion does not displace the design rules of the original system—it only defines a new layer of visible information. In contrast, porting creates a whole new "apex" of visible information.

Porting also creates links between systems in the form of translator modules. With porting, it is no longer possible to delineate strict boundaries between systems, for hidden modules may look to many different places for visible information and design rules.

Unix and C

The early development of Unix illustrates how porting works. Unix's inventors, Ritchie and Thompson, wanted to make the system portable (within constraints imposed

---

[28] In 1976, Dennis Ritchie and Steve Johnson proposed to their boss that Bell Labs purchase a non-DEC machine to test the idea of porting Unix. An Interdata 8132 was acquired in the spring of 1977. "Proposing a UNIX Portability Experiment," Lucent Technologies, Inc., 1998; viewed 1 May 1999; available at http://cm.bell-labs.com/cm/cs/who/dms/firstport.html.

by the technology of the time) both for aesthetic reasons and because they had no desire to be tied to DEC hardware. But the original design was expressed in terms of a DEC hardware instruction set.

Then, as now, there were two ways to write code. Programmers could use the commands built into the processor's instruction set,[29] or they could write code in a high-level language, like FORTRAN, and use a machine-specific *compiler* to translate that code into "object code" that was understandable by the machine.[30] At the time of Unix's creation, FORTRAN, a creation of the 1950s, was the most widely-used programming language. However, FORTRAN's structure did not make it easy to write good systems programs—that is, operating systems, compilers, and editors.[31] And other high-level languages of the time were not much better for this purpose.

What was needed then was a new language that could do justice to the structure of Unix. This language was created by Ritchie and Thompson between 1969 and 1973. It was a painful bootstrapping process. Dennis Ritchie described the genesis of the language and its connection to Unix in the following way:

B [the predecessor of C] started out as system FORTRAN. . . . It took [Ken Thompson] about a day to realize he didn't want to do a FORTRAN compiler. So he did this very simple language, called it B. . . . A few system programs were written on it, not the operating system itself, but the utilities. It was fairly slow, because it was an interpreter.[32]

[The] first phase of C was really two phases in short succession of, first, some language changes in B, . . . and doing the compiler. [The] second phase was slower. . . . It stemmed from the first attempt to write Unix in C. . . . There were sort of two things that went wrong. [One was that Ken] couldn't figure out . . . how to switch control from one process to another, the

---

29 There is a unique correspondence between each element in the instruction set of a processor and the electrical events that occur within the particular processor. However, the instruction set defines a virtual machine, which can be implemented in hardware in different ways. A range of processors with the same instruction set is a family. In theory, families can run the same software with minor modifications. Bell and Newell (1971, Chapters 1–3); Hennessy and Patterson (1990).

30 Object code is the string of bits (0s and 1s) that reside in the computer's memory and that the machine actually processes. High-level languages, like FORTRAN, include many standard commands that need six or eight lines of machine-language code to express. *Compilers* are software programs that translate programs written in a high-level language into object code for a specific machine.

31 Kernighan and Plauger (1976).

32 An interpreter translates commands into an intermediate representation that is not specific to a particular machine. The intermediate form instructions are then translated into machine instructions as the program executes. This can slow down processing substantially.

relationship inside the kernel of different processes. . . . The second thing . . . was the difficulty of getting the proper data structure. The original version of C did not have structures. So to make tables of objects—process tables and file tables, and that tables and this tables—was really fairly painful. . . . It was clumsy. . . .

The combination of things caused Ken to give up that summer [1972]. Over the year, I added structures and probably made the compiler somewhat better—better code—and so over the next summer, we made the concerted effort and actually did redo the whole operating system in C.[33]

In their first article, Ritchie and Thompson reported that one of the "major programs available under Unix" was "a compiler for a language . . . with types and structures (C)." They went on to say:

Early versions of the operating system were written in assembly language, but during the summer of 1973 it was rewritten in C. The size of the new system is about one-third greater than the old. Since the new system is not only much easier to understand and to modify but also includes many functional improvements, . . . we considered this increase in size to be quite acceptable.[34]

A New Design Hierarchy

The paragraph just quoted modestly notes one of the most significant facts about Unix. The invention of C, and the rewriting of Unix in this new language, led to a dramatic change in the design hierarchy of computers and in the relationship of the operating system to the hardware.

The new design hierarchy appears in figure 13.3. A noticeable difference from figure 13.2 is that the hardware instruction set is no longer the highest "apex" of visible information. There are instead two "apexes"—one for hardware and one for software. At the apex of software is the C programming language. It is visible to all programs written in the language, including the Unix operating system.

C is also a module, albeit a very high-level architectural module. Our theory of design evolution thus predicts that C would change very slowly (because costs of visibility are high) and that changes were likely to be backward-compatible with previous versions of C. (In other words, syntactically correct programs written in previous versions of C should be correct in later versions of the language.)

---

[33] Dennis Ritchie, quoted in "The Unix Oral History Project: Release.0, The Beginning," ed. M. S. Mahoney, available at http://gardens.princeton.edu/~mike/h398/h398.html, also in Salus (1994, pp. 48–49).

[34] Ritchie and Thompson (1974).

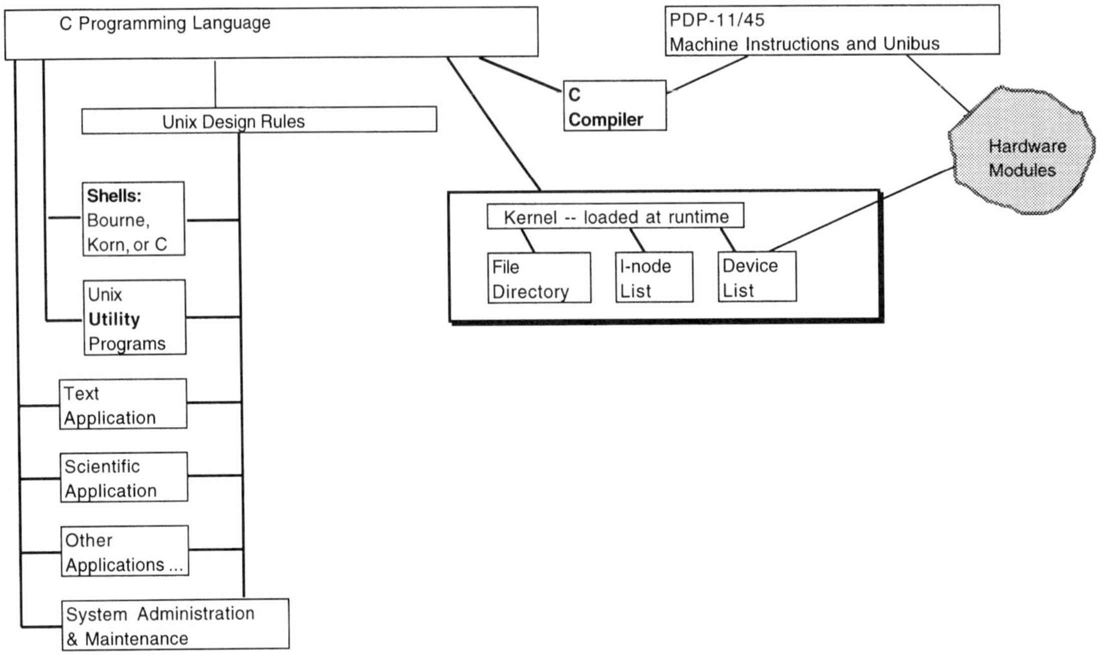

Figure 13.3 Design hierarchy for C and the PDP-11 after the porting of Unix to C.

The Unix design rules are also visible to programs designed to run under Unix, including the Unix Shells and Utilities. Knowing C is not enough—to operate under Unix, and to write good (efficient) programs, the programmer must know how the operating system is structured (tree-structured directory, programs are files, recursive Shell, standard input and output), as well as commands and utilities that are specific to it. But Unix is modular and thus much of Unix is not visible. Programmers do not have to know about the i-node list, or the details of the device files. They do not have to know how the kernel manages the relationship between processes, or whether a particular utility resides in primary memory.

Between C and the hardware is a translator module—the C compiler. The compiler translates C commands into machine language commands automatically. But other hidden modules do not have to know how the translation process works—only that it is correct.

Finally, between Unix and the hardware are the device files, which tell Unix about the particular hardware setup it is managing. Dynamically, the device files are visible information: they are part of the kernel that is loaded when the system boots. But from the perspective of applications designers, the device files are another hidden translator module. The user can be sure that "all programs [will] be usable with any file or device as input or output." However, the fact that this takes placc via a set of device files in the kernel is not something that the user needs to know.

The Value of Porting

Porting is like inversion in that it promotes a common solution in a wide range of contexts—this reduces the costs of design (the module does not have to be redesigned from scratch in every system), but may diminish gains from subsequent experimentation. Portable modules and subsystems also are not "trapped" by the design rules of a particular system; in a sense they are "free to roam" from system to system. Therefore investments in porting and portability are often defensive investments, aimed at deterring an architect's opportunistic behavior at a later stage in the evolution of modular design.

The components of the net option value of porting are the same in all contexts, even though their magnitudes will differ. Thus let us consider the opportunity to develop a portable operating system and compare it with the opportunity to design a separate operating systems for *M* different hardware systems. We assume that each separate operating system has a net option value as shown in equation (13.1). Thus for each of the *M* separate operating systems, there is an option value gained that depends on the modular structure of the system, a cost of design, a lost option value from reduced experimentation, a cost savings from the same source, and a cost of visibility (cost of redesigning hidden modules to be compatible with the operating system).

Let NOV_i denote the net option value of one of the operating systems; *i* is an index running from 1 to *M*. For simplicity, we assume that, on average, the functionality of the portable operating system is the same as that of the separate operating systems. The sum of the NOV_is is realized under either alternative, and thus nets out of the valuation of the portable operating system.

Four elements then remain and are depicted in equation (13.2). The first is the option value (to the designer of the operating system) of being able to switch among *M* systems at a switching cost *S*. The second is the cost savings associated with not having to redesign each operating system from scratch. The third is the cost of designing appropriate translator modules. And the fourth element is the "cost of bootstrapping"—that is, the cost of finding a representation of the portable module that is independent of any system.

$$\underset{}{\text{NOV}_{\text{Port}}} = \underset{\substack{\text{Value of}\\\text{option to}\\\text{switch}}}{V(M;S)} + \underset{\substack{\text{Cost savings}\\\text{in operating}\\\text{systems}}}{(M-1)C_{\text{OS}}} - \underset{\substack{\text{Cost of}\\\text{translator}\\\text{modules}}}{MC_{\text{Translator}}} - \underset{\substack{\text{Cost of}\\\text{bootstrapping}}}{C_{\text{Bootstrap}}} \qquad (13.2)$$

Equation (13.2) shows that, as with inversion, the economic motives for porting are complex. The net option value may come out positive, indicating that the port is worthwhile, or negative, indicating that separately designed systems are more cost-effective.

The last three "cost" terms are likely to net out to a negative number if the "porting template"—the prototype of the design to be ported—has not been split, and has a system-dependent representation.

If the porting template has not been split, hence is not modular, the translation function will be interdependent with the design itself, instead of being isolated in separate translator modules. Designing *M* translator "modules" will then be tantamount to designing *M* new operating systems—hence the net cost savings will be very small.

If the porting template has a system-dependent representation, then a system-independent mode of representation must be selected or invented. This is an exercise in setting design goals and specifying design rules for a *new* modular family—the portable subsystem. In the case of Unix, we can "see" and even measure the cost of this architectural endeavor in terms of the time and effort that Ritchie and Thompson put into the development of C and the writing of the first C compiler. And we can see how ignorance constrains such architectural efforts in the difficulty Thompson had expressing the dynamic process relationships of Unix (which he had already programmed in PDP-11 assembly language) in terms of the new language.

Quite often the "pure costs" associated with porting net out to a negative number. In this case, the value of applying the operator depends on the value of the option to switch. This option clearly has some value independent of the ownership of the module designs in the "grand system" (made up of the portable system and the *M* "host" systems). However, the value of this option goes up dramatically if the system to be ported and the host system are owned by different enterprises.

In the previous chapter, when discussing the exclusion operator, we described the intertemporal paradox and embedded conflicts of interest that arise when separate economic actors *coinvest* in a particular modular architecture. Once the architecture becomes established, investors in hidden modules may see their economic gains expropriated

- if the architect enters their market;
- if the architect changes its pricing structure;

- if the architect changes critical design rules to give one set of hidden modules an advantage over another set.

These actions are not mutually exclusive, and thus profitable hidden-module investors may see themselves attacked on all three fronts at once.

Faced with this long-term threat, an effective defense is to make the hidden module portable. In effect, the value of the switching option to a coinvestor who is not the architect is higher than the value of the option to the architect. For the coinvestor, the value of the option includes the value of changing the architect's incentives and controlling its actions after the fact.[35] The owner of a portable module's design can respond to adverse moves by the architect by switching to another host system. As with many games of this type, knowledge that the switch is possible may be sufficient to deter the architect from taking the threatened action in the first place.[36]

What this means is that it was no accident that Unix was the first, and for a long time the only, portable operating system of any significance. Soon after PDP-11's introduction, DEC moved to supply its own operating systems for that family. And through the rest of the 1970s, except in the very smallest "kit" computers, operating systems became part of the basic set of modules that computer systems manufacturers were expected to provide.

Indeed, manufacturers had no interest in making their proprietary operating systems portable. Quite the opposite—since virtually all applications software "looked to" the operating system, proprietary operating systems became one way that manufacturers increased users' switching costs and locked them in to a particular family of computers. From the manufacturers' standpoint, therefore, the more tightly the operating system was interconnected with the hardware instruction set, the better.

---

35 Investing in a "best alternative" before the fact is a classic way of guarding against expropriation after the fact. Absent a very complete long-term contract, a coinvestor's profit stream may be subject to expropriation by the architect (more precisely, the owner of an architectural module) after the fact. Porting the hidden module is an effective defense against this threat of expropriation. Grossman and Hart (1986) show that vulnerable parties will "overinvest" in defensive mechanisms (in this case, the switching option), relative to the social welfare optimum.

36 Making the module portable is *not* a defense against the architect's entering the market with its own version of the hidden module. However, entry necessarily increases competition in the hidden module's market, and for that reason is generally not as attractive to the architect as increasing the price of the architectural module or cutting out the hidden module via a strategic change in the design rules.

Thus, for at least fifteen years, Unix was sui generis. It was a proof of the concept of portability for operating systems, and a proof that an operating system could form the apex of a design hierarchy. But there was no other operating system like it in the world.

An Overview of All the Operators

We have now come to the end of our analysis of the six modular operators. Table 13.1 lists them, and indicates the effects of each on the overall design. As a group, the operators are a set of tools that assists the designers' imaginations. In a large, complex design, they define a set of potential "next steps" in the designers' search for value. Although there are only six, the operators can be applied at many points in a large design, and in many combinations and sequences. Hence they define a very large number of potential paths, which can be used to explore a very large space of designs.

Taken all together, the operators, applied locally in different combinations and sequences, are capable of generating designs of great variety and complexity. (See box 13.1.) They impose no limits on the number of modules in the system, or on the functions the modules might perform. There are no restrictions on the depth of a design hierarchy, on the number of apexes that hidden modules may look to, or on the number of systems to which a hidden module may be ported.

In earlier chapters we described some of the most critical episodes in the development of computer designs. The intellectual lineage of all computer designs today can be traced back to the first ENIAC machine at the University of Pennsylvania, and to von Neumann's conceptual splitting apart of the artifact's design. Other important events in the history of these designs included the initial modularization of System/

Table 13.1 The six modular operators and their outcomes

Operator	Outcome
Split	Hidden modules; visible design rules
Substitute	Improved hidden modules
Augment	Modules performing new functions
Exclude	New systems; minimal systems
Invert	New levels in the design hierarchy; architectural modules
Port	New visible apexes; translator modules; overlapping systems

Box 13.1 Operator Moves and Sequences in the Value Landscape

A move is the application of an operator at a particular point in a modular design. A sequence is a series of moves. For example, IBM's System/360 sequence was split-substitute; DEC's PDP-11 sequence was split-exclude-augment-substitute; Ritchie and Thompson's Unix sequence was invert-split-port-augment-substitute. The feasibility of particular moves and sequences depends on the state of the design and the state of knowledge at a particular time. In addition, economic value determines whether a particular move or sequence is attractive. Thus it is fair to say that a design creates a "supply" of options in the form of moves and sequences, but economic value determines the "demand" for such options.

We assume that the world is peopled with economic actors who are "seeing and seeking" economic value. Because these actors are value-seekers, moves and sequences with high perceived economic value are more likely to be chosen than those perceived to have low value. However, the economic actors generally have imperfect foresight, hence they will proceed in a trial-and-error fashion. Some moves therefore will succeed in creating value, while others will fail through bad luck or a misunderstanding of the circumstances.

Moves and sequences that are actually implemented determine the future state of the design (how many modules, how many layers, how many systems, etc.). Hence, by influencing the moves and sequences designers choose, value exerts a "pull" on the design as a whole. Over time, subject to the constraints of imperfect foresight, we would expect designs (of both systems and modules) to change in ways that increase value. In terms of the "value landscape" we would expect to see movement toward higher "ground," and a tendency for designs to stabilize at high "peaks."

360, the introduction of DEC's minimal PDP-11 family of minicomputers, and the creation and porting of the Unix operating system.

These are only a few of the most salient episodes in the evolution of computer designs. The whole story is vastly more dense and complicated. It spans millions of large and small experiments past and present, in hundreds of thousands of workgroups, at tens of thousands of enterprises. Thus, practically speaking, it is impossible to grasp the whole history of computer designs, just as it is impossible to catalog all the modules in all the systems, and all the linkages among systems that exist today.

This explosion in the sheer number of designs is a manifestation of the power of modularity. The combinatorial possibilities inherent in the six modular operators vastly multiply the options embedded in a large design. In these four chapters, we have shown in some detail how the *multiplication of options* can dramatically increase the economic value of the whole system.

Decentralization of control over options is the second important consequence of modularity. A modular architecture partitions information, and allocates certain parameters to decision-makers working on the so-called hidden modules of the system. The design and task structures are intentionally set up so that hidden-module designers do not have to have detailed knowledge about the whole system under development. Hidden-module designers have only to master the design rules, and then bring their specific knowledge to bear on a subset of the design tasks.

With respect to their own modules, then, hidden-module designers can act independently of both the central architects and one another. In this fashion, modularity allows human beings to divide effort and knowledge to create very complex artifacts, which function in an integrated way, even though no one person comprehends the whole.

In the last two chapters of this volume, we focus on the consequences of the decentralization of options for the organization of firms and markets in the computer industry. Between 1967 and 1980, the computer industry split into sixteen different subindustries, hundreds of new firms entered the industry, and market values in the industry came to be increasingly dispersed across modules and firms. These trends continued through the 1980s and 1990s until in 1996, almost one thousand firms were members of the "greater" computer industry. Today that number continues to grow.

Without modularity in the underlying artifact designs, the division of effort and knowledge across so many firms would have been impossible. Thus modularity was a necessary precondition for the industry structure we have today. But, in principle, there was no reason why the six modular operators could not have been contained in one or a few big firms. Hence modularity alone does not explain why the structure of the computer industry changed during the 1970s, nor why economic value came to be distributed across so many separate enterprises. In the next two chapters, we turn our attention to those questions.

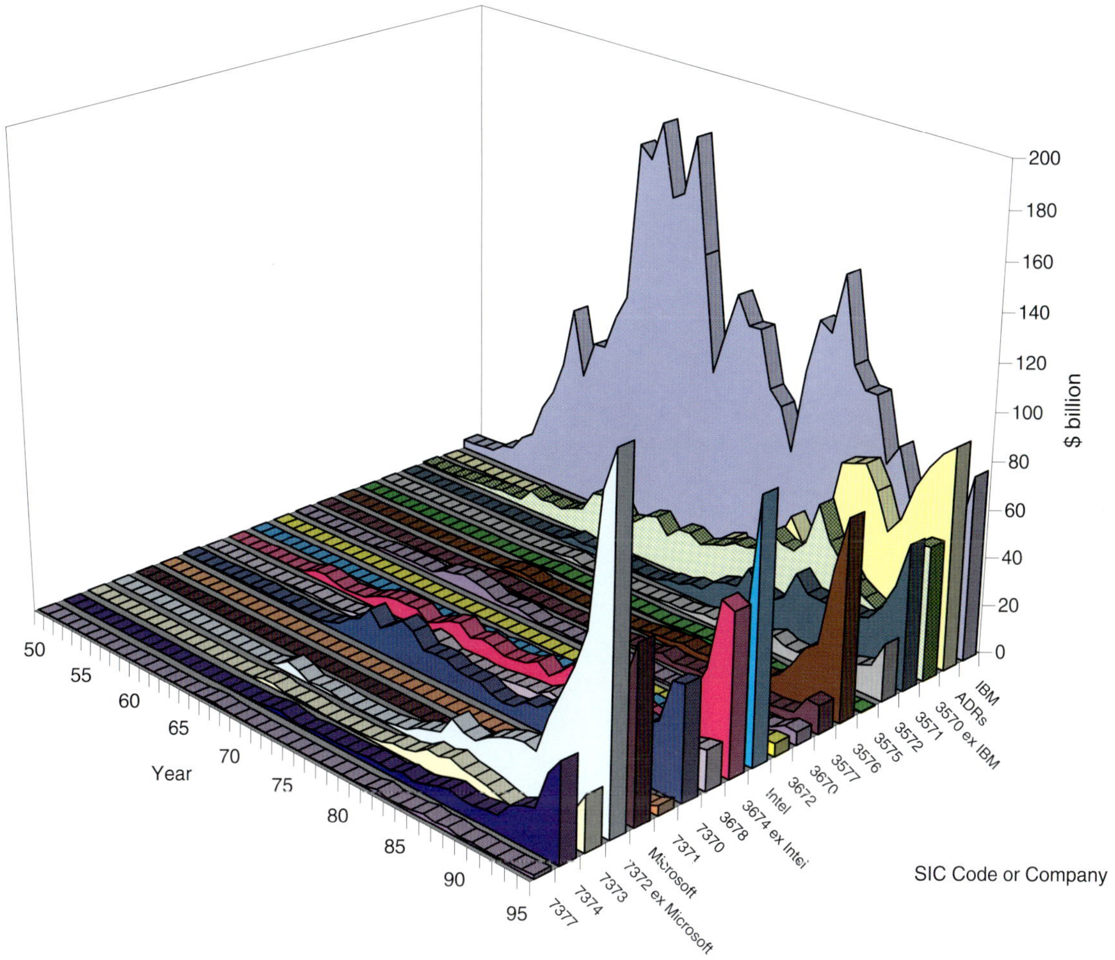

Plate 1.1 Market capitalization of sectors of the computer industry in constant 1996 dollars.

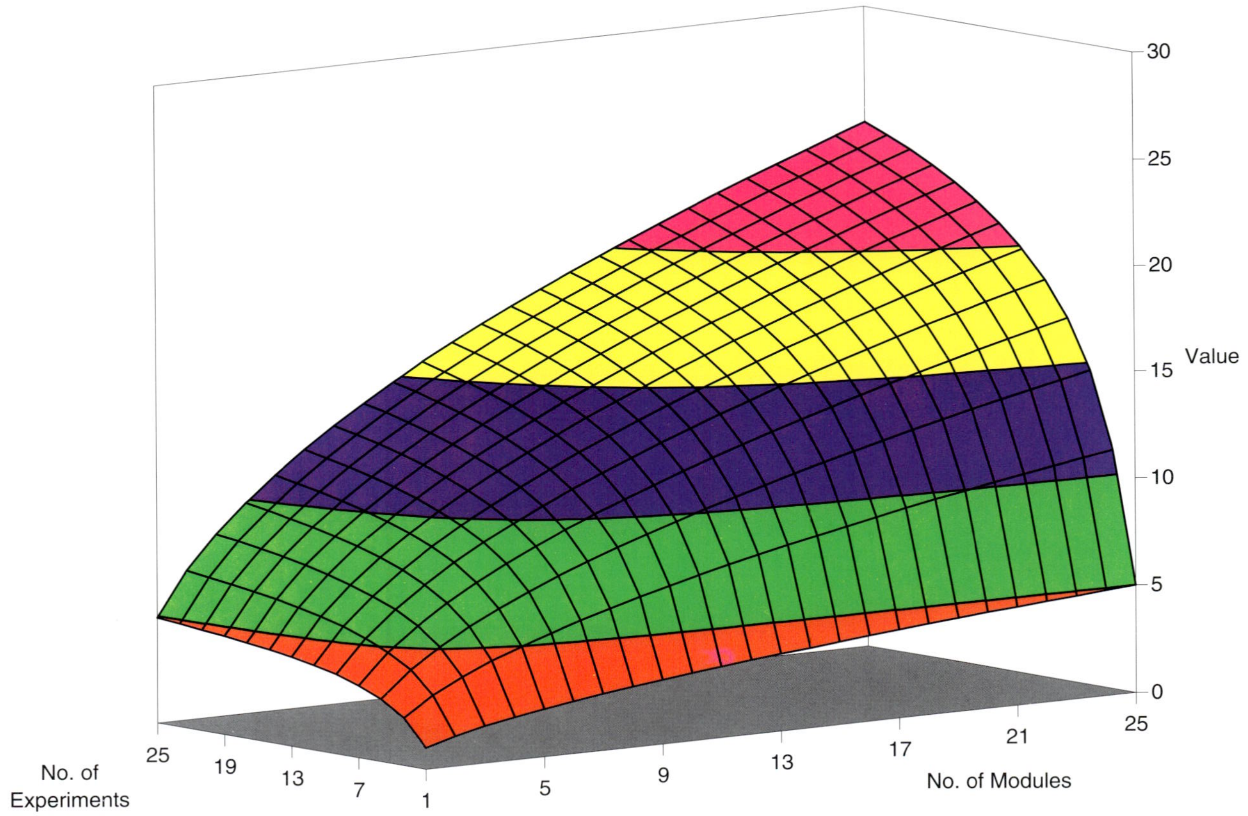

Plate 10.1 The value of splitting and substitution: 3-dimensional view.

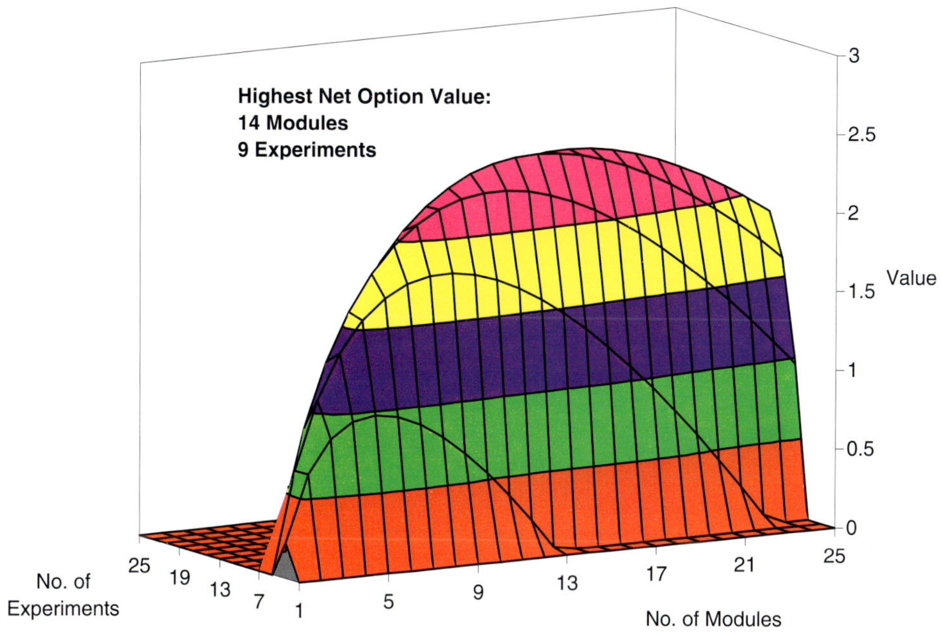

Plate 10.2 The net option value of splitting and substitution: 3-dimensional view.

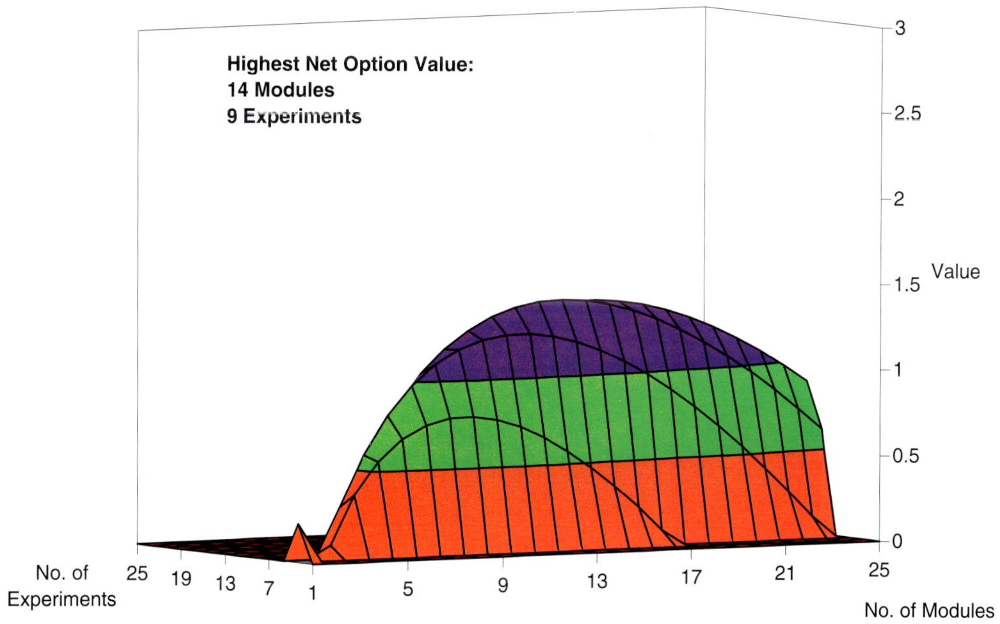

Plate 10.3 The net option value of splitting and substitution: a multi-peaked value landscape.

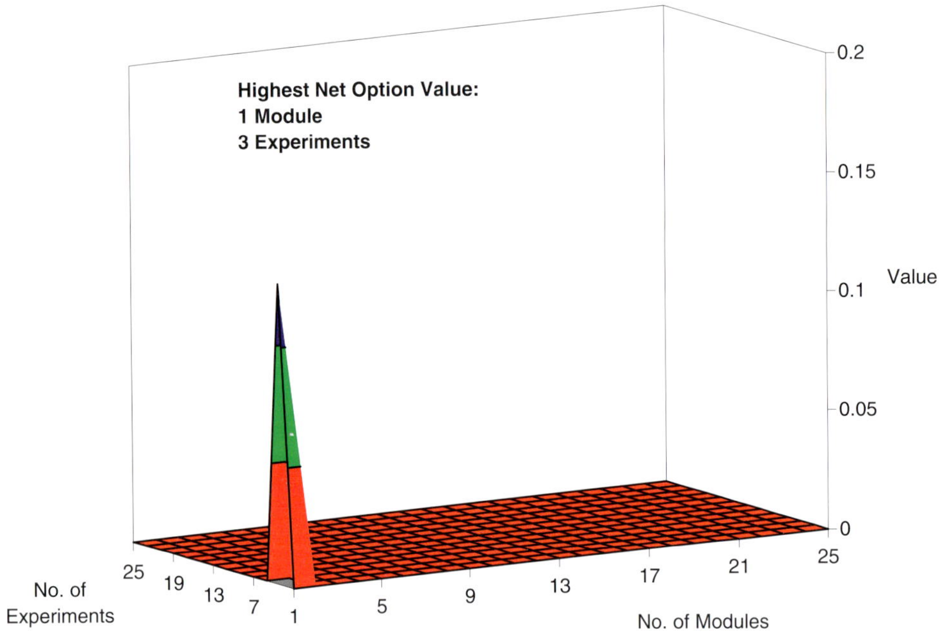

Plate 10.4a The net option value of splitting and substitution with system-level tests: cost per test = 10% V(1,1).

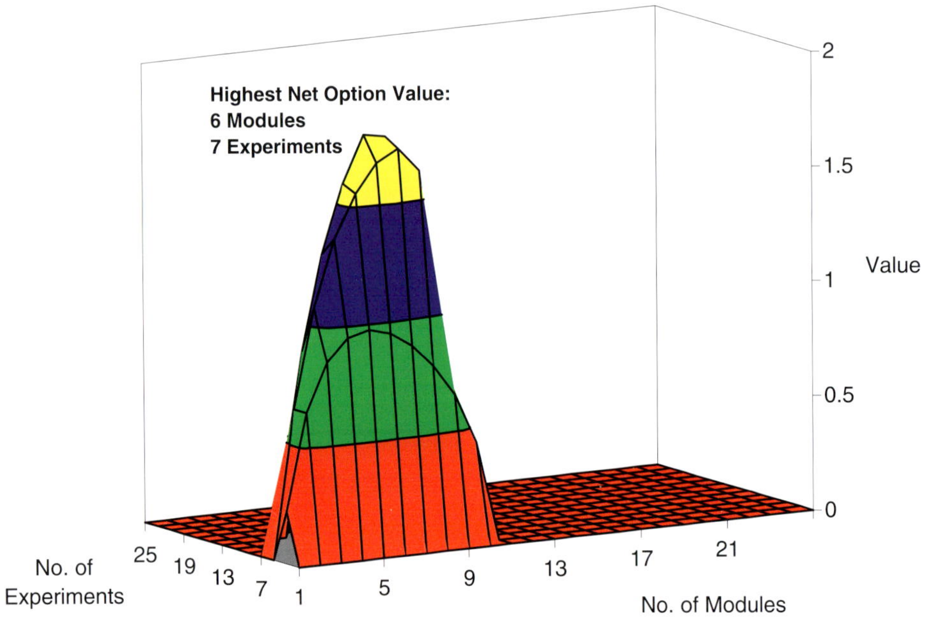

Plate 10.4b The net option value of splitting and substitution with system-level tests: cost per test = 0.00001% V(1,1).

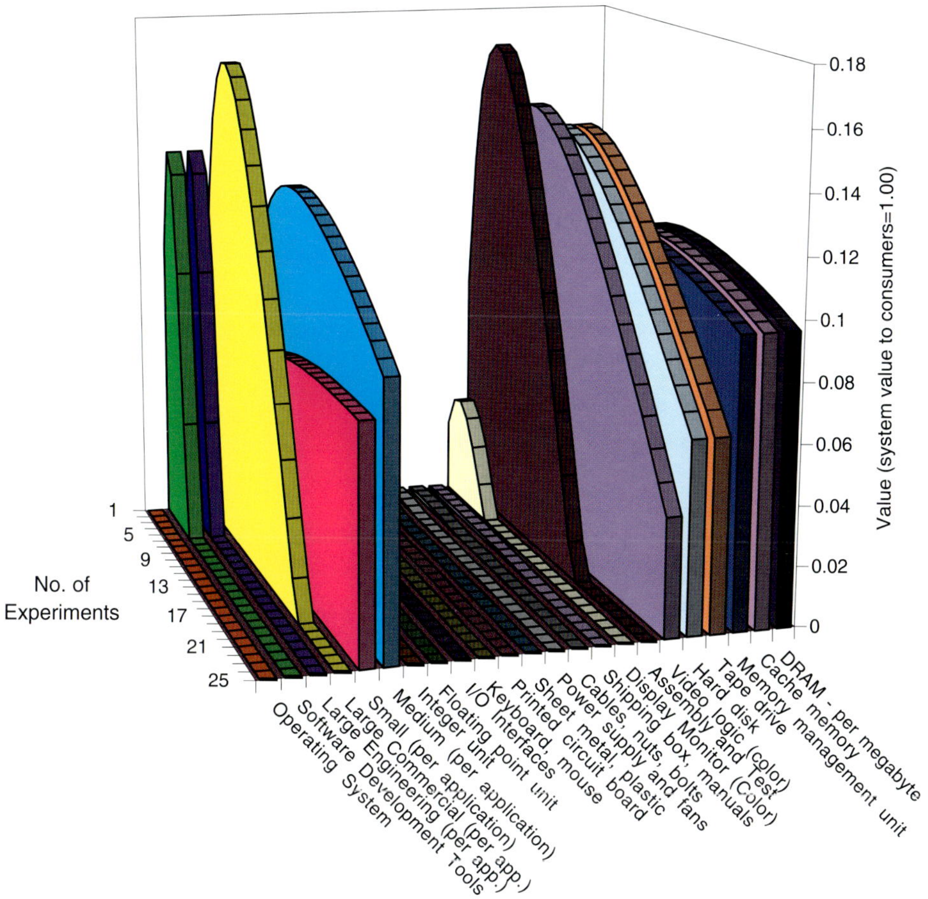

Plate 11.1 Value by module of a computer workstation.

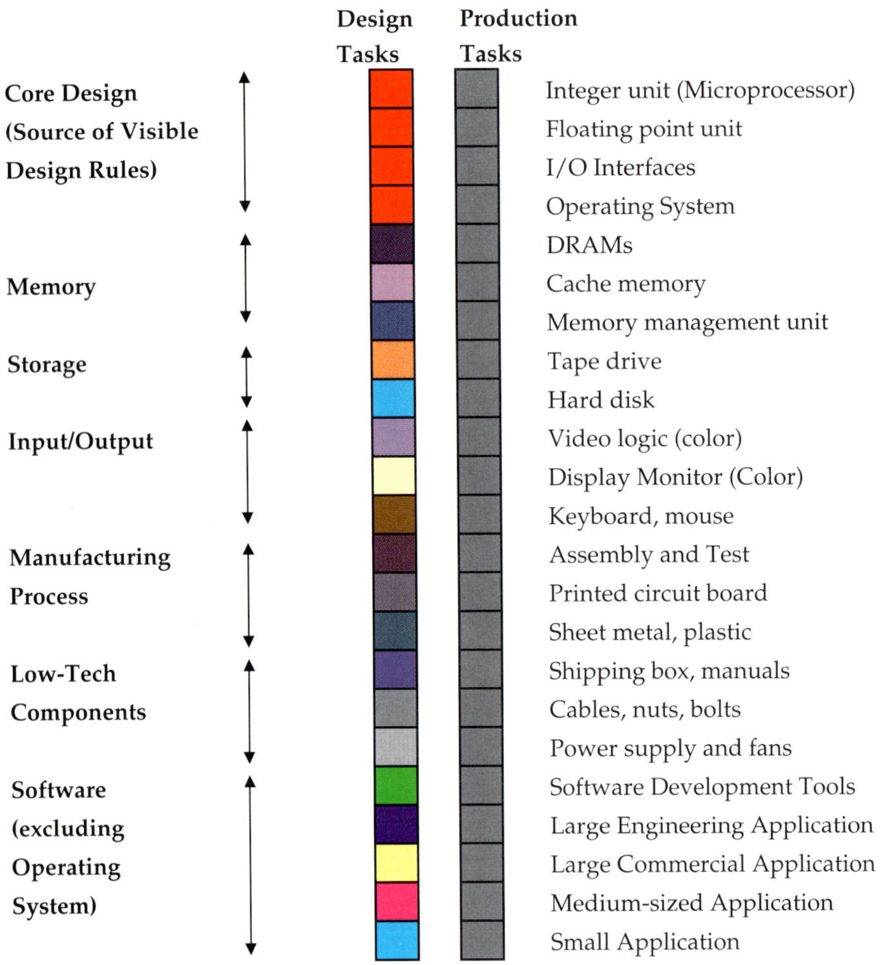

Plate 14.1 The workgroups needed to design and produce a modern computer system. Modules and colors correspond to modules and colors of plate 11.1.

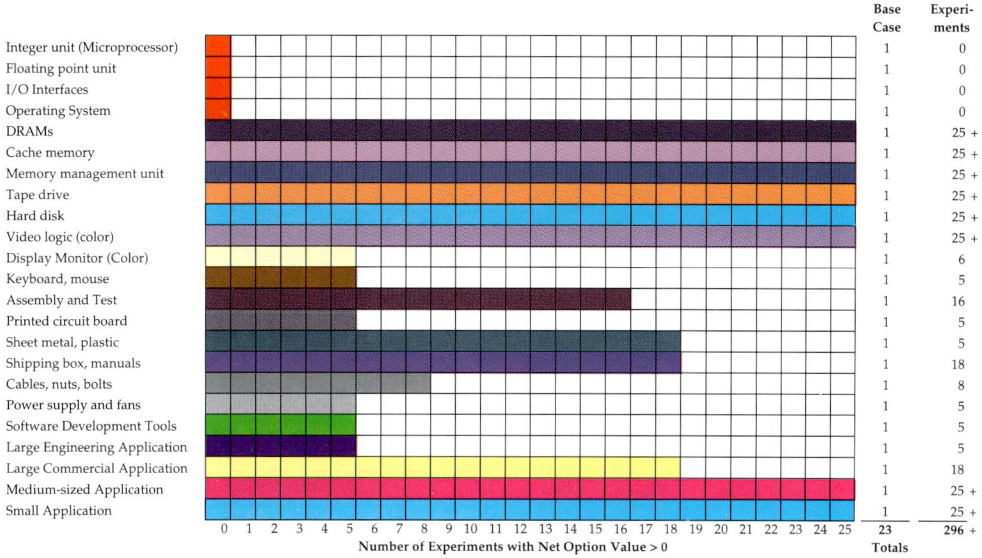

Plate 14.2 The valuable design projects in a modular computer system. Modules and colors correspond to modules and colors of plate 11.1.

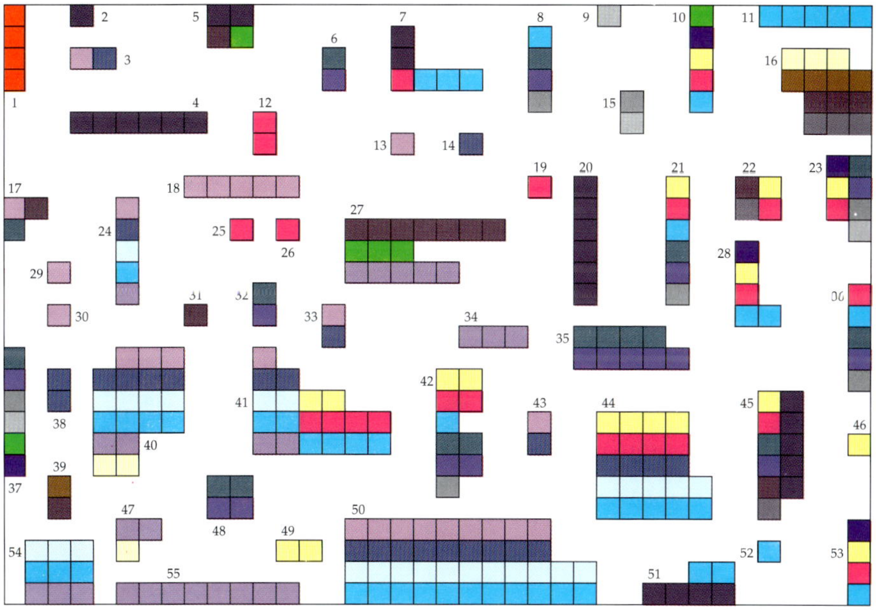

Plate 14.3 A cluster configuration of a modular computer design (contiguous blocks indicate separate work-groups within one corporation). Modules and colors correspond to modules and colors of plate 11.1.

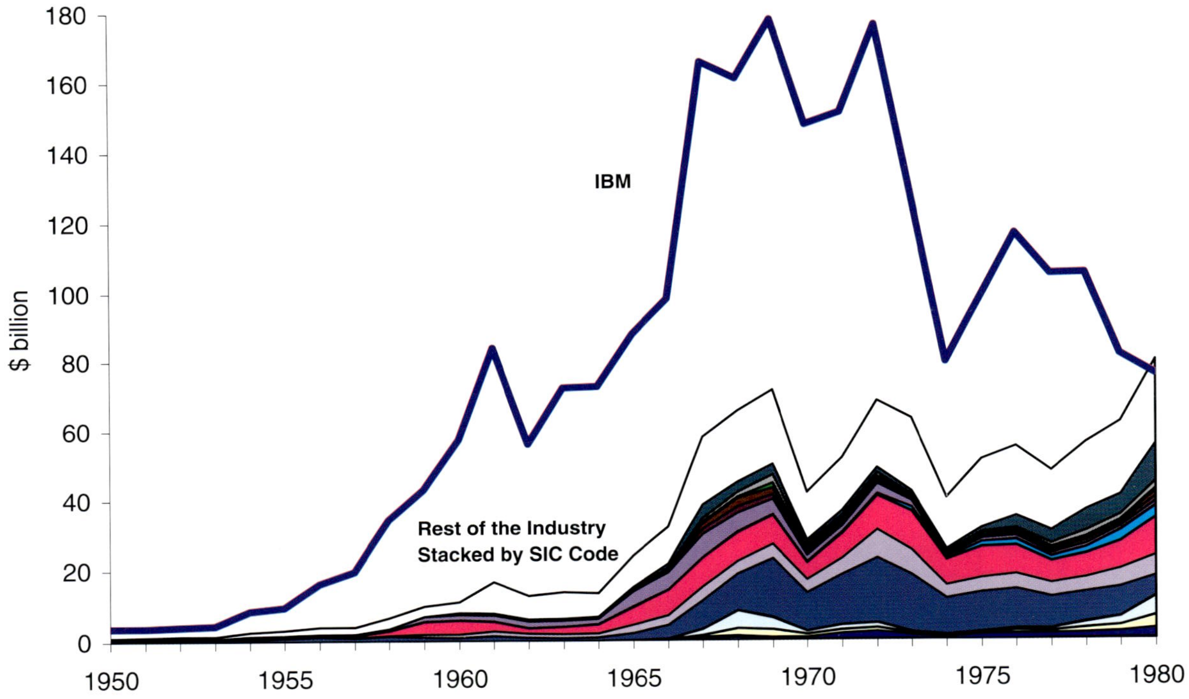

Plate 14.4 The market value of IBM and the rest of the computer industry in constant 1996 U.S. dollars, 1950–1980. Areas and colors correspond to areas and colors of plate 1.1.

Part IV Modular Clusters

14 The Emergence of Modular Clusters

What happens when a newly modular design emerges in an advanced market economy? A modularization, as we have explained, multiplies and decentralizes valuable design options. Market economies are decentralized complex adaptive systems in their own right: their members have the ability to act unilaterally in response to local calculations of value.[1] Hence it is interesting to investigate what happens when a design laden with valuable, decentralized options "hits" an economy that permits and, indeed, rewards decentralized action and initiative.

A modular design makes possible decentralized *design evolution.* In the presence of advanced capital markets, a modular design also makes possible decentralized *industry evolution.* In other words, when an artifact with a modular design is created in an economy with advanced capital markets, subindustries of firms and markets organized around modules may emerge and evolve in parallel with the module designs themselves.

The creation of System/360, which we described in Chapter 7, was a case in point. In addition to being a pathbreaking design, System/360 constituted a large natural experiment in the larger economic system. Following its introduction, people in many different parts of the economy had to grapple with the question of what this design meant for them and for their enterprises.

As we shall see, what arose was a *modular cluster*—a group of firms and markets (for goods, labor, and capital) that played host to the evolution of modular computer designs. The artifacts designed and produced in such modular clusters are, by definition, systems of modules. The individual modules in turn have very little stand-alone value, but they are valuable parts of larger systems. For example, disk drives,

---

[1] Hayek (1945).

spreadsheet programs, and Web pages have no value except in the context of larger computer systems and networks. Given an infrastructure of other computer-related artifacts, these things are very valuable; without that infrastructure, they would have no reason to exist.

Characteristics of a Modular Cluster

For a modular cluster to form, the design rules of a particular system (its architecture, interfaces, and tests) must be known to a number of firms, the hidden-module tasks must be truly independent, and hidden-module tests must be "good enough" to support arm's-length transactions between the module suppliers and system integrators.

When a cluster does form, the modular design will come to be "mirrored" in the cluster's own structure.[2] Individual modules will become the products of specific firms, and markets will arise at the key interfaces established by the design. However, the correspondence of modules to firms, and interfaces to markets, will not be perfect. One firm can always make several modules, or internalize several interfaces within its boundaries.

Modular clusters may take two generic forms, corresponding to two different design structures. In a "one-apex" cluster there is a single set of design rules promulgated by a central architect. These design rules will be visible to designers in many firms, operating in different hidden-module product markets. In a "multi-apex" cluster, there are several sets of design rules, and thus hidden-module designers must look to several sources of visible information. As in the one-apex cluster, however, there will be many firms and many markets, including, possibly, markets for design rules.

One-apex modular clusters arise in conjunction with one-apex design hierarchies. In chapters 10 and 12, we saw that a one-apex design hierarchy can be created in two ways: (1) via splitting and substitution, as in the case of IBM's System/360; or (2) via exclusion and augmentation, as in the case of Digital Equipment Corporation's PDP-11.

Then in chapter 13, we saw that one-apex design hierarchies can be converted into multi-apex hierarchies through application of the inversion and porting operators. Thus, over time, it is logically possible, and indeed reasonable, for single-apex clusters to evolve into overlapping multi-apex clusters, and even into a very large "cluster

2 Henderson and Clark (1990).

of clusters." That is what we will argue occurred in the U.S. computer industry between 1970 and the present.

In presenting this argument, our first task is to explain what enabled the original, single-apex clusters to form. This task will carry us through the year 1980 and to the end of this volume. In volume 2, we will seek to understand how computer designs and the industry evolved after 1980, by which time modular clusters were an established mode of doing business. Within the clusters, new institutions were required to support transactions and protect property rights; new strategies arose as firms sought to capture value at different points in the evolving modular designs; and new valuation technologies were needed to calibrate the worth of new modular products and the enterprises that made them. The analysis of institutions, strategies, and valuation technologies will be our organizing framework in volume 2. However, before we look at economic behavior in "mature" modular clusters, we must consider how and why modular clusters emerged in the computer industry during the 1970s.

Form and Forces

In the introduction to his book, *On Growth and Form,* the biologist D'Arcy Thompson observed that "the form of an object is a 'diagram of forces' . . . that are acting or have acted on it."[3] The bones in an elephant's leg must be thick, because of the action of the force of gravity on its mass; by the same token, the surface areas of birds' wings must increase in proportion to their weight.[4] Thus, to understand why an organism has a particular shape or arrangement of parts, one should study the relationship between the organism's form and the forces operating on it in the environment.

Our problem is like Thompson's but translated into the realm of economics. The object of our study is the computer industry; its form is the way it is configured into firms and markets. We know the form of the industry has changed. In the 1960s its revenues and value were highly concentrated in a single firm (IBM); today its revenues and value are widely dispersed over about one thousand firms. To understand and explain this transformation, Thompson would advise us to look at the "forces" that act on firms and markets.

We have said that the primary force operating in an advanced market economy is the force of value. Human beings have the capacity to "see and seek" value.

[3] Thompson (1942, p. 16).

[4] Ibid., p. 204 ff.

Moreover, among all types of value, capital market value creates the highest degree of consensus, hence provides the strongest basis for collective action. Thus, it makes sense to begin by asking, What capital market values are inherent in the form, and how have those values changed?

From 1964 onward, in parallel with the creation of System/360, the "principles of modularity," that is, the body of scientific knowledge needed to design a modular computer system, were codified, clarified, and disseminated to computer designers via technical articles and books. Gordon Bell and Allen Newell's historic book, *Computer Structures,* which was published in 1971, played a key role in this process. Thus, from about 1970 onward, virtually all new computer designs were based on the principles of modularity and exploited the option values inherent in modular structures.

We know that in relation to an interdependent design of similar scale and complexity, a modular design multiplies valuable options and decentralizes design-parameter decisions. Both of these aspects of modularity—multiplication and decentralization—change the way in which "the force of value" operates on firms and markets in the economic system. The *multiplication* of design options calls for the formation of many more workgroups. The *decentralization* of decision making makes it possible for the workgroups to be dispersed across many independent firms.

In this chapter, we deepen the argument behind these propositions and consider their implications. The multiplication of workgroups, we will show, follows immediately from the multiplication of design options described in part III (chapters 9–13) of this volume.

The argument that workgroups can be dispersed across many firms requires additional analytic apparatus, however. We must first introduce the concepts of transactions and agency costs, and show how these costs impinge on the contract structures of an industry. We then show how a modularization changes transactions and agency costs by eliminating coordinating links within a given task structure. The absence of those links makes possible a dispersal of effort across separate firms. Hence the modularization of a design makes it possible for a modular cluster to form where none would have been possible before.

Following the analytic sections of this chapter, we look again at how the computer industry changed following the general modularization of computer designs. We consider data on entry, the distribution of firms by subindustry, and changes in market values between 1956 and 1980. Following that, in chapter 15, we look at the dynamics of hidden-module competition. This new type of competition emerged in the wake of System/360, and was another force that shaped the industry's structure during the decade of the 1970s.

The Multiplication of Options Increases the Number of Workgroups Involved in Both Design and Production

In this section, we consider what the modularization of a large, complex design—such as System/360—implies for the formation of workgroups in an industry. To do this, we draw on the theoretical apparatus developed earlier in chapters 10 and 11. We argue that in an advanced market economy, a large, newly modularized design will cause many new workgroups to be organized. The workgroups will be needed to seek the option values inherent in the design.

Revisiting the Thought Experiment

To see why this is so, we need to think about how value depends on the existence of workgroups. Let us go back to the workstation thought experiment of chapter 11, which looked at the design options embedded in a system of heterogeneous modules. Readers may recall that we constructed an example of a computer workstation made up of twenty-three modules.[5] We then estimated the cost of experimentation, the technical potential, and the visibility of each module, and plugged those estimates into the net option value formulas [see equations (11.1) and (11.2)]. That allowed us to calculate illustrative values for each module as a function of the number of experiments on that module. The results of those calculations were summarized in plate 11.1 and table 11.2.

We measured the value of the modular system in relation to the base case value of an interdependent system with a net present value equal to zero.[6] To anchor our thought experiment, we constructed the base case so that value-seeking designers, operating in a market economy, would create one and only one interdependent whole system.

Now we ask: How many workgroups would be involved in the design of that interdependent system? In one sense, because of the design's interdependencies, all people working on the system would be part of one workgroup. But the artifact itself would still need to have a full complement of functions, and the tasks of providing

---

[5] Our decomposition was based on a list of the purchased components for a computer workstation provided by Andreas Bechtolscheim of Sun Microsystems to John Hennessy and David Patterson. The list was published in Hennessy and Patterson (1990, figure 2.13, p. 63).

[6] This assumption was not necessary, but it simplified the exposition and the graphs.

those functions, in turn, would have to be assigned to individuals and subgroups within the larger group.

Hence the interdependent computer design would still need to have all the components of a von Neumann computer: arithmetic unit, control, memory, storage, input, output, and software. All these components would need to be designed and manufactured in order to make a functional whole computer system.

Plate 14.1 depicts a mapping of the tasks of designing and producing components to workgroups. Each colored square in the figure denotes a set of design tasks, and each of the corresponding gray blocks denotes a set of production tasks. All of these tasks must be performed in order to offer a real computer system for sale. In this example, we have twenty-three separate components in the overall computer design—that means there are twenty-three sets of design and production tasks. Therefore, something on the order of twenty-three "groups" would be needed to perform the tasks of design, and twenty-three other "groups" would be needed to perform the tasks of production.[7]

However, because of the interdependencies in the design, the tasks involved in creating it would be tied together by many coordinating links. In these kinds of designs, it is common for individuals to cross over task-group boundaries, and work on different parts of the design at different times. Hence the boundaries of the "component workgroups" would not be very clear.

What then would change if the design were modularized? For example, what if the design were split into twenty-three modules, each having a cost of experimentation, technical potential, and degree of visibility as shown in table 11.1? In that case, as we have already shown, the hidden modules of the system would come to have option values, which could be exploited by mounting additional design experiments. The detailed breakdown of option values would be as shown in plate 11.1 and table 11.2.

Plate 14.2 indicates the number and the nature of projects that would have positive net option value under the new regime of modularity. In the figure, if a square is colored, then the corresponding design effort has positive net present value; if a square is uncolored, then it has negative value. We simulated the value of up to

---

[7] We are being vague here about the relationship between the tasks of design and production. As we saw in chapter 6, modularization of design and production tasks is necessary to obtain the continuous-flow economies of mass production, and the task structure of production does not have to correspond to the task structure of the design process. However, all the components of a whole computer system must be produced as well as designed, hence the scale and complexity of the production processes will be proportional to the scale and complexity of the overall system design.

twenty-five experiments per module: the plus signs in the column labeled "Experiments" signify that more than twenty-five experiments had positive net present value under the assumptions set forth in table 11.1.

Within the new, modular architecture, the twenty-three original design workgroups are still necessary. But, *in the hidden modules of the system,* at least 296 additional design projects now have positive value. Hence, the people and resources "demanded" in the design effort might be more than ten times greater than were demanded before the modularization.[8]

We must emphasize that the cost and quality of design efforts here are no "better" than before. What has changed is the modularity of the system: the designers can now mix and match design solutions at the module level.

Some Caveats

There are factors omitted from the simulation that would tend to reduce the number of profitable experiments. First, as the number of experiments increased, design resources (e.g., qualified people) might become scarce, causing the costs of experimentation to increase in a nonlinear fashion. This effect, we think, would be especially pronounced right after a modularization, as the larger economic system had to adjust to new levels of demand for design resources.

Second, at any given time, theories about how to improve a particular set of designs will depend on the existing state of knowledge, and thus will be related. As a result, experimental approaches and outcomes are likely to be correlated. In our model, correlation across experiments would cause a reduction of technical potential, which in turn would lead to a reduction in the number of profitable experiments.

Third, the structure of product and labor markets will affect the way in which modularity gets translated into actual behavior. If the product markets are monopolies or the labor markets monopsonies, then the monopolists or monopsonists will not

---

[8] Detail of the calculation: (23 + 296)/23 = 13.9. In fact, under our assumptions, some of the smaller hidden modules, for example, DRAMS and small software applications, can support many more than twenty-six experiments, which is all our grid permits. Thus 23 + 296 = 319 is less than the full set of projects with positive value. However, for shifts in resource utilization of this magnitude, the assumption of linear factor costs is clearly not valid. Nonlinear factor costs would drive the total number of worthwhile projects down. The point of the analysis is not to find "the number," but to show that the change in value is large enough (1) to draw substantially more resources into the design effort; and (2) to perturb the surrounding product and factor markets.

want to conduct as many experiments as have positive net present value. However, in a competitive capital market, when design-parameter decisions are not subject to central control, it is difficult to prevent others from mounting independent design efforts. (We revisit this issue when we look at dynamic patterns of hidden-module competition in Chapter 15.)

Fourth, the impact of a design modularization on the number of *production workgroups* depends on two additional dimensions of modularity: that within the production process itself, and that between the production and design process for each module.

At one extreme, the production process for a given module may be fully independent of the design process, highly synchronized,[9] and capital-intensive. That is in fact the nature of chip fabrication production processes, as we saw in Chapter 3.[10] In cases like this, the multiplication of design options will have little impact on production efforts.

At the other extreme, the production process for a module might be independent of the production of other modules, but interdependent with the module's own design. That is the nature of disk drive production processes. In these cases, then, every module design effort would "carry with it" a corresponding new production effort.[11] The multiplication of options following a modularization would then multiply production workgroups by the same factor as design workgroups.

Among the modules of a computer, these two extreme relationships between design and production both arise, and many intermediate patterns can be observed as well. Therefore, a modularization will increase the number of workgroups engaged in production, but by less than it increases the number of workgroups engaged in design.

In summary, the multiplication of options associated with a large-scale modularization calls for a great expansion in the number of workgroups employed in making that system of artifacts. Modules are heterogeneous, and thus it is impossible to predict the precise number of new workgroups, but our analysis points to the fact that a

---

9 Highly synchronized production processes often have a high degree of interdependence stage to stage, hence are difficult to break apart.

10 Recall that Mead and Conway said "pattern independence" was "the most powerful attribute" of the planar process of chip fabrication (Mead and Conway 1980, p. 47). Pattern independence is a form of modularity that obtains a clean interface between the designs of artifacts and the production systems used to make them.

11 This is the essence of initiatives aimed at "design for manufacturability" (DFM). See Nevins and Whitney (1989); Shirley (1990); Suzue and Kohdate (1990); Ulrich and Eppinger (1995).

very large number of new workgroups will be needed to exploit the option values inherent in the design.

Such workgroups are in turn the basic structural units of an industry. Each workgroup is fundamentally—isomorphically—involved in carrying out the tasks of design or production, or both, for a specific module experiment. In this fashion, the modular form of the artifact begins to be visible in the form of the organizations that design and produce it. The artifact's task structure, in effect, determines industry structure up through the level of workgroups.[12]

However, nothing we have said thus far sheds light on how these multifarious workgroups will be organized into firms, or what markets will arise to mediate transactions among them. Will all workgroups be gathered into "one big firm," or will each workgroup make up a separate firm? Will it be advantageous to have several workgroups within the boundaries of a single firm, and if so, which ones, and how many? Conversely, what information and material need to flow between workgroups, and what types of markets can facilitate these exchanges? The answers to these questions will determine the overall structure of the industry and of the subindustries within it. We now begin to address them.

Transactions Costs and Agency Costs

In chapters 2 and 3, we argued that the "microstructure" of designs affects the economics of design processes in deep and unavoidable ways. In particular, the many coordinating links that are needed to implement an interdependent design process will have a profound impact on the costs of getting things done under different contractual regimes. For this reason, the interdependencies in a design and task structure will influence the optimal boundaries of firms and the location of markets in the surrounding industry.

In this section, we will make this reasoning more precise, arguing as follows: In the presence of moderate transactions and agency costs, putting interconnected tasks within a single corporation is a uniquely efficient way to "package" the task structure.

To make this argument, we must first define what we mean by "transactions and agency costs." For our purposes, *transactions costs* are the costs associated with a formal transfer of property.[13] They can in turn be broken down into

[12] Henderson and Clark (1990).

[13] This is a narrower definition than is commonly found in the field known as "transactions-cost economics." Oliver Williamson and others working in this field include certain other

- *information costs*—the costs to either party of verifying the quality of the property being transferred, as well as the costs (also to either party) of verifying the value of the payment being made for the property;
- *operational costs*—the costs of effecting the transfer once the parties have agreed to it (including transport costs, clearing costs, etc.).

Agency costs are the costs of delegating tasks to other human beings, whose objectives are not identical with those of the delegator (called the "principal").[14] For their part, agency costs can be broken down as follows:

- *direct costs*—the cost of the work not being done as the principal would like it to be;
- *bonding costs*—the costs of changing the nature or timing of the agent's incentives so as to reduce direct agency costs (e.g., the agent may post a bond, or the principal may provide a reward in proportion to the quality of the job done);
- *monitoring costs*—the costs of watching and supervising the performance of the tasks so as to reduce direct agency costs. (e.g., the principal may require that the work be audited or evaluated by a third party).[15]

An Interdependent Design and Task Structure

What transactions or agency costs are incurred in the course of carrying out an interdependent design process? To fix ideas, consider figure 14.1, which depicts a generic task structure matrix for an interdependent design process.

This matrix is representative of the design processes used to create early computers like the IBM 7070 or 1410.[16] As we described in chapter 2, the off-diagonal **x**'s

---

costs, for example, the costs of ex post opportunistic behavior, under the general rubric of transactions costs. Our definitions are closer to those used in financial economics. However, even within financial economics, the classification of costs is not wholly consistent. Since naming practice differs across subfields of economics, we are here using what we believe is a reasonable compromise. (See Williamson 1975, 1985).

[14] Jensen and Meckling (1976).

[15] This categorization is due to Jensen and Meckling (1976). We have stretched their definitions a little by including the costs of setting up incentive compensation mechanisms in the category of bonding costs.

[16] We base the assertion that this task structure was representative of all pre-1970 computer design processes except System/360 on Bell and Newell's classic text *Computer Structures,* which analyzed "all interesting" computer designs up to that point in time. (Bell and Newell 1971.)

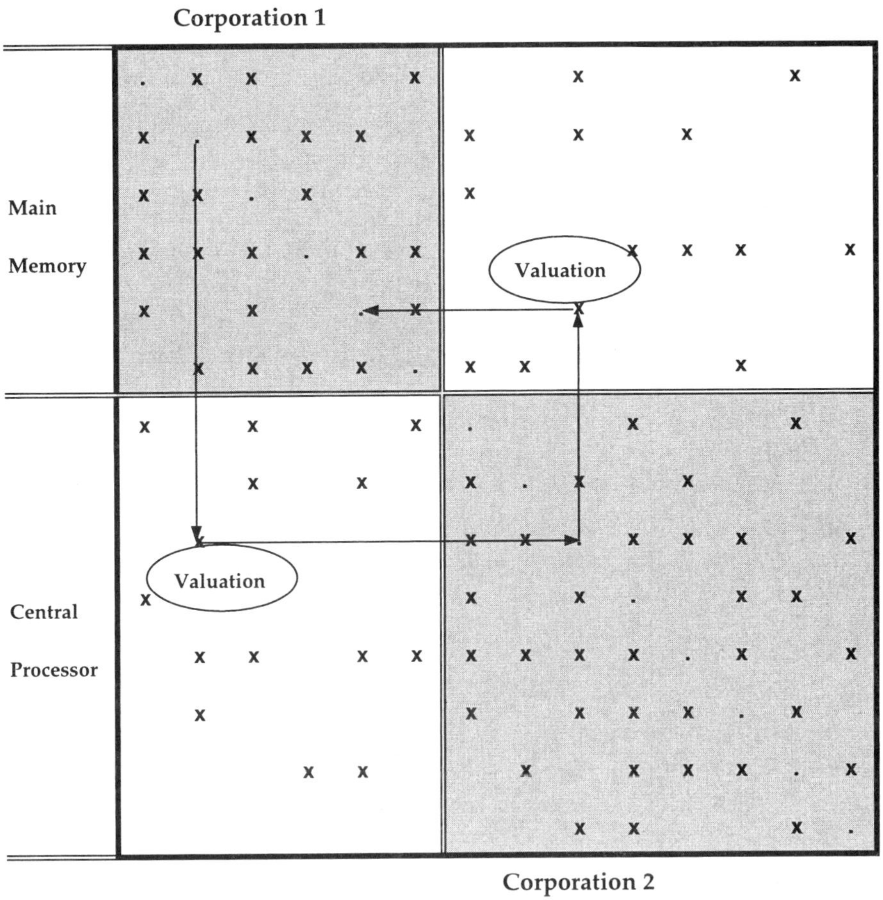

Figure 14.1 A generic fragment of an interconnected task structure with tasks distributed across two corporations.

in the TSM indicate the information transfers and coordinated decisions needed to ensure that the computer designed by this process works, is manufacturable, and is useful. Then, as now, there were many ways to organize day-to-day work in such a matrix. However, all ways of organizing interconnected tasks rely heavily on cycling and iteration. As an example of cycling, the arrows in the figure show a problem-solving path, which originates in the memory unit, proceeds to the central processor unit and from there goes back to the memory unit.

Suppose for the sake of argument that the tasks of memory and central processor design were spread across two corporations, whose boundaries were defined by the gray boxes in the figure. The tasks and links inside each of the gray boxes would then occur within the boundary of the respective corporation. However, as the diagram shows, a considerable number of links in this task structure lie outside those corporate boundaries. Moreover, there are links both above and below the main diagonal, hence there is no natural supplier-customer relationship within this task structure.

When the problem-solving path moves out of the boundary of one corporation, work on the design "passes" from a designer at one firm to a designer at the other. At that point, either property (the design) or effort must "cross over" from one corporation to another. In the path shown in the figure, at the first crossover point, property rights to the "design in progress" may be transferred from Corporation 1 to Corporation 2. Or Corporation 1 may retain those rights, and enter into a subcontracting relationship with Corporation 2, which then would delegate one of its employees to work on the design.[17]

The first type of action (the transfer of property) triggers a transaction with transaction costs. The second type of action (subcontracting) creates an agency relationship with agency costs. We will consider the transfer of property and transactions costs first, and subcontracting and agency costs second.

Transfers of Property and Transactions Costs

When a design is passed from designers in one corporation to designers in another, the legal rights to the design may need to be transferred as well. Corporation 1 must realize the value of the work it has done on the design up to the point of transfer; Corporation 2 must establish its rights to the design in the future in order to reap the benefits of any improvements it may make.

But what *is* the value of the design-in-process? Clearly that depends on how much work needs to be done to complete it. The work remaining in turn will depend on what the designers at Corporation 1 did to the design (and learned in the process), as well as what the designers at Corporation 2 might do, given the design's state at the time of the transfer. Thus the value of the design-in-process depends on the separate specific knowledge of designers at both corporations. And if their compensation de-

---

[17] One might ask, is it possible to do neither of those things? It is possible, but in that event the relationship between the two parties with respect to the value created in the design process will be unspecified. The parties may then disagree about the appropriate division of value after the fact. This is such a common occurrence that, with even a modicum of foresight, both parties should be eager to establish the terms of their business relationship ahead of time.

pends on the valuation of the design at the time of transfer, then designers working for the seller will have incentives to overstate its value, while those working for the buyer will want to understate its value.

Faced with this valuation impasse, the managers of the two corporations could hire an independent third party to evaluate the design and set a price on it. But ascertaining "the true value" of the design at this stage may be truly difficult—so much so that the shortest and cheapest way to verify its value may be to finish it and introduce it in the market.[18] Hence third-party valuation, though feasible, is costly, especially if the design must go back and forth between the two corporations many times.

An obvious way to address this problem is to make all payments to both corporations contingent on the final value of the completed design:

$$C_1\text{'s financial claim} = \alpha V, \text{ and}$$

$$C_2\text{'s financial claim} = \beta V,$$

where V equals the present value of the surplus (profit) generated by the completed design, and α and β are fractions determined by contract ($\alpha + \beta < 1$). However, this action converts the two separate corporations into one—a joint venture—for the purpose of completing the design. Hence the only way to avoid interim valuations, and the attendant transactions costs, is to collapse the two corporations into one economic entity.

Subcontracting and Agency Costs

A second way to implement a design process across two corporations is to have ownership of the design-in-process remain with one of them, and have that company subcontract with the other to supply the required services. This contrivance avoids the transactions costs associated with transferring title to the design-in-process, but it creates a different set of "agency" costs, which may be as large as or larger than the transactions costs.

---

[18] The technology of financial valuation rests on the application of powerful algorithms to estimates of primitive quantities (inflows and outflows at different points in time). Strictly speaking, the algorithms work only if the primitive estimates are reasonably precise. When the artifact is new (designs are, by definition, new artifacts), estimation of the primitives takes place by analogy, and can be very imprecise. It is always possible to generate low-cost, low-precision estimates of value (a blind guess, for example), but these will be subject to dispute, and hence will not support a frictionless transfer of title.

Agency costs arise when a decision-maker (the agent) acts on behalf of another party (the principal), but does not in all instances act as the principal would wish. In a subcontracting relationship, there are in fact two "levels" of agency: the company that owns the design-in-process is the principal; the subcontractor is the first-tier agent, and the subcontractor's employees are the second-tier agents.

Agency costs arise at both levels, but we are interested in those caused by conflicts of interest between the subcontractor (the first-tier agent) and the principal. Because design outcomes are dependent on the quality and timeliness of the designers' decisions, the principal must be especially concerned with who gets assigned to perform the tasks, and when the effort takes place. But with respect to these two management decisions, the subcontractor's interests will generally be in conflict with the principal's.

For example, suppose the design cycles through the subcontractor's "shop" more than once. (This is practically inevitable if the design structure is interdependent.) On the second pass, is the subcontractor obligated to supply the same designer as before? Or can it simply assign any designer to the process? Clearly, there are instances when dealing with multiple persons will greatly increase the cost of completing the design. In particular, if a designer gains critical specific knowledge on the first round, then, from the principal's point of view, he or she will no longer be interchangeable with other designers in the subcontractor's labor pool.

Foreseeing this eventuality, the principal may ask (or require) that the subcontractor provide continuity of service (the same designer each time). However, this is an expensive request to fulfill. Given uncertain demand, any labor pool can be managed more efficiently (less downtime per employee) if its members are interchangeable.[19] It costs more to supply a particular individual each time a request is made: these costs will be apparent from the fact that some of the subcontractor's employees will have waiting lines for their services while others are idle.

The possibility that queues will form raises another contractual question: what kind of priority does the principal require? Does the principal want a specific designer to drop everything whenever there is a call for his or her decision-making capabilities? Or can the principal manage with a nonpreemptive call? Again, there are instances when having a specific designer available on short notice will be very

---

[19] Merton (1973b). Providing continuity of service is the same as selling call options on the time of specific employees vs. call options on the "portfolio" of employees. To be meaningful, the contract for continuity of service must include the stipulation of a response time; otherwise, the subcontractor could just tack each new request onto the end of a potentially infinite queue.

valuable to the principal, and other instances when it will be less critical. Symmetrically, it is more costly for the subcontractor to provide a "first call" on a designer than a last call. Hence, if a quick response is needed, then that, too, needs to be recognized in the initial contract between the two companies.[20]

Finally, in addition to worrying about who gets assigned to the design process and when, the principal must consider how the subcontractor plans to measure the performance of its employees and compensate them. There will, necessarily, be a principal-to-subcontractor compensation formula and a subcontractor-to-designer formula. Formally, the principal has a problem of optimal mechanism design with respect to the subcontractor (the first-tier agents) and the subcontractor's employees (the second-tier agents). And if the value of the design is affected by continuity of service, by performance measurement systems, and by incentives and rewards, then the principal will want to have a hand in designing the subcontractor's employment policies, as well as its job assignment policies.

The agency costs introduced by a subcontracting relationship are likely to be proportional to the number of times the subcontractor's services are used. Essentially, each time the principal calls upon the subcontractor to supply a designer, there is the potential (a probability) that the subcontractor will not act as the principal would have wished. Usage of the subcontractor, in turn, will be related to the number of "out of block" **x**s in the TSM of the design process. The path to a final design must pass through each of those coordinating links at least once, and with cycling and iteration some links may be revisited over and over again.

It follows that a subcontracting relationship is most likely to be unsatisfactory (inefficient) when the underlying design is interdependent, and the task structure interconnected. The more the principal needs the services of the subcontractor, and values continuity of service and preemptive calls on the time of designers, the more it will need to have long-term, exclusive contractual relationships with specific individuals. These will be the functional equivalent of employment contracts even if they are called by another name.

The principal can achieve the "functional equivalent" of employment contracts in one of two ways: it can cut out the subcontractor and hire the designers directly; or it can align the interests of the subcontractor with its own by giving the subcontractor

---

[20] Moreover, a dishonest subcontractor might sell a designer's time twice, playing the odds that no conflict would arise. Therefore *to be really certain* of a designer's availability on an as-needed basis, the principal would want an "exclusive" contract—that is, an agreement that a specific designer would work on the principal's project and no others until the design was completed.

shares in the final design. However, as with the joint venture solution to the transactions costs problem, these actions "solve" the problem of agency costs by converting two corporations into one for the purpose of completing the design. In the first and simpler case, the principal assumes directly the employment role of the subcontractor; in the second, more complicated case, the principal continues to work through the subcontractor, but the erstwhile agent becomes a principal (that is, a shareholder) with respect to the final value of the design.

The Advantages of Having a Single Corporation Span an Interconnected Set of Tasks

Our arguments about the incidence of transactions and agency costs in an interdependent design process can be encapsulated in a valuation equation. Let the fundamental value of the design process apart from transactions and agency costs be denoted V_0. Let the design tasks be distributed across two corporations as, for example, in figure 14.1, and let l denote the number of across-boundary linkages (the number of **x**s in the white blocks of the figure). Let the average number of iterations through any link be t, and let the cost per transaction or agency event be c.

The net present value of the design[21] is then

$$\text{NPV} = V_0 - clt \qquad (14.1)$$

Obviously, for c of any noticeable magnitude, this NPV deteriorates rapidly with both l, the number of across-boundary linkages, and t, the number of iterations through each link. Given the known tendency of interdependent design processes to cycle, it is easy to imagine such "frictional" transactions and agency costs overwhelming the innate value of the design.

However, this fact is itself a design problem in the domain of contract structures. If, as we assumed, the links and iterations are fundamental to the design process, then a focus of effort will be to devise a contract structure that brings the cost of links and iterations down. That is exactly what the contract structure of a single firm or corporation does.

Since the publication of Ronald Coase's seminal paper "The Nature of the Firm," it has been recognized that firms (generally) and corporations (specifically) are prearranged contract structures, whose function in part is to reduce transactions and

---

[21] As in previous chapters, for simplicity, we do not show discount factors for time or risk. Such factors complicate the expression, but would not change the conclusions in any way. Readers may think of them as being subsumed in V_0 and c.

agency costs.[22] By setting up a firm or corporation, it is possible, not to eliminate transactions or agency costs, but to reduce the amount of friction these costs introduce into the flow of day-to-day work.

The reduction of frictional transactions and agency costs may be accomplished in two ways (which are not mutually exclusive):

- A group of transactions may be standardized and regularized, so that their variable cost approaches zero.
- Claims on the ending value of a collective effort may be created and used to compensate agents for their work.

For example, an entrepreneur might create a standardized set of employment contracts with designers that included a reward proportional to the value of the final product. By the act of standardizing the terms of employment, the entrepreneur would be using Coasian logic to create a firm.

Moreover, in a modern, market-based economy, that firm can quite easily become a corporation in the sense we defined in chapter 4. There we said that in functional terms, a corporation is a collective enterprise that has rights

- to hold property;
- to transact and to enter contractual arrangements;
- to issue claims that are themselves recognized as property.

Hence, for a Coasian firm to become a corporation, all that needs to happen is for that firm's various transactions, contracts, and claims to be recognized as valid transactions, contracts, and claims within the larger economic system in which the firm operates.[23]

---

[22] Coase (1937). His paper is framed as a search for a definition of a "firm" that would be at once "manageable" and "realistic." Coase's initial insights have been extended by Alchian and Demsetz (1972); Williamson (1975, 1985); Jensen and Meckling (1976); Holmstrom (1979, 1982); Milgrom and Roberts (1995); and many others. Unfortunately, one of the things missing in these formulations is a taxonomy of all the different types of transactions and agency costs that may influence real contract structures. Because we lack an exhaustive taxonomy, it is difficult to know if one has identified all possible directions of search in the space of complex contract designs. We believe this is a reason to seek out new analytic methods that are better suited to the hyperdimensionality of task and contract structure spaces.

[23] This can occur through a legal event (the act of incorporation), or simply as a matter of prevailing practice. The degree to which corporations are formally recognized within legal systems varies across economies even today. Absence of a legal framework is an impediment, but not an absolute barrier to the formation of corporate enterprises.

Substantively, what happens when a firm or corporation is formed is that a large number of questions involving who owns what, who shows up, what work gets done, and how pay and other rewards are divided get taken out of the ongoing flow of decisions and negotiations. It is not that these questions disappear—they just go "off the table" for some of the time that work is being done. Either they have been contracted for in advance, or there is an agreement (and a formula) for settling them later on.

The Boundary of a Corporation

The *contract-structure boundary* of a corporation refers to an invisible but economically significant boundary that separates the "inside" from the "outside." Property and actions that arise inside the boundary can be rearranged using the corporation's internally sanctioned procedures. However, property and actions outside the boundary are not under the corporation's direct control. When property (or action) crosses the corporate boundary, an event occurs in the larger economic system. For example, the sale of a product, the issuance of a security, and the hiring of an employee are all boundary-crossing events.

Thus a corporation is a social artifact that provides for a reduction in frictional transactions and agency costs within prearranged contractual boundaries. But what should those boundaries be?

Figure 14.1 and equation (14.1) give us ways to think about this question. We have posited that links within the corporations will take place at a low variable cost, and links across corporations will take place at a high variable cost. In other words, **x**s in the gray boxes are low-cost connections, and those in the white boxes are high-cost connections. Clearly then the lowest-cost structure is to have no white-box **x**s at all. But that is the same as having all the tasks take place in what Coase called "one big firm."[24]

Thus in the presence of frictional transactions and agency costs, an interconnected task structure is best packaged within one corporation. Contract structures that spread interconnected tasks over two or more corporations are suboptimal in proportion to the **x**s left in the "white spaces" outside the boundaries of any of them.

In summary, a *corporate contract structure* is a way to economize on variable transactions and agency costs when tasks are interdependent, workgroups are fluid, and the flow of work is not predetermined. Often, given an interdependent design

[24] Coase (1937).

structure, the most cost-effective contract structure is to have a single corporation, which holds title to the design-in-process, and enters into standardized, long-term employment contracts with individual designers. For example, in our thought experiment, a good enterprise design for the premodular design and task structure would be an organization made up of:

- *one* workgroup;
- *twenty-three* subgroups; and
- *one* corporation;

The corporation's *contract-structure boundary* ideally would encircle all the interdependent tasks.

The economic advantages of having a single corporation obviously depend on the differentials in transactions and agency costs. If transactions or agency costs differentials are very low, then even with a high degree of interdependency, the advantage of having a single corporation will be slight. Moreover, there are always ex ante costs associated with forming a corporation, thus in some cases, the savings obtained by reducing frictional transactions and agency costs may not justify the cost of setting up the necessary contracts in the first place.[25]

Nevertheless, in the presence of frictional transactions and agency costs, the best way to package an interconnected set of tasks is generally within a single firm or corporation. However, this reasoning does not hold for a modular task structure, as we demonstrate in the next section.

The Decentralization of Design-Parameter Decisions Reduces Opportunities to Incur Transactions and Agency Costs

By its very nature, a modularization decentralizes decision-making and eliminates the need for ongoing connections and links between designers working on different modules. If the design structure is modular, then, we have said, the hidden-module

[25] This was Stigler's (1951) essential argument: that activities necessary to production would be conducted within a firm until the market grew enough to warrant the setup costs of vertical separation. Contracting technology, the legal system, and financial institutions may all affect setup costs substantially. Hence the number of corporations formed and their scope may vary dramatically from one economy to another. See Khanna and Palepu (1996, 1997) for further discussion of this issue, as well as empirical evidence.

designers only need to know the design rules of the system, and have access to a systems integrator. For their part, systems integrators only need to know the design rules, and have the ability to test modules "at their own level."[26] Thus, after a modularization, the forces giving rise to frictional transactions and agency costs will be greatly weakened.

To see this more clearly, recall that a modular task structure consists of a nested hierarchy of tasks, represented by a boxes-within-boxes configuration of the TSM. Figure 14.2, which reproduces figure 3.4, depicts a generic modular task structure for the design of a computer system. Each block in this matrix represents an interconnected sets of tasks,[27] performed by a corresponding workgroup. Work inside each group is highly interconnected, requiring frequent communication among group members, and cooperative problem-solving efforts.

From the analysis of the previous section, in the presence of transactions and agency costs, we would expect each of the interconnected task blocks in the matrix to be "packaged" within a single corporation. But what about the task structure as a whole? Is it necessary for one corporation to span all the tasks, or can the individual blocks be allocated to different corporations? To answer these questions, we must look more carefully at the different stages of the design process.

Recall that a modular task structure always has three distinct stages:

1. the design rules stage
2. the parallel work stage
3. the system integration and testing stage

Let us look first at the middle stage (parallel work). Blocks corresponding to this stage are contained within the heavy black border in the inside of the matrix in figure 14.2.

---

[26] We do not mean to imply that systems integration and testing tasks are insignificant, but only that in the presense of good module tests, the integrators do not need to know how the modules themselves were constructed. Systems integrators do need to know the design rules, including the architecture, interfaces, and tests that pertain to the modules of the system. In volume 2, we will see that if the architecture and interfaces are public knowledge, as they are for some systems, then systems integrators may compete on the basis of superior tests and the ability to integrate (Iansiti 1997b).

[27] In general, hidden-module task structures themselves may be interconnected, sequential, or modular to any depth.

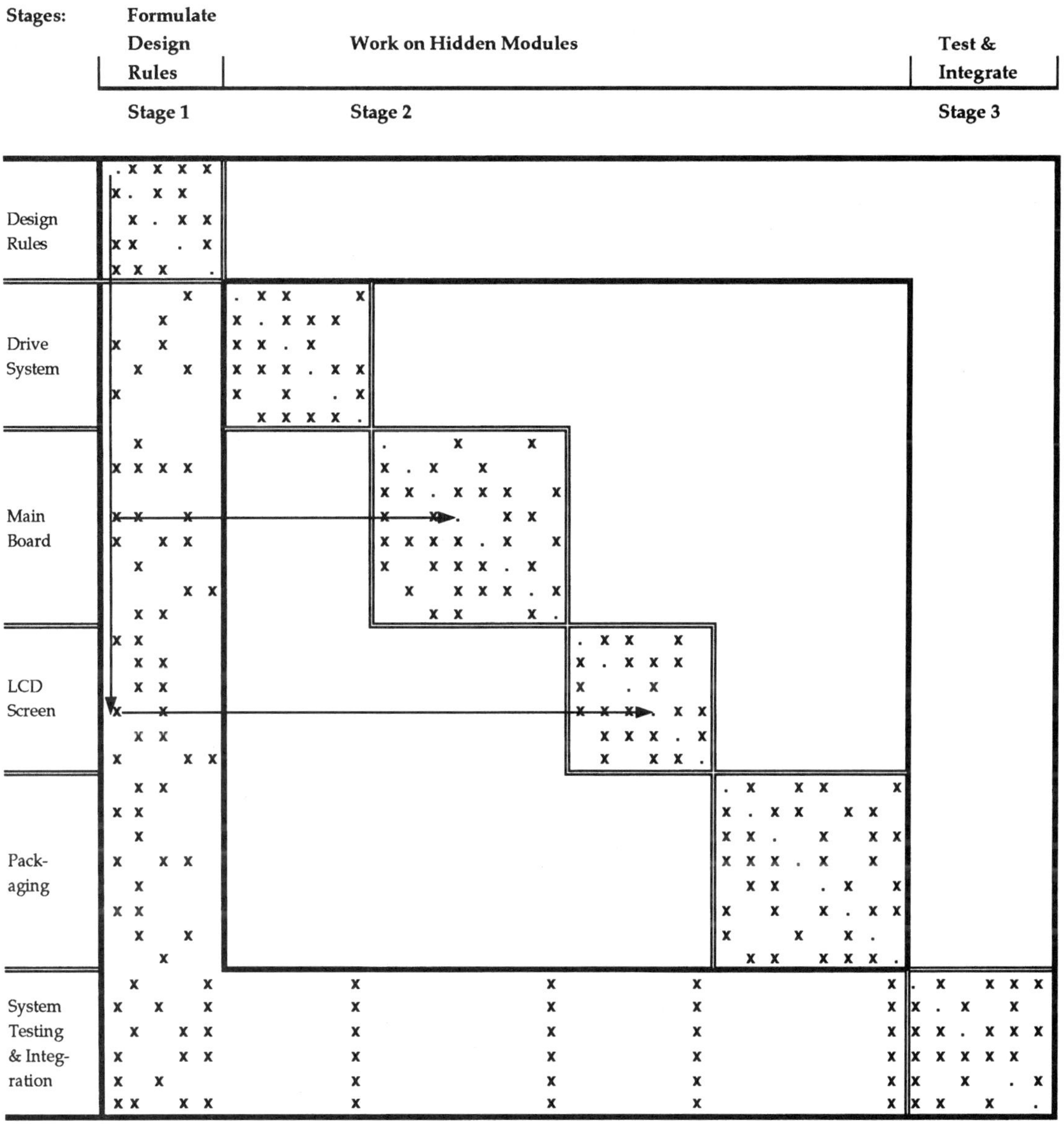

Figure 14.2 A generic modular task structure matrix for a computer system.

Hidden-Module Blocks Are Independent

By construction, in the parallel work stage of a modular design process, there are no interdependencies across task blocks. Work within each block can proceed independently of work in every other block—this is the defining characteristic of a modular task structure. In theory, everything any designer needs to know about work going on in other modules has been codified in the design rules. Moreover, all designers are bound to obey the design rules, and limit their actions to changing their own "hidden" design parameters.

In the absence of interdependencies across blocks, the transactions-and-agency-costs rationale for having just one corporation vanishes. Mathematically, the *clt* term in equation (14.1) becomes zero, not because *c*, the cost, has disappeared, but because *l*, the number of links, is now zero. We then must fall back on Franco Modigliani and Merton Miller's fundamental irrelevancy proposition: as long as the value of the project (V_0) is unaffected by the contract structure, all ways of packaging the enterprise are equally good.[28] Thus we surmise that the different task blocks in the parallel work phase might be implemented equally well by separate corporations, by divisions within a single corporation, or in other ways (e.g., via joint ventures). Given what we know, as long as the interconnected blocks are not themselves divided, a large number of contract structures would work equally well for this stage.

The "given what we know" caveat is important, however. At this point, our "diagram of forces" includes only the following:

- technological opportunities (the value of the base case, and the option values inherent in the modular operators);
- technological costs (the cost of creating, communicating, and enforcing design rules, of implementing experiments, and of testing outcomes);
- "frictional" transactions and agency costs.

These elements, we have argued, create or destroy value in the design process. Thus, our theory predicts that designers (who understand these elements) will attempt to structure their ventures in ways that maximize the value of opportunities

---

[28] Modigliani and Miller (1958). Strictly speaking, their irrelevancy proposition applies to a world in which there are no transactions or agency costs. We are considering a world in which there are such costs, but, because of the independence of the task blocks, they are not affected by the contract structure. The irrelevancy proposition will then hold within that limited domain.

and minimize the costs. This prediction follows from our axiom that designers "see and seek value" in their designs.[29]

However, there are other elements, such as market power, property rights, and organizational design factors which we have not yet incorporated into our "diagram of forces." These, too, affect value, and thus have the capacity to influence the configuration of firms and markets in an industry. These omitted elements are a primary focus of our attention in Volume 2; for now, however, we simply note that they exist, and place them to one side for later systematic consideration.

The Three Stages Are Vertically Linked

Let us now turn back to the generic, modular TSM (see figure 14.2), and look at the relation of task blocks *across different stages* of the design process. Although they are not linked to one another, all the hidden-module task blocks have backward connections to the design rules and forward connections to the system integration and testing tasks. The design rules are inputs to the hidden modules, and the hidden-module parameters plus the design rules are inputs to the system integration and testing stage of the design process. Thus if we isolate one hidden module and look at all *its* linkages, we would see that the underlying task structure was block-hierarchical, in the sense defined in Chapter 2 (cf. Independent, Sequential, Hierarchical, and Hybrid Designs and Task Structures, and figure 2.8).

Figure 14.3 shows a subset of the rows and columns of the generic, modular TSM (see figure 14.2) in a way that highlights these hierarchical relationships.

This hierarchical task structure corresponds exactly to the "upstream-and-downstream" relationships that are at the core of the dual concepts of vertical integration and supply chains. The existing theoretical and empirical literature on vertical integration and supply chains says, in effect, that these relationships are *not determinative* of a uniquely efficient contract structure.[30] Instead, a large number of factors—scale, asset specificity, asset durability, the cost of bilateral bargaining, the

---

[29] As we have said before, the designers' "understanding" does not have to be complete, universal, or certain. In an economy with an advanced capital market, a guess followed by a bet—that is, an investment—can lead to the formation of a new firm, and thereby change the structure of the system.

[30] As Oliver Williamson, one of the major theorists in this area, puts it: "technology does not imply a unique organizational form" (Williamson 1985).

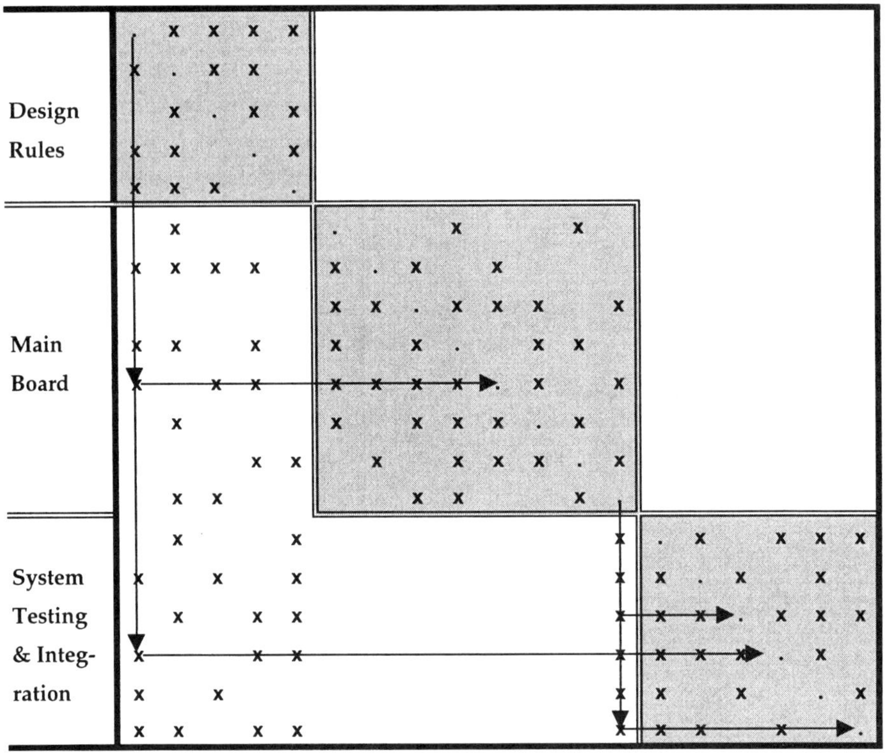

Figure 14.3 A subset of the generic modular task structure matrix highlighting the vertical relations between stages.

repetitiveness of trade, the potential for adverse selection, the potential cost of ex post opportunism, the completeness and the enforceability of formal contracts—all play a role in determining the optimal contract structure for this generic task structure.[31]

This indeterminacy plus the independence of the parallel work blocks leads us to conclude that a modular task structure can be implemented by one, a few, or many firms. The constraints that frictional transactions and agency costs impose are fairly

[31] Stigler (1951); Klein et al. (1978); Williamson (1985); Grossman and Hart (1986); Hart and Moore (1990); Holmstrom and Tirole (1991); Baker et al. (1997); Rajan and Zingales (1998 a, b).

weak; simply, each of the interconnected task blocks within the modular structure should be implemented by a single firm or corporation.

Characteristically in advanced market economies, it costs very little to set up a new corporation.[32] This means that there are many, many ways to organize a modular task structure across different corporations, all of which may yield approximately equal value to investors.

For example, plate 14.3 shows one possible configuration for the modular task structure of our thought experiment (the workgroups depicted in plate 14.2). To construct this configuration, we randomly broke apart the modules and experiments to arrive at a set of 53 "corporations" ranging in "size" from one workgroup (firm #2) to forty (firm #50). We kept all elements of the visible design rules together in one corporation (firm #1), even though history shows that such a high degree of integration is unnecessary. (In the most famous example of "split design rules," the original IBM personal computer included a microprocessor designed by Intel, and an operating system supplied by Microsoft. IBM provided only the basic architecture and the I/O interfaces of the system.[33])

In fact, there are more than 10^{500} ways to organize 300 workgroups into 1, 2, 3, 4, . . . up to 300 corporations.[34] Asking how the set of 300+ workgroups will be organized within the larger economic system is equivalent to asking which of the 10^{500} configurations will be most highly valued. Frankly, it is daunting just to count the alternatives, much less choose between them!

The Origins of Modular Clusters in the Computer Industry

A class of artifacts whose designs are fundamentally modular poses three basic challenges for the surrounding economic system. The system, whether it takes the form of a few large firms or a cluster of smaller ones must provide

---

[32] A new firm can be organized informally in the course of a conversation. Turning that firm into a corporation usually requires a formal registration and the filing of legal documents. In the U.S., for example, any legal resident of a state can apply to incorporate an enterprise. The procedure itself generally costs less than $1000.

[33] Ferguson and Morris (1993).

[34] This is approximately the number of subsets of all sizes that can be formed from three hundred different elements (n=300). We attempted to calculate the figure more precisely, using a simple recursive algorithm, but our 32-bit computer gave out at n=208.

- a sufficient number of workgroups to realize the option values inherent in the designs;
- mechanisms for coordination (communication, agreement, and obedience) across workgroups on design rules, product definitions, and performance standards;
- a way to accommodate ongoing decentralized change in the underlying artifact designs.

It is precisely by achieving a delicate balance of coordination and decentralization that a modular design creates value through the multiplication of design options. Continual change in the artifact designs—what we have called design evolution—is the result: it is visible evidence of the value being created.

However, the realization of option values calls for a complex and ever-changing set of collective enterprises, whose members are cognizant of the design rules, and yet also aware of and responsive to local changes in value. Such a complex configuration is inherently difficult to organize and manage, particularly from a position of central control.

Given these organizational challenges, which are inherent in modular designs, it is small wonder that as these designs became widespread in the computer industry, the structure of the industry itself began to change. As we shall see in the next section, in each year from 1965 to 1980, the number of publicly traded firms in the computer industry increased. Over the same time period, capital market value shifted from IBM to the expanding modular clusters of firms that made up the rest of the industry. We review these developments in the next section, and then close the chapter with a brief recapitulation of the "natural history" of a module.

What Actually Happened: The Changing Structure of the Computer Industry from 1965 to 1980

As we indicated in chapter 1, between 1967 and 1974, ten new computer subindustries emerged. Throughout the 1970s, new firms entered these subindustries at a steady rate. Approximately twice as many entered as exited, and thus, from 1970 to 1980, the number of publicly traded corporations in the "greater" computer industry grew—from 108 in 1970, to 298 in 1980.

Table 14.1 lists the sixteen four-digit SIC categories that we consider to be part of the greater computer industry, and shows the number of firms in each subindustry in the years 1960, 1970, and 1980, respectively. The products of the firms in these cate-

Table 14.1 Number of firms making computers and computer modules by SIC code: 1960, 1970, 1980

Code	Category definition	1960	1970	1980
3570	Computer and office equipment	5	2	9
3571	Electronic computers	1	8	29
3572	*Computer storage devices*	1	6	36
3575	*Computer terminals*	2	5	23
3576	*Computer communication equipment*	1	1	10
3577	*Computer peripheral devices, n.e.c.*	3	5	12
3670	*Electronic components and accessories*	11	7	11
3672	*Printed circuit boards*	2	19	39
3674	*Semiconductors and related devices*	8	4	10
3678	*Electronic connectors*	5	15	16
7370	*Computer programming, data processing, and other services*	1	9	26
7371	*Computer programming services*	0	2	12
7372	*Prepackaged software*	0	7	13
7373	Computer integrated systems design	1	3	16
7374	*Computer processing, data preparation, and processing*	0	5	29
7377	*Computer leasing*	0	10	7
		41	108	298

Firms in italicized subindustries make modules of larger computer systems.

Firms making modules =	34	95	244
Percent of total =	83%	88%	82%

gories are distinguished by the fact that all are "descended" from the original von Neumann design for a computer.[35] The data show that between 1970 and 1980, the vast majority of firms entering the industry made hidden modules of computers—that is, they made storage devices, terminals, communication equipment, electronic components, and software. This empirical fact is consistent with the results of the valuation analysis in chapter 11. There we saw that in a modular design, most of

[35] Refer back to the description of design "descent" in chapter 5. The Compustat data set broken down into four-digit SIC code classifications is the closest thing we have to an observable map of the whole computer industry divided into meaningful submarkets. Needless to say, this map is very imperfect, but for now it is the best we can offer.

the valuable options were associated with experiments on the hidden modules of the system.

Plate 14.4 then graphs the market value of IBM against the sum of the market values of the other firms in constant 1996 dollars from 1950 to 1980. During these three decades, IBM dominated the industry: it accounted for over two thirds of total industry value until 1978.[36] The chart shows how IBM's capital market value grew in the mid-1960s, as it began to ship System/360. However, in 1973–1974, IBM's value dropped precipitously. The value of the rest of the industry also fell, but not as much. (The entire economy was in difficulty at this time, struggling to cope with inflation, recession, and the first oil shock.) Then in 1977–1980, IBM's value fell again, but this time the rest of the industry grew in aggregate market value.[37]

Figure 14.4 shows the same data in a different form: it graphs the "market value share" of IBM vs. the rest of the industry.[38] Somewhat surprisingly, the introduction of System/360 did not cause an increase in the concentration of market values in the industry as a whole. Instead, even as IBM's own value was skyrocketing, it was losing ground in terms of "market value share" to all other firms in the industry.[39] With minor exceptions, that trend continued through the rest of the 1960s and the 1970s. Then in 1980, for the first time, the 297 publicly traded corporations that made up the rest of the computer industry were worth more in aggregate than IBM.

From these data, we can infer that in the years following the introduction of System/360 a large number of new workgroups formed, whose business was to design and produce modules of computer systems. Many of the new workgroups, in turn, were organized as small, new firms.

In this fashion, the tasks of designing and producing computers came to be distributed over many separate firms. Von Neumann's mental decomposition of the design of a computer became a true economic division in which effort, knowledge, and value were spread across literally hundreds of independent enterprises.

---

[36] IBM's position is somewhat overstated, because large diversified firms like General Electric, which were active in the industry until the early 1970s, are not represented in the sample.

[37] The growth in the aggregate market value of the rest of the industry was caused by the entry of new firms into the database, not increases in the value of existing firms.

[38] Slywotsky (1996).

[39] Our data are not so robust that the starting dates of this trend can be identified with confidence. However, the overall change in market value shares is large, as the chart shows. Moreover, press reports and the technical literature of the time also provide evidence that entry into the industry accelerated in the late 1960s, and that said entry was concentrated in the "hidden module" subindustries.

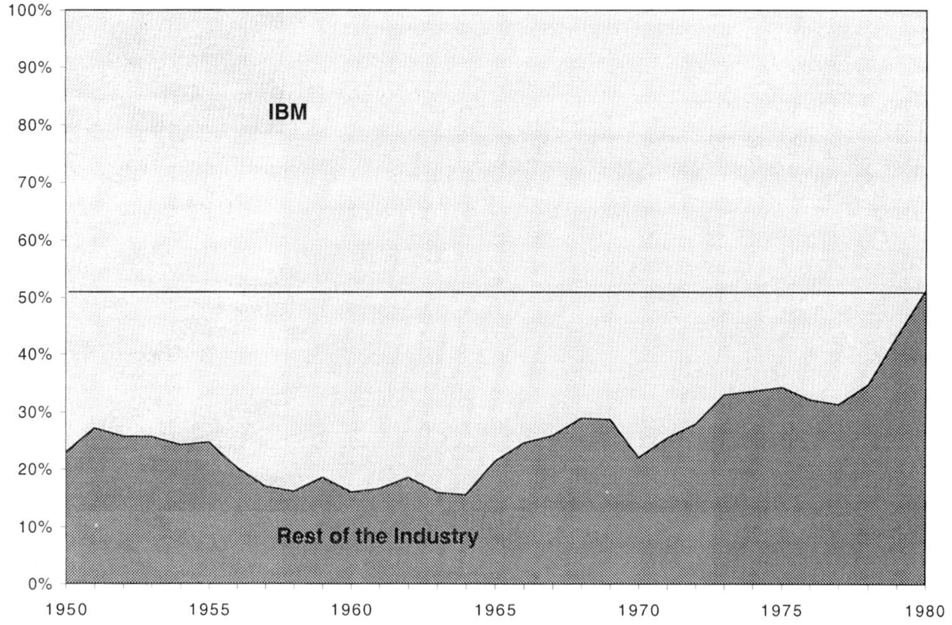

Figure 14.4 The market value share of IBM and the rest of the computer industry 1950–1980.

The "Natural History" of a Module

If we now step back and envision changes in a system over many design generations, we can see an outline of what we might call the "natural history" of a computer module. It begins as an unnamed element—a knot of structure within a generally interdependent system design. In this early stage, it is still tied to the system by many parameter interdependencies, but designers can "see" that it is a separate "organ" within the "body" of the system.[40] At this point, the component is given a name, which often reflects its role within the system.

As it comes into focus through having a name and a role, the component becomes an object of study. Specialized knowledge begins to accumulate around it, even

[40] This was von Neumann's metaphor for the computer in his preliminary memorandum (von Neumann 1987).

though it is still highly interdependent and connected with the rest of the system. Designers specializing in the component may appear at this time, and become highly sought-after members of design teams. For example, over the fifty-plus years of history of computer designs, there have been experts in the designs of core memory, chip memory (RAM and ROM types), tape drives, disk drives, microprocessors, operating systems, graphical user interfaces, Web applications, and many other components.

The component may stay in this protomodular state for a long time. However, the possibility also exists for one or more architects to construct a modular design and task structure in which the component is a module. This was essentially what occurred on a very large scale with System/360. A modular design and task structure severs the coordinating links between hidden modules, and replaces them with design rules and an ex post integration and testing phase. At this point, component specialists can begin to work separately from one another, with each team of specialists focusing on one hidden module. By definition, these "hidden-module designers" (as we now call them) do not need to be in day-to-day contact with others who are working in parallel on other modules.

But the story isn't over yet. Now that the module is separate, the groups working on its design need to know what their objectives are. One objective, clearly, is for the module to "work" within the system; another is for the system to work better with this team's module than that team's. However, it is tedious and expensive to have to try out provisional designs in the larger system each and every time.[41] Thus hidden-module designers will want to spend some of their time and resources developing module-level tests.

The appearance of module-level tests has an important economic side effect. The testable, verifiable dimensions of the module are the foundation that supports arm's-length contracts and market transactions—without tests, there is no way to know what is being bought or sold. Thus once a module's performance is understood well enough to be summarized via benchmarks and test scores, the module can be traded. It may then come to have its own markets and industry structure.[42] At this point the

---

[41] Indeed, the larger system may not be available when a particular module design is ready to be tried out. This is a consistent theme in the historical record—see, for example, the records on IBM System/360, AST 4000, and Windows NT. In each of these cases, designers cobbled together a test system because the target system was not ready to play host to the modules being designed (Pugh et al. 1991; Bauer et al. 1992; Zachary 1994.)

[42] Sutton (1998).

module has "come of age," becoming a separate and distinct product in the larger economic system.[43]

In the last chapter of this volume, we show how, in the presence of an evolving modular design, competition among companies making hidden modules gives rise to a decentralized process of *industry evolution.* Design evolution and industry evolution interact, a process known as coevolution. We end with a brief Afterword in which we look ahead to volume 2.

---

[43] Stigler (1951).

15 Competition among Hidden Modules and Industry Evolution

In the previous chapter, we argued that a modular design and corresponding task structure make possible a particular type of industry structure, called a "modular cluster." In a modular cluster, the tasks of designing and producing the different parts (modules) of a complex artifact are spread over many different firms. The firms' activities are coordinated by design rules that govern key parameters of the artifacts' design. Conformance with design rules is meant to ensure that modules produced by different firms in the modular cluster will work together as functioning systems.

Modular clusters, by definition, "play host" to modular design evolution. Hence, unless and until an artifact design has been modularized, there is no cause for a modular cluster to form.[1] Following a modularization, we have seen, there will be a concomitant multiplication and decentralization of design options. The number of workgroups engaged in design (and production) will go up, while, simultaneously, the forces of transactions and agency costs that tend to bind workgroups together into firms will diminish. Depending on the balance of these and other forces acting on the industry, a modular cluster (or clusters) may then emerge as a viable form for the industry to take.

As well as by design rules, the firms in a modular cluster are connected to one another and to the larger economic system by a set of mediating markets for goods, labor, and capital. Subsets of firms in a modular cluster must cooperate in setting up markets to facilitate transfers of goods, services, and information among themselves and with the external environment.

The first modular computer design to be realized was IBM's System/360, whose creation we described in chapter 7. By 1970, System/360 as a whole was worth

---

[1] There may be other types of clusters, for example, geographic clusters of retail stores. However, we are specifically interested in the nature and organization of modular clusters.

around $170 billion in today's dollars. It contained approximately twenty-five distinct hardware modules,[2] each of which could evolve independently of the others. Each module also had a set of designers within IBM who were familiar with System/360's design rules. Although we do not have any way to reconstruct the cost of experimentation, technical potential, and degrees of visibility of System/360's modules, we believe that their option value potential was similar to that of the hypothetical twenty-three-module system analyzed in Chapters 11 and 14. If that assumption is correct, then the option values embedded in System/360's modular design would have justified something like 300 new design efforts focused solely on the hidden modules of that system.

Those hidden modules in turn could be designed (and for the most part, produced) by groups working independently of one another, within or outside IBM's corporate boundaries. Thus in theory it was possible for new enterprises to capture significant amounts of value by supplying hidden modules to IBM's customers. And that is in fact what happened. Shortly after the introduction of System/360, there emerged numerous companies that offered IBM "plug-compatible" peripherals, including disk drives, tape drives, terminals, printers, and memory devices. In effect, these new companies used the modular operator *substitution* as a means of gaining entry to the computer marketplace.

IBM attempted to deny these hidden-module competitors access to its system's design rules and its customers, but was unsuccessful in this effort. In the late 1960s and early 1970s, IBM's managers discovered that it was far easier to claim to control a modular design than to maintain control in the face of a series of challenges aimed at its hidden modules.

Indeed, what is most striking about the computer industry from 1970 onward is its highly dynamic pattern of entry, competition, and exit by firms, combined with rapid turnover in products and product definitions. These competitive dynamics, we believe, were instigated by the values inherent in the underlying, evolving modular designs. Competition eventually led to an industry made up of almost a thousand firms, organized as a multi-apex modular cluster. By far the majority of firms in this cluster make hidden modules. Thus understanding the nature of competition among the makers of hidden modules is crucial to understanding the path of development of the computer industry from 1970 through today.

In this final chapter of our first volume, we look at what happens when multiple firms compete by offering different designs of the same hidden module within a

---

[2] Software modules did not become an important source of value until the late 1970s.

modular cluster. We begin by looking briefly at the modular cluster that formed around System/360. As an example of a hidden module subindustry within that cluster, we take the case of disk drive makers. After briefly describing how the disk drive subindustry emerged in the late 1960s and early 1970s, we use our tools to analyze the structure of this type of industry (a hidden module subindustry within a modular cluster).

The model we present is a formalization and extension of a model of the disk drive industry's behavior, which was put forward by William Sahlman and Howard Stevenson in 1985.[3] In our version of this model, we combine elements of the theory of design evolution, developed in previous chapters, with game-theoretical methods of industry analysis, in particular, the concept of a rational expectations equilibrium, and John Sutton's analysis of endogenous sunk costs.[4]

We argue that there are a large number of so-called rational expectations equilibria in a hidden-module subindustry, each of which in turn corresponds to a different industry configuration. Thus from a static game-theoretical perspective, assuming no repetition of play, almost any industry structure is possible as an equilibrium for a particular subindustry. The disk drive subindustry, for example, might become a monopoly; it might be competitive (in the design phase, the production phase, or both); or it might be something in between.

We then consider the possibility of dynamic movement within the set of rational expectations equilibria. Against the backdrop of rapidly evolving artifact designs, movement from one subindustry equilibrium to another can be instigated by value-seeking behavior on the part of portfolio investors and enterprise designers operating in the capital markets. Specifically, financiers and entrepreneurs can use financial operators, such as

- the founding of corporations;
- the funding of corporations;
- mergers and acquisitions;
- changes in ownership and management;

to create new firms, and restructure older ones. The existence of these reconfiguration opportunities in turn creates nonstationarity, and therefore the possibility of evolution in the firms and markets of the hidden-module subindustries.

---

3 Sahlman and Stevenson (1985).

4 Sutton (1991).

Historical Overview

Defections from IBM

We pick up the story of System/360 where we left off at the end of chapter 8. In December 1967, less than a year after IBM began to ship System/360 in volume, twelve key employees left IBM's San Jose laboratory (home of disk drive R&D) to set up shop in a new company called Information Storage Systems (ISS). The twelve—known within IBM as the "dirty dozen"—took with them state-of-the-art knowledge about how to design high-capacity disk drives.[5] They also knew System/360's design rules—its architecture, interfaces, and tests. (See box 15.1.)

As these twelve people left IBM, there was a great deal of ferment taking place in the disk drive market.[6] Memorex, Potter Instruments, Telex, Marshall Labs, and other firms had announced IBM-compatible disk drive products. The defecting engineers quickly struck a deal to provide drives to Telex. Within six months, they not only developed an IBM-compatible design, but had made it faster and less expensive than the disk drives IBM was offering.[7]

In the months and years that followed, a stream of experienced designers and engineers left IBM to go to work for manufacturers of plug-compatible peripherals. With such an inflow of talent, these companies were able to quickly launch products with superior performance at lower prices than IBM's. Often they implemented concepts that had been under development at IBM's R&D centers.

By offering products with superior performance and lower prices, plug-compatible peripheral manufacturers were able to make substantial inroads into IBM's markets between 1968 and 1970. By early 1970, the plug-compatible manufacturers as a group had achieved an 11% market share in tape drives and a 4% share in disk drives.[8]

---

5 Pugh et al. (1991, pp. 490–495).

6 Christensen (1993).

7 Pugh et al. (1991).

8 Brock (1975, p. 113); see also Fisher, McGowan, and Greenwood (1983, pp. 310–340); DeLamarter (1986, pp. 121–218); Pugh et al. (1991, pp. 489–553). Most of the facts contained in these sources were drawn from documents released in legal proceedings against IBM during the 1970s. Each book has a different point of view: Brock was an academic; Franklin Fisher was an expert witness retained by IBM; DeLamarter was an economist who worked for the Department of Justice on the U.S. v. IBM case; Pugh et al. is IBM's officially sanctioned technical history.

Box 15.1 Interfaces: What Disk Drive Designers Need to Know

When a particular type of module is defined to lie within the *architecture* of a larger computer system, then the system's design rules must describe the *interface* through which the module will interact with the rest of the system. Interfaces are like treaties or contracts between subparts of a system. As such, they must specify not only general principles, but also some very detailed protocols (who says what, when, and in what language). In computer systems, one important subclass of interfaces involves storage devices like disk drives. Storage device interfaces differ significantly in their details, but each must answer a set of questions about how the device will be connected to and communicate with the system. For example, the key elements of a disk-drive interface are the following:

- *Logical block addressing scheme:* The interface must specify how data will be addressed by the controller. By convention, location of data on a disk is determined by three numbers: the *cylinder*—the surface formed by the combination of identical tracks lined up on vertically stacked disks; the *head*—which of the several tracks the data are on; and the *sector*—the specific location on the track (tracks usually are divided into seventeen or twenty-five sectors). For example, the controller in the first IBM PC used a twenty-bit addressing scheme (ten to address the cylinder count, four to select the head, and six to select the sector on the disk).
- *Data transfer mode:* The interface must indicate how the CPU will format and send the data to the drive (and vice versa). The size of the data path (e.g., eight-bit, sixteen-bit, etc.) is crucial, but the interface must also specify the methods used to read and write data. In the System/360, for example, IBM used what was called a "data channel" (essentially a processor dedicated to moving data between memory and peripherals) to provide direct links between the disk drive and the system. In some of DEC's early machines, however, the task of managing the data flow was handled by the CPU using an interrupt scheme.
- *Physical cabling and pinout:* The interface must establish the type of cable, the number of pins in the connector, and the electrical signal attached to each pin. The IDE standard interface, for example, uses a flat, forty-pin cable, and reserves pins 3–18 for data transfer. (The other pins are used for electrical grounding and for control information.) The eight-bit SCSI interface uses a fifty-pin cable with pins 26–34 handling data.
- *Software commands:* The interface includes a set of commands that defines the language through which the drive communicates (sends messages/receives instructions) from the CPU. Each command has a specific syntax (in SCSI, the READ command comes in five bytes that specify the operation, the address, and the transfer length) defined according to conventions established in the interface language.

In principle, every modular computer system or family could have a different set of interfaces. However, such redundancy offers a clear opportunity to apply the *inversion*

(continued)

operator in order to standardize the interfaces across a number of systems (see chapter 13). Thus, since the mid-1960s, storage companies have cooperated with one another and with standards committees in the development of several disk-drive interface standards, for example, the IPI (intelligent peripheral interface) used in mainframe computers; SCSI (small computer systems interface) introduced by Apple Computer with the Macintosh family; and most recently, PCI (peripheral connect interface), which has been sponsored by Intel.

Source: Bodo, M. (1994). *Hard Drive Bible,* 7th ed. Santa Clara, Calif.: Corporate Systems Center.

IBM's Countermoves

In 1970, IBM began to take actions aimed specifically at these competitors. First, it changed the physical interface between disk drives and processors, by moving the controller of the disk drive into the processor's box.[9] Next, it lowered the price on its newest disk drive product (called the 2319) by 46%.[10] The plug-compatible manufacturers responded with their own new designs and matching price cuts. Their growth did not slacken, and by April 1971 plug-compatible drive installations were growing at a rate of about 20% per month.[11]

In May 1971, IBM struck again, more effectively. It announced a new set of lease terms, known as the fixed term plan (FTP). Before the fixed term plan went into effect, standard IBM leases were cancellable at will with one month's notice. The FTP allowed customers to obtain one- or two-year leases on some pieces of peripheral equipment at rentals 20% to 25% below standard rates. However, there were large penalties attached to premature cancellations of FTP leases. This move put IBM's prices below those of the plug-compatible manufacturers, and also switched many IBM users away from one-month leases, which had made it contractually easy for customers to swap IBM equipment for plug-compatible equipment on short notice.

Plug-compatible manufacturers dropped their prices once again, but this action caused their earnings to disappear. (Because of IBM's manufacturing efficiencies, derived from the modular structure of its production processes, plug-compatible

9 Changing the interface amounted to a change in design rules.

10 Brock (1975, p. 114).

11 Ibid., pp. 119–120.

manufacturers had variable costs that were substantially higher than IBM's. Hence they were competitive only so long as IBM maintained its own high margins.) Plug-compatible orders fell off as customers began to question whether these companies would be able to survive and support their products in the future.

A year later, in the summer of 1972, IBM introduced a new top-end disk drive in which the controller was split between the drive and the processor, and dropped prices again.[12] IBM took aggressive action on memory as well. It dropped memory prices significantly, but it also "unmodularized" memory, by bundling it with the processor on popular models of System/370 (the successor of System/360).

Plug-compatible manufacturers met each of IBM's moves with lawsuits alleging unfair competition. IBM countersued charging theft of trade secrets. During the 1970s, these suits, plus a large antitrust suit, which the U.S. Department of Justice filed in 1969, slowly wended their way through the legal system. In the end, the courts rejected most of the plug-compatible manufacturers' complaints of monopoly, and in 1982 the government withdrew its suit.[13] Thus, by the early 1980s, there was little doubt that, with respect to System/360 and its descendants, plug-compatible manufacturers only existed on the sufferance of IBM.

However, by that time, hundreds of hidden-module designers had left IBM. Partly as a result of those departures, IBM lost control of the evolving design trajectories of most of the hidden modules of a computer system. By 1980, the most advanced and capable designers of disk drives, tape drives, memory devices, processors, and applications software often did not work for IBM. Instead they worked for the growing numbers of firms that focused on the design and production of their specific hidden modules.

In the 1970s, as we have said, minicomputer firms like Digital Equipment, Data General, Prime, and Wang used the exclusion operator to design stripped-down modular systems, and invited other firms to coinvest in their hidden modules. The chip designers, peripherals designers, and software designers who had initially sought to supply parts for System/360 found new opportunities to create chips, disk drives, tape drives, printers, terminals, and specialized software for these new systems. In this fashion, the minicomputer architects and the independent hidden-module designers began to invest in decentralized, yet coordinated ways. They developed new ways of communicating and enforcing design rules across their own corporate

---

12. Ibid., pp. 124–125.

13. DeLamarter (1986, pp. 365–366).

boundaries, their systems worked and gained market acceptance, their designs coevolved, and each group fueled the others' growth.[14]

The Economics of Hidden Modules—Static Analysis

To construct a sensible model of competitive forces in a hidden-module subindustry, we must be concerned not only with competition in the product markets for hidden modules but also with competition in the market for new designs. Our theory of design evolution via modular operators indicates that a modularization creates valuable design options. One of the operators, hence one class of valuable design options, is *substitution.*

After a design has been modularized (split), hidden-module designers can substitute new designs for old ones without changing the overall system, as long as they adhere to the system's design rules. In Chapters 10 and 11, we showed how the value of these module-level substitutions justifies increased levels of experimentation and investment in hidden-module designs (including designs of production processes).

To fix ideas, let us assume that we are looking at a complex artifact, which has already been split into a system of modules. As value-seeking designers and investors, we would like to know what reward we can expect if we introduce a new design for one of the hidden modules of the system. For example, in late 1967, what gain might we expect from the effort of designing a new disk drive that was compatible with System/360?

Figure 15.1 shows the characteristic net option value function for parallel experiments mounted on one hidden module in a larger system. We obtained this figure by taking one "slice" of plate 11.1 (which graphed the values of parallel experiments for all the modules of a computer workstation). The function in figure 15.1 shows the expected payoff minus the cost of running different numbers of experiments to improve the performance of the module in question. Values have been scaled relative to the cost of a single experiment—thus the rate of return on the *first* experiment for this module is around 200%. (All other assumptions are as described in Chapter 11.)

A similar graph may be constructed for any hidden module in any modular system of artifacts. Hence, with respect to von Neumann–type computers, the module in question could be a storage device (like a disk or tape drive), a memory device, an I/O device, or any one of a vast set of application programs. Such graphs are usually

---

[14] Christensen (1993).

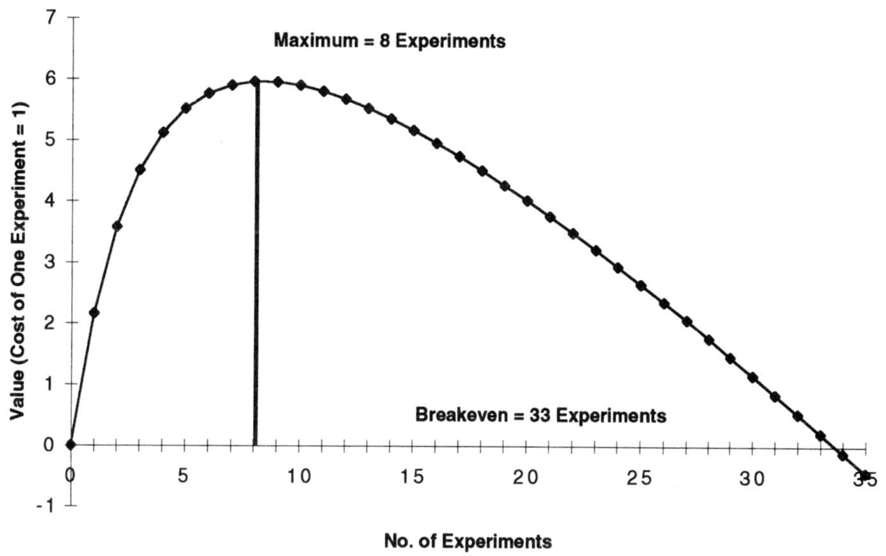

Figure 15.1 The net option value of experimentation on one hidden module.

single-peaked functions,[15] but the height, breadth, and the location of the peak will vary from module to module. Modules with high technical potential have high peaks because there is much room for improvement. Large modules have sharp, narrow peaks because the cost per experiment is high. Small modules have flat, wide peaks because the cost per experiment is low.

Up to this point, we have defined the value of a module in terms of the users' incremental willingness to pay for a better module within a particular system. Thus the function graphed in figure 15.1 reflects the net present value to users of different levels of investment in the particular module's design. This in turn is the total value that may be divided among all participants, including

- the users;
- the architects of the modular design;
- the hidden-module designers;
- the systems integrators and testers;

[15] Concave expected returns and convex costs are sufficient to obtain a single-peaked value function. In Chapter 10, we discussed factors that can sometimes lead to multipeaked value functions.

- the producers of modules;
- the investors, who fund all of the above.

Our task at this juncture is to figure out how many separate enterprises may be involved in the design and production of hidden modules.

Rational Expectations Equilibrium

Consider a two-stage investment process as follows. In stage 1, "firms" or "corporations" invest in experiments that generate new disk drive designs.[16] In stage 2, some of the new designs will be converted into new products and sold to users. The rewards to investment will then be a function of the number of firms that enter the search for new designs, and the number of firms and the nature of competition in the final product market.

We assume that potential entrants make investments "rationally"—that is, they base their decisions on assessments of the value they can capture in the final product market. Their assessments in turn are based on expectations about how completed designs and artifacts will be priced.[17] And those prices will be affected by the state of competition in the final product market.

Under these assumptions, investments in new designs have the characteristics of "endogenous sunk costs." John Sutton has argued that in industries with endogenous sunk costs, weaker anticipated competition in the final product market will induce more initial investment.[18] In other words, the expectation of high returns in the future may induce more firms to enter—but if they survive, their presence in the product market later on will intensify competition and serve to drive prices down.

A *rational expectations equilibrium* exists when what is expected and what occurs are the same. Technically, in the language of finance, one says that "the distribution of forecast values is consistent with the distribution of outcomes after the fact." In other words, a rational expectations equilibrium is that set of expectations that elicits

---

[16] In an advanced market economy, firms above a certain size will almost certainly be organized as corporations to economize on ongoing transactions and contracting costs. Hence, in this section, we use the terms "firm" and "corporation" interchangeably.

[17] Actually, we are concerned with the *value* of a completed design, which is the product of the price of the artifact times the quantity sold, less the variable and fixed costs of making it. We subsume those issues by treating a design as one artifact with a zero marginal cost. This greatly simplifies the discussion, and does not affect the conclusions in any significant way.

[18] Sutton (1991).

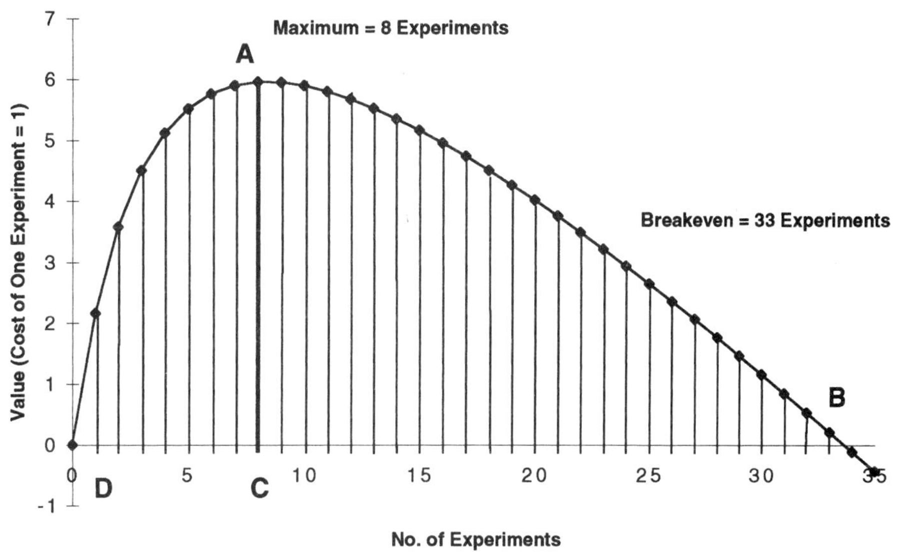

Figure 15.2 Rational expectations equilibria for experimentation on one hidden module.

just enough entry so that the initial expectations turn out to be true. This is a useful organizing concept because it identifies patterns of behavior that are sustainable over time in both the product and the capital markets.[19]

Figure 15.2 shows the rational expectations equilibria that arise with respect to investments in hidden modules with the value function we have posited. Every point on each vertical line in the figure corresponds to a different rational expectations equilibrium. In other words, each point on any vertical line represents a feasible investment policy and a rule for sharing the gains that (in expectations) adequately compensates investors. As the figure clearly shows, there are an infinite number of such equilibria associated with the two-stage design investment process we have outlined.

Interesting Extreme Cases in the Set of Rational Expectations Equilibria

There are also four interesting extreme cases, which are indicated in figure 15.2, summarized in figure 15.3, and described briefly below. (Readers seeking greater detail may consult the appendix to this chapter.)

[19] Fama (1965, 1970); Sutton (1991).

Market for New Design Experiments is: \ Final Product Market is:	Monopoly	Competitive (Bertrand)
Monopoly	**A** NOV to investors	**D** No investment
Competitive	**B** Winner-Take-All	**C** Design Competition

Figure 15.3 Extreme types of market structure for hidden module designs.

The first extreme case, marked *A* in the figure, is the case in which both the experimental process and the final product markets are controlled by a monopolist. In that case, the full net option value (NOV) associated with design improvements will be captured by the investors and employees in the monopolistic enterprise.[20]

The second case, marked *B* in the figure, is one in which there is competitive entry into the search for new designs, but the final product market is a monopoly. The first-stage investment process then has the characteristics of a "winner-take-all" contest—also known as a tournament. For this structure to arise, there must be limited entry into the second stage. This can occur if there are economies of scale in the production process, or "network economies" in the use of the module.

The third case, marked *C*, has competitive entry into the market for new designs, followed by "tough" competition in the final product market. The most stringent form of competition consistent with rational expectations is a form of Bertrand price competition. Under this regime, the equilibrium price for a module in the final product market would be its cost (assumed to be the same for all competitors) plus the difference in value (quality) between the first-best and second-best designs. The first-best design would make a small profit (in expectations); the second-best would survive on the margin; the rest would exit. Most of the net option value in this case would be captured by the users of the hidden modules. This type of equilibrium can occur if entry to the second stage occurs through the mechanism of design competitions. (See box 15.2.)

In the fourth case, *D*, no firms invest in the design at all. This case arises when designs can be copied at will, with no compensation paid to the originators of the design.

[20] There may be substantial overlap between the investors and the employees of such a firm.

Box 15.2 Design Competitions

Design competitions have been common in the computer industry since the mid-1970s. For example, Gordon Moore, a founder of Intel Corporation, recounted that in 1979, "we set a goal of achieving two thousand design wins" for the sixteen-bit microprocessors, the 8086 and the 8088.[1] At this time, Intel's primary business was designing and making chips that were built into larger systems, including computers, calculators, consumer electronics products, and controllers. Most of its revenue was generated by "design wins" like the two thousand it aimed for in microprocessors. (Unlike today, microprocessors accounted for a relatively small part of Intel's revenue and profit in 1979. The company was then primarily known for its memory chips.)

A design competition requires a more completely modularized architecture than an *exclusive* subcontracting relationship. The architecture must include functional definitions of all modules being outsourced, a complete list of design rules, precise conformance standards, and the means to test many different versions of a module at a very low cost per item. Moreover, on an expected value basis, the rewards to hidden-module designers must be commensurate with their investments in the new design.

The fact that Intel, and a growing number of firms like it, could survive and prosper as design subcontractors in the mid-1970s is evidence that computer design rules and module tests had reached the point where they could support modular task structures and arm's-length contracts between systems architects and hidden-module designers.

[1] Moore (1996). In fact, the company exceeded this goal. What is more: "[B]uried among these two thousand wins was one that proved crucial to Intel and important to the electronics industry. A small IBM group at Boca Raton, Florida, chose the 8088 for their first personal computer." Intel's market value and market power today can be traced back to that one, relatively obscure, design win.

The Range of Possibilities

Cases *A* through *D* outline the region of rational expectations equilibria in terms of value captured by investors and number of design experiments mounted. Speaking generally, in figure 15.2, points on the horizontal axis correspond to equilibria in which the net option value realized by investors is zero. Such outcomes are consistent with a competitive capital market or inadequate protection of intellectual property, or both. Points on the value function itself (like *A*) are attained when the full value surplus flows to investors (who may be the designers). For this to happen, the final product market must be a monopoly or, equivalently, a perfect cartel.[21] However, the

[21] Strictly speaking, the final product market has to be a price-discriminating monopoly for the full surplus to accrue to the investors (unless the investors are also the users).

profits of the monopoly may be reduced by either underinvestment (points to the left of the maximum) or overinvestment (points to the right of the maximum) in experimental substitutes.

Points in the interior of the set are those in which the net option value is split between the investors and other parties (e.g., systems integrators, users, or both). Such outcomes are consistent with imperfect competition in both the product and capital markets. In contrast to points *B* and *D*, these outcomes generate a surplus for society (in expectations). However, the fraction of the surplus expected to flow to investors is less than in a full-scale monopoly.

Points on the heavy vertical line in figure 15.2 maximize the total pecuniary value gained by all parties. In our example, the maximum (in expectations) is attained when the number of experiments equals eight.

Rational expectations equilibria correspond to points on the lighter vertical lines in the diagram. They do not provide as large a total pecuniary surplus as points on the heavy line, but they do provide enough to pay for the cost of mounting the experiments. Thus all informed, rational investors (including the designers themselves) can be induced to finance design efforts that correspond to points on any of these vertical lines.

Mistakes

To arrive at a rational expectations equilibrium, wherein it makes sense for all parties to make a particular investment, investors must have consistent beliefs. That is, before they invest, they must know (and know that all other investors know) the terms of entry and competition both in the financing stage and in the final product market. If the investors do not have such mutually consistent beliefs, then someone's (perhaps everyone's) expectations will be confounded by what actually happens. After the fact, the basis on which the investments were made will turn out to be untrue, and investors will be called upon to adjust their beliefs.

Given time to act, the forces that will lead investors to a mutually consistent set of beliefs are quite powerful. Hence, over some period of time, we would expect an industry to settle within the set of rational expectations equilibria—that is, on one of the vertical lines of figure 15.2. However, that in itself is a hypothesis, which requires formal specification and testing on industry data.

Moreover, mutually consistent beliefs will not arise costlessly at the exact moment that a new product creates the possibility of a new market and a new industry. The coordination capacity of the economic system is not vast enough to convey to all potential entrants knowledge of what all other potential entrants are likely to do.

Meanwhile, before the consistency-checking process concludes, investors may make mistakes. Misunderstanding the game, any single investor or enterprise might underinvest, and thus leave potentially valuable designs "on the table." Or, investors may invest too much, hence in aggregate lose money. Inconsistent beliefs leading to mistakes in turn raise another set of questions: If misperceptions are possible, how serious will the resulting misallocations be? How soon will they be corrected? And what will happen in the meantime?

The Economics of Hidden-Module Design—Dynamic Analysis

These questions cause us to want to understand better the *dynamics* of the hidden-module design and investment processes. We are especially interested in knowing how investors' learning about industry structure and rewards interacts with the evolution of the designs and artifacts themselves. To get at these issues, we would now like to nest the static, single-play analysis of a hidden-module subindustry in a dynamic framework that includes other parts of the larger economic system.

The larger system encompasses, on the one hand, decentralized design evolution and multiple modular architectures, and on the other, value-seeking enterprise designers operating in an advanced capital market. What difference do these contextual factors make?

The fact that the underlying modular designs are evolving in a decentralized fashion means that there are many repetitions of the basic investment game, but always in different settings. That is, both within and across different types of modules, investors are routinely confronted with design investment opportunities yielding value functions and competitive assessments like those we described above. For each one, they must ask, How valuable is this opportunity? How many others are likely to mount design experiments? How many designs will survive in the ultimate product market? How much value can one enterprise expect to capture from this design effort?

Figuratively, we can think of a set or collection of value functions like the one in figure 15.2, arrayed one on top of another. Investors move through time from one value function to another, making the relevant assessments and "bets" with respect to each opportunity. The collective bets of all investors, combined with the outcomes of the individual design processes, result in a particular set of design outcomes, a particular product market configuration, and a particular disposition of rewards to the investors. Those outcomes are in turn data, which investors can use to perform basic consistency checks on their investment decisions: Is the game we just played the one we thought we were in?

From outside this complex, dynamic process, what would we observe? Let us look at several rounds of the investment process for one type of hidden module. In a world of "perfectly consistent" expectations, at the beginning of a round, all actors would know what game they were in. In terms of figures 15.1 and 15.2, they would know *k*, the number of experiments mounted by all participants, and *V*, the total expected reward.[22] Hence, with perfectly consistent rational expectations, the state of the game at the beginning of the round would correspond to a point on the chart.

In reality, however, this high degree of consistency and precision places too great a burden on the channels of communication and coordination in the economic system. It is simply not worthwhile (not cost-effective) for the investors to expend the resources needed to be so consistent and so precise.

Investors' Expectations Ranges

In advanced market economies, it is extremely easy for investors—both individuals and firms—to enter a staged investment game like the one we have described. All the investor needs to "play" is something of value to contribute to an endeavor—the "thing" may be money, time, organization, knowledge, contacts, or a reputation. Then, to justify the investment (to make it an investment and not a gift), the investor must perceive a potential reward. The perception may be fuzzy, and does not have to be right—that is, the investor *may* believe something that, upon further analysis, would turn out to be untrue.

Any further refinements of the investor's valuation of the opportunity will themselves be investments that demand effort. For example, it is possible to translate rough perceptions of costs and benefits into a formal range of outcomes; to map the range of outcomes onto a probability distribution; to condition the probability distribution on the behavior of other investors; and to apply consistency checks after the fact. These refinements all require knowledge and effort, hence they are costly to implement.[23] Because they are costly, such refinements are justifiable if and only if the investor perceives that they will result in better ex ante decisions.

---

[22] Actors would also know the distribution of the total reward and how it would be divided among participants, conditional on outcomes. This would allow them to infer their own, individual expected rewards, V_i.

[23] These refinements and consistency checks make up a large part of the curriculum of finance and strategy courses offered in business schools. Thus they require not only effort but advanced education on the part of the decision analyst (who may be different from the decision-maker).

This means that the investors' own model of the game is in fact endogenous to the investment process.[24] On the one hand, if large errors can be avoided through a better understanding of the underlying market structure, then it will pay investors to invest in more knowledge. But, conversely, if more precise knowlege of the market structure does not allow large errors to be avoided, then the cost incurred in gaining that knowledge will only detract from the final payoffs. Investors will not then find it worthwhile to do the extra work to improve their own understanding of the structure of the game—they will be content to "roll the dice."

We can depict an investor's range of expectations as a gray area within the larger area defined by the overall value function. An investor who has some knowledge of the competitive structure of the game, and understands how entry affects expected rewards, will have an "expectations range" within the subset of rational expectations equilibriums. What's more, the range of value captured, V_i, conditional on the number of entrants, k, should be downward-sloping in k. Figure 15.4 presents an example of this "expectations range" for one investor, who might be an individual or a corporation. (In what follows, we assume the investor is a corporation.)

Within the gray area, we have placed a point that represents the observed outcomes of V_i and k on the last round of play of this game. This "last round of play" may be the previous generation of design for a particular type of module, or some other competitive interaction which the investor deems relevant. A new entrant (or an investor that did not learn from past experience) would have no last round of play recorded.[25]

Placing the last round of play within the expectations range says something about the investor's view of the stationarity of the process: if the structure of the game is changing rapidly, and the investor understands this, then the outcome of the last round of play may lie outside the investor's range of expectations for the next round of play.

From the figure, we can infer that the investor whose expectations are depicted did very well on the last round of play, and is relatively optimistic about its prospects on the next round.

The location of the expectations range indicates that the investor expects some entry leading to deteriorating market conditions; this is evident from the fact that

---

24 Massa (1998).

25 Capturing and storing information about prior competitive rounds is costly, hence if the system is very dynamic, it may not be cost-effective even to try to learn from experience. There may be other, better ways for investors to form an expectations range.

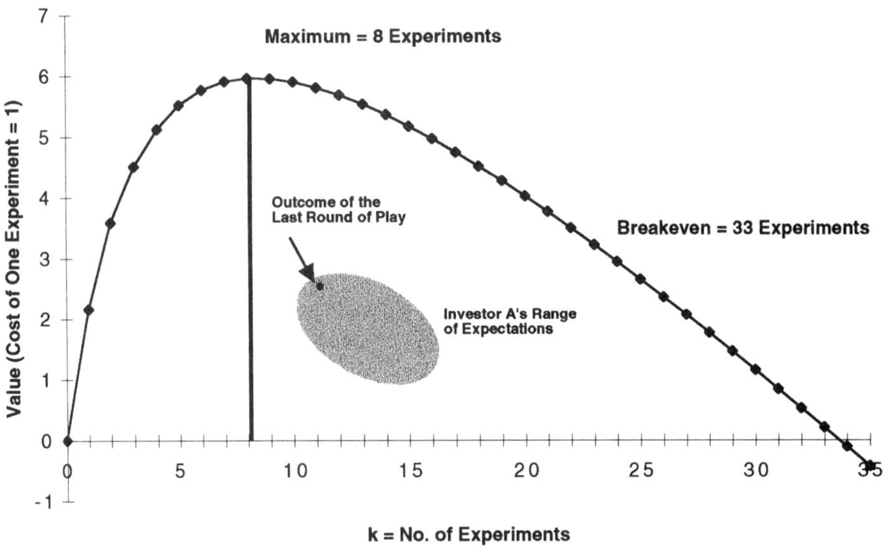

Figure 15.4 The expectations range of one investor in relation to that investor's outcome on the last round of play.

most of the expectations range lies below (lower V_i) and to the right of (higher k) the last round of play. But all points in the expectations range lie above the horizontal axis, indicating that, for any level of entry in the expectations range, this investor expects to make money.[26]

For this investor, whom we label investor A, investing in the next round of play is a proverbial "no brainer." But other investors may have different models of reality, different past experiences, and thus different expectations ranges. For example, figure 15.5 adds to figure 15.4 the expectations range of another investor (investor B), who lost money on the last round of play, and who does not expect to do well on the

[26] There is a subtlety here with respect to expectations. We are only showing the expectations range as a function of different levels of entry into the design-investment process. There may be other dimensions of expectations, for example, technological variables, which, if the outcomes were unfavorable, would cause the investor to expect to lose money. Technically, the expectations range should be conditioned on an assessment of "average" or "likely outcomes" in these other dimensions. However, this assumes that the investor is putting in the effort needed to make his, her, or its expectations range perfectly consistent with probability theory. Most investors in our experience are not so rigorous, but they invest anyway!

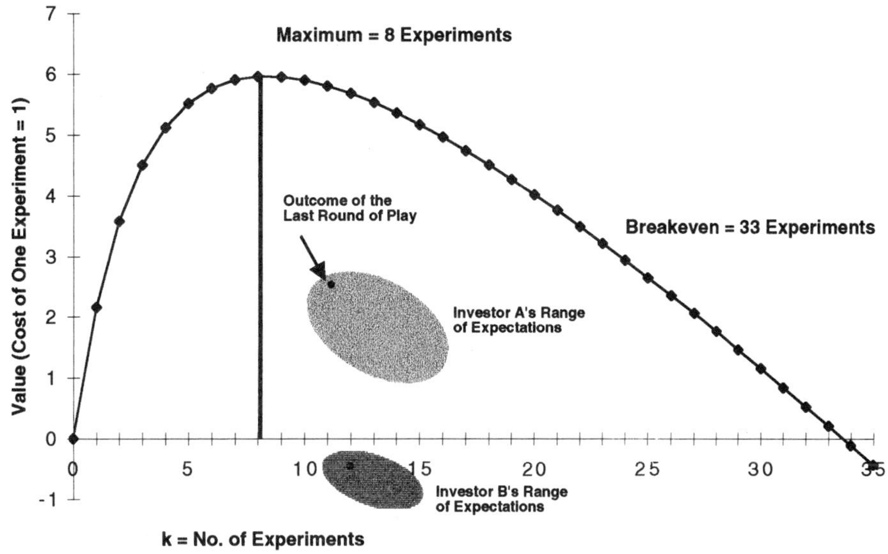

Figure 15.5 The expectations range of two investors in relation to their outcomes on the last round of play.

next round. In contrast to investor A, investor B has a clear mandate *not to invest* in the next round of the design investment process.

Obviously it is possible for an investor's expectations range to span the horizontal axis, meaning that the investor knows it is possible to lose money, but is not sure to do so. In that case, it might be worthwhile for the investor to spend money (or other resources) to narrow its expectations range. However, even after making the expectations range more precise, it often happens that the range still spans the axis, meaning that the decision is not clear-cut. In that case, the decision to invest or not will depend on the investor's specific decision rule.[27]

But who are these investors A and B? Thus far we have assumed that they are "firms" or "corporations" operating as discrete, independent enterprises within the subindustry defined by the value function. And that is, indeed, one kind of investment

---

[27] If the investor is an individual, his or her decision rule may be evaluated in terms of its rationality. However, if the investor is a collective enterprise, for example, a corporation, attributions of rationality become very tricky. For example, if the proximate decision-maker is an agent, then the decision rule may be "rational" from the agent's perspective, but not from the principal's.

that takes place within this complex, dynamic system. Existing firms assess their opportunities and competition, form expectations, and decide whether or not to participate in the next round of investment in a particular class of designs. (The designs themselves may be of hidden modules, architectural modules, or system architectures. However, it is in the nature of modular structures that the opportunities to design new hidden modules far outnumber the opportunities to design new architectural modules or new system architectures.)

Portfolio Investors and Entry into the Submarket

In an advanced capital market, however, there is yet another level of investor: specifically, there are investors who hold portfolios of corporations, and allocate capital across sectors in the greater economic system. This type of investor does not have information as detailed as that of the subindustry participants, like investors A and B. But a portfolio investor *can* observe aggregate or average returns, and allocate funds based on that information.

Moreover, enterprise designers can structure contracts that give them access to the financial resources of these portfolio investors. The enterprise designers can then use those resources in conjunction with financial operators to restructure industries and subindustries in ways that create or capture value for the portfolio investors. For example, using the resources of portfolio investors, enterprise designers can instigate value-seeking moves such as

- starting new firms;
- designing new employment contracts;
- shutting down enterprises;
- merging two enterprises;
- having one enterprise acquire another.

In this fashion, the enterprise designers can change some of the parameters of a subindustry equilibrium, even as the design-investment process and the product-market game are being played out.

How then will the corporate level and the portfolio level of investment interact? First, to portfolio investors, any above-market expected rate of return is an incentive to invest. Thus, if it happens that all firms in an industry or subindustry are experiencing above-normal profits, portfolio investors will want to invest in that sector. Their investments in publicly-traded stocks will drive up the market value of existing corporations in that subindustry. The high market values of those enterprises will then create opportunities for entrepreneurs to assemble a credible set of contracts and start new firms of that type.[28]

Entry by new firms into the subindustry, fueled by the funds of portfolio investors, will continue until all opportunities to earn above-normal returns disappear. Corporations already in these product markets and subindustries may not be able to stop this entry, for as long as the next investment opportunity is attractive, the portfolio investors and the enterprise designers will have every reason to snap up the opportunities.

Entry Deterrence

Incumbent firms can garner above-normal returns and simultaneously deter entry only if there is a large discontinuity in the value function—as, for example, the break between $k=8$ and $k=9$ shown in figure 15.6. Such breaks occur for low k (few entrants) when product or process designs are large, complex, and interdependent.[29] In those cases, the first system of a given type may have a high positive net present value, while the next has a large negative net present value.[30]

However, this is not the characteristic pattern of the value functions of most hidden modules. For most of the hypothetical value functions shown in plate 11.1, the value function forms a high peak and slopes gently downward as k, the number of experiments, increases. In other words, there is much room for additional entrants to prosper.

In this case, unless the incumbent has incontestable property rights to all new hidden-module designs, there is little that it can do to stop investment and the influx of entrants.[31] Absent such property rights, an incumbent might as well try to hold back the tide as stem the inflow of capital into the hidden-module submarkets of a

---

[28] See our discussion of this process in Chapter 4. This is the type of "arbitrage" envisioned by Modigliani and Miller (1958) in their initial paper on capital structure.

[29] In the literature on strategic positioning and entry barrier, this is called "lumpy" investment. Lumpy investments are large, irreversible, and indivisible, hence give rise to discontinuous value functions, with NPV < 0 for small k.

[30] Chandler (1977); Sutton (1991).

[31] It is conceivable that an architect who created a modular design might be deemed to have rights to the design of any hidden module that could be hooked into the system. Such rights would have to be vigorously enforced because they run counter to the natural structure of a modular system, which is highly decentralized. This is an example of John Sutton's equilibrium "law of stability" (Sutton 1998).

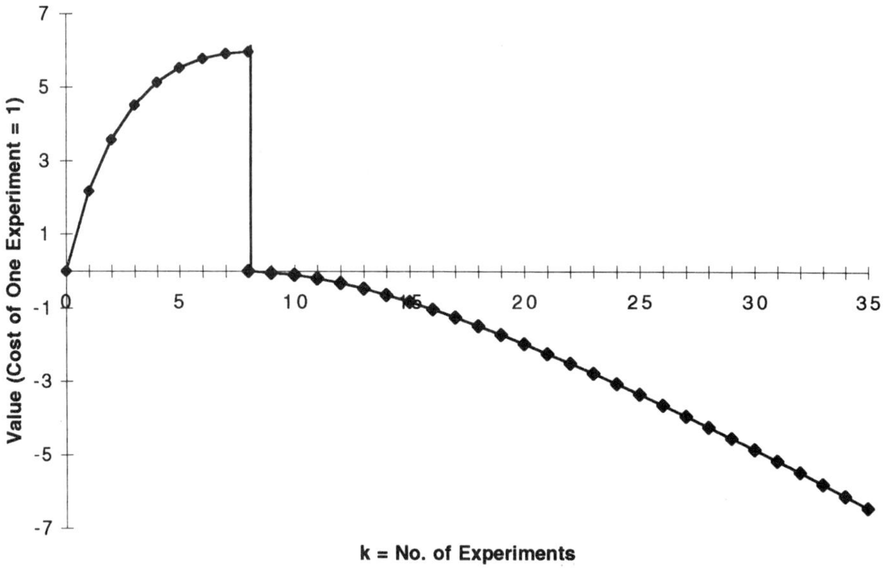

Figure 15.6 A value function with an entry-deterring discontinuity between $k=8$ and $k=9$.

modular design. The question is not, Will entry occur? but, How fast? and, When will it stop?

The Dynamics of Entry

The speed of entry in turn will be affected by the cost structure of experimentation, particularly the cost of forming new workgroups and enterprises that are capable of creating the designs in question. If the supply of qualified designers of a particular hidden module is elastic, and if the cost of forming and disbanding new corporations is low (as it is in the United States), then many new firms will be able to enter the design phase in parallel very rapidly. Conversely, if the supply of designers is inelastic, or the cost of forming new corporations is high, then experimentation will take place in series, and the number of entrants will be correspondingly lower.

How fast and furiously entry occurs in the product market (the second stage of this two-stage investment process) depends of course on how many firms enter the design phase. However, entry into the product market will also be affected by the technology of the experimental process itself, especially its lead-times. If being first to market confers an advantage in competitions among substitute designs, then the "winners" in the product market will be firms that excel in speed as much as quality.

It is even possible, in theory and in practice, to capture value via a strategy of "transient monopoly." That is, a hidden-module designer may bring its designs to market quickly, charge high prices while it faces no competition, and then exit when subsequent entrants cause prices to drop. (During the 1970s and 1980s, Intel Corporation was renowned for its ability to execute this strategy.)

Exit and Consolidation

Entry funded by portfolio investors will stop only when above-normal returns disappear. How long this takes depends on the time it takes to complete the relevant design and production processes. Investors cannot know industry-level profitability until the entire process is complete and the returns are received. Only then can they check their rewards against their initial expectations.

However, at the point when it is clear that the subindustry is no longer profitable in aggregate, three things will happen which will serve to bring the subindustry's returns back into line with average returns in the greater economy. First, a certain number of firms in the industry will have lost money in the last round of play. Many of them, like investor B, will see no prospect of making money on the next round, and should voluntarily exit.

Second, portfolio investors, who are conscious of aggregate or average performance, will note that the subindustry's returns are below normal. As they become aware of this fact, they will cease to fund new investments. They will also pull funds out of existing investments, thereby driving the stock market values of firms in the subindustry down. The net inflow of new funds to the industry will then dry up, and the value to enterprise designers of starting new firms will disappear.[32]

At that point, and not before, it will become profitable for incumbents to begin to merge and acquire one another, in order to consolidate their design processes and product markets. It does not make sense to consolidate a subindustry while external capital is still flowing in, because then any reduction in the number of competitors will most likely be offset by entry financed with external funds. But if external capital is not flowing into the industry, and if the weakest firms (B-type investors) are exiting voluntarily, or going bankrupt, then the strong and middling firms can make themselves stronger, and capture value for their investors, by combining with one another. Moreover, these firms will be "assisted" in their restructuring efforts by the

---

[32] Figuratively, as stock market values fall, the value landscape sinks. Enterprise designs that were previously value peaks will drop below sea level, becoming negative NPV projects.

same enterprise designers and financiers who, a short time before, were bringing new entrants into the subindustry![33]

Cycles of Expansion and Contraction in the Hidden-Module Subindustries

The process of design evolution at the artifact level of the economic system is capable of generating a rich series of opportunities to invest in new hidden-module designs. If investors do not have perfectly consistent expectations, then design evolution in hidden modules will give rise to a dynamic cycle of entry and exit into the hidden-module submarket, as shown in figure 15.7. (Readers should imagine the process taking place over a series of four different value functions, although to avoid clutter we show only one.)

The numbered points in figure 15.7 represent the portfolio investor's perceptions of the subindustry's average expected returns. The cycle begins at point 1—here the subindustry has a small number of firms, which in aggregate mount only ten experiments. The total value-surplus created by these experiments is quite large.[34] In the figure, most of the surplus is captured by users, but even so, the anticipated return on investment is above the required rate of return (defined by the horizontal axis).

This level of anticipated return acts as a beacon to portfolio investors, who, if the appropriate financial conduits exist, will fund new entrants (formed by entrepreneurs). Hence in the next round, the subindustry resides at point 2. Industry participants and even portfolio investors know that much entry has taken place (there are now twenty-five independent experiments underway) but the product market for hidden modules does not appear to be overly competitive. Expected returns are still high (though lower than before), and thus entry continues.

---

[33] In the U.S. economy, entry into attractive industries (like the greater computer cluster) and the restructuring of underperforming industries (like oil and railroads) increased dramatically in the 1980s. Both entry and exit were assisted by specialized financial firms—venture capitalists, leveraged buyout (LBO) organizations, and others—who in turn obtained their funds from large, professionally managed investment pools (portfolio investors) (see Chandler 1990; Jensen 1986, 1988, 1989).

[34] The expected value surplus at point 1 is around 200% of the cost of one experiment, giving an average expected rate of return of 20% above the cost of capital to each experiment. This is consistent with target rates of return for many venture capital firms. Of course, not all venture capitalists earn their target rates of return. Venture capitalists in turn get much of their money from large pension funds that seek to earn only a few points above the cost of capital, but would like to do so consistently. Again, not all do.

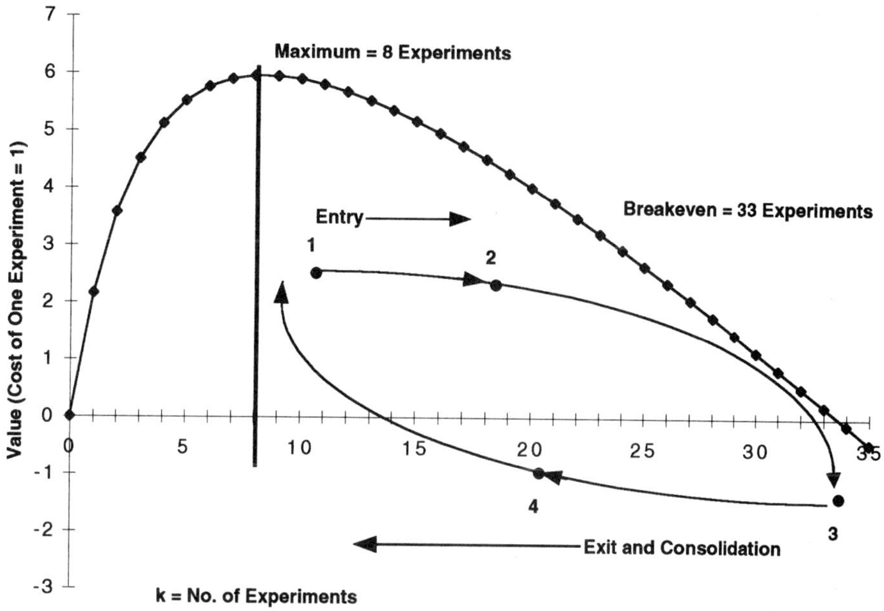

Figure 15.7 A dynamic cycle of entry, followed by exit and consolidation in a subindustry devoted to hidden-module design.

In the next round, the industry starts at point 3. The number of industry participants has grown larger, and the product market has turned sour. Even to the relatively uninformed portfolio investors, expected returns are now decidedly negative. At this point, the external funds spigot will be shut off, entry will stop, and exit and restructuring may begin.

At the next round, at point 4, some corporations will have exited, and others will have reduced their commitments to new design experiments. Hence, the number of independent design experiments will be much lower than before. But, as we have drawn the figure, portfolio investors are still very pessimistic about average returns. This could happen for a number of reasons: for instance, portfolio investors in general will not know the scope of the current opportunity (represented here by the height and breadth of the value function), nor how much net exit has actually occurred, nor how the reduced number of design experiments will affect pricing and profits in the product market. Thus exits and restructuring moves (mergers and acquisitions) will continue, leading the subindustry back to a point close to point 1.

A Spiral Pattern of Entry and Exit in Hidden-Module Subindustries

Thus we hypothesize that a cycle of entry and exit in a hidden-module subindustry can arise from the interaction of design evolution at the artifact level with decentralized value-seeking at the capital market level of the economic system. However, in understanding the nature of such cycles, and in interpreting figure 15.7, it is important for readers to bear in mind that the underlying space in which designers and portfolio investors are searching for value is much more complex than the subspace depicted in the figure. The cycle drawn in the figure is a projection on the two-dimensional subspace (defined by k,V) of a series of four "moves" in a complex value landscape.[35]

Viewed across time, the four points define a segment of a spiral process, wherein the subindustry progresses through a sequence of module designs (generations), and oscillates between high and low profitability. The particular shape, slope, and amplitude of the spiral will in turn depend on parameters like design lead-times, variation across value functions, and costs of entry, exit, and coordination in both the product and the capital markets. We hypothesize that such spirals exist, and suggest that they may prove to be a fruitful area for future modeling and empirical investigation.[36]

Industry Evolution

The dynamic process we have described is based on decentralized value-seeking by many persons, who act on the basis of local calculations of advantage (How do I and my enterprise benefit from this move?). This process is decentralized, and thus it conforms to our definition of an "evolutionary" adaptive process, as described in Chapter 9. However, in contrast to design evolution, which occurs at the artifact, design, and task levels of the economic system, *this* process takes place at the firm and market level. Moves in this process are the entry, exit, and restructuring (e.g.,

---

[35] In view of the complexity of the underlying value landscape, we have drawn the cycle as an open loop that does not return to its starting point. In the terminology of chaos theory, the cycles will "look and behave" like strange attractors. However, in our setting the "wobbling" is caused by uncertainty and imperfect coordination in a complex adaptive system, not by deterministic nonlinear dynamics.

[36] Merton and Bodie (1995) suggest that a similar spiral pattern marks the evolution of structures (financial institutions *and* markets) in the financial sector of the economic system.

by mergers and acquisition) of corporations. Over time, combinations of moves by different players in response to local calculations of value give rise to new industries and subindustries like those listed in chapters 1 and 14. For this reason, we label this process *industry evolution.*

In the computer industry, industry evolution overlays the process of design evolution, which we described in detail in previous chapters. Every new design does not give rise to a new firm, much less a new industry. However, "industry-level operators," involving entry, exit, and restructurings, do permit new firms and new product markets to be formed and old ones to be shut down when local conditions warrant it.

We do not have space in what remains of this volume to develop a theory of industry evolution in as much depth as we have developed our theory of design evolution. However, at least in the computer industry, there are notable parallels between industry evolution and design evolution, which may point in the direction of a general theory.

Like design evolution, industry evolution is a decentralized adaptive process that relies on human actors who see and seek value in a complex space of designs. Their search for value encompasses units of selection, a selection criterion, sources of variation, and selection mechanisms.

The units of selection in industry evolution are collective enterprises (both firms and markets); the selection criterion is economic value (specifically, capital market value); the sources of variation are actions (operators) that can be taken with respect to collective enterprises. The selection mechanisms in this process are the various methods used to evaluate and choose actions (e.g., the calculation of the net present value, resource allocation decisions by managers, or voting by shareholders). As in the case of new module-level tests, the selection mechanisms necessarily evolve in parallel with the enterprises themselves.

Finally, in the larger economic system there are strict design rules that divide property rights and thus partition effort and knowledge in a modular fashion. These rules govern

- the ownership of property;
- exchanges of property (transactions);
- exchanges of promises that have the status of property (contracts).

Anything, including knowledge, that is not property may flow quite freely among different parties in the economic system. Anything, including knowledge, that is property may not be so loosely passed around.

We cannot get significant dynamic behavior at the firm and industry level of an economy unless both the knowledge and the rights to form, disband, buy, or sell enterprises are themselves decentralized. Hence, for industries to evolve in significant ways, competitive capital markets and investors outside the "local group" are needed to give impetus to both entry and exit into submarkets. Industry evolution (under our definition) does not occur in centrally planned economies or in industries that are cozy oligopolies, with a small number of firms. As with design evolution, decentralized actions by many, independent, value-seeking actors are needed for the process of industry evolution to take hold.

Finally, when the submarkets of a modular cluster are subject to the forces of industry evolution, they will display rates of entry and exit in proportion to the opportunities generated by the underlying artifact design. Thus hidden-module subindustries will be "turbulent," displaying high rates of entry and exit. Architectural-module subindustries will be less turbulent, but changes in them will affect many firms focused on the design and production of hidden modules. Finally, global design rules will change very slowly indeed, but have vast implications for all the enterprises in the relevant modular cluster.

Appendix: Discussion of the Extreme Points in the Set of Rational Expectations Equilibriums

Case A—Monopoly/Monopoly

In the first instance **A**, a single firm, controls both investment opportunities and the product market. This firm maximizes its own net option value by mounting eight design experiments: when the experimental results are in, it will select the best (most valuable) design, and charge users as much as they are willing to pay for it. The surplus over the cost of experimentation flows into the hands of investors.

A value-seeking monopolist will not undertake as many design experiments as have positive net option values, because each new experiment above eight *cannibalizes* part of the value of those that preceded it. Thus the ninth experiment's *incremental value to the firm that owns the other eight* does not justify its cost.

Case B—Competition/Monopoly

But stopping at eight experiments leaves potentially valuable investments on the table. If other firms have access to the technology, *they* will have incentives to invest. For the sake of argument, assume eight experiments are underway, and the incumbent and potential entrants are equally good at generating designs. A potential entrant's expected net option value, given a winner-take-all payoff, is the value of winning times the probability of winning, where "winning" means having the best design. If the entrant is technologically equal to the incumbent, this expected payoff is the total net option value divided by the number of experiments.[37]

Because the net option value function is concave, *the potential entrant's expected payoff is always higher than the incumbent monopolist's payoff.* The monopolist takes account of the effect of the next experiment on those already underway, whereas an entrant has no such concerns. The concavity of the net option value function creates an interdependency across experiments: part of the "cost" of the next experiment is that it decreases the (average) value of those that went before it. The incumbent monopolist bears this "cost of cannibalization," whereas an entrant does not.

Therefore, an entrant will see valuable opportunities where an incumbent monopolist sees none. This relationship, which is typical of winner-take-all contests, may be summarized as follows:

$$\begin{array}{l}\text{Payoff to}\\ \text{entrant on}\\ k^{\text{th}}\text{ experiment}\end{array} = \frac{1}{k}\text{NOV}(k) > \text{NOV}(k) - \text{NOV}(k-1) = \begin{array}{l}\text{Payoff to}\\ \text{incumbent on}\\ k^{\text{th}}\text{ experiment}\end{array} \quad (15.1)$$

The value of "the next" experiment is positive as long as the net option value (NOV) function itself is positive. If the market for investment opportunities is competitive (case **B**), new firms will enter until the net option value function crosses the horizontal axis. In the particular case shown in figure 15.2, up to *thirty-three* firms will enter the search for a new design.

Because case **B** assumes a winner-take-all final product market, when the search is over, thirty-two of the firms will drop out. The last one (the one with the best design) will stay in and charge users for the full value added of the design. In this instance, the final product market will have one firm that will appear to be very profitable. But the investment process as a whole will be a breakeven proposition—what the investors make on the winning firm, they will lose on the rest of the experiments that fall short of the best.

In case **B**, as in **A**, the users pay "full value"—that is, as much as they are willing to pay—for the finished design. The design itself will be better than in case **A,** for it will be the best of thirty-three experimental design efforts, not eight. There is no surplus, however, for the value created by the design is completely offset by the cost of mounting the experiments.

Case C—Competition/Competition

Now suppose that there is free entry into the search for new designs, followed by competition in the product market, and that all potential entrants know that this is the case. In a rational expectations equilibrium, investors and potential entrants must make some assumption about what type of price competition will prevail in the product market and see what kind of return on investment such competition yields.

---

[37] In contrast, the payoff to the incumbent monopolist is the difference in the net option values function obtained by increasing the number of experiments by one.

The most stringent form of price competition is Bertrand pricing with no variable costs or quantity constraints. To see how Bertrand pricing works, suppose for a moment that thirty-three firms enter the search for new designs. After the fact, the designs can be ranked from highest to lowest value: $\{x_1, x_2, \ldots x_{33}\}$. With no variable cost, and no quantity (capacity) constraints, firms sponsoring designs have incentives to cut price until all but the top two contenders are forced to exit.

The equilibrium price will then be the difference between the best and the second-best design: $x_1 - x_2$. Investors who expect Bertrand-type competition in the aftermarket can forecast these prices. We can approximate those forecasts by using order-statistic distributions to compute the expectation of the highest and second highest draws in a sample of size k, and taking the difference of the two expectations. We can then calculate the net option value *captured by investors* for different numbers of entrants (experiments).

As it happens, if all firms are technically equal, the equilibrium number of experiments funded will be the same as in case **A**—eight. But the economic payoff to investors is quite different. Under Bertrand competition, the total surplus is transferred to users via lower prices.

Case D—Monopoly/Competition

In case **D**, no firm has exclusive rights to the design in the final product market. If the originator of new designs has no property rights that allow it to earn back the cost of experiments, then investors will have no incentives to invest in the design technology. The equilibrium number of experiments in that case is zero. This is a perverse outcome, because valuable opportunities will go begging. However, it is a realistic case when designs do not have the status of protected intellectual property.

Afterword

The year 1980 marks the end of a natural epoch in the history of computer designs and the computer industry. By this date, von Neumann's fundamental concept of the artifact—a composite system, made up of complementary "organs" and capable of running general programs—had been realized. Indeed, as we have shown, von Neumann's concept proved capable of being elaborated and extended in unexpected ways.

The principles of modularity, which are the focus of this work, emerged early in the history of computers, and were a recurring theme in technical discussions. Briefly stated, the principles of modularity are as follows:

- Create nested, regular, hierarchical structures in a complex system.
- Define independent components within an integrated architecture.
- Establish and maintain rigorous partitions of design information into hidden and visible subsets.
- Invest in clean interfaces and "good" module tests.

These principles gave designers a means of coping with the constantly growing complexity of computers. Conscious, well-founded divisions of effort and knowledge made it possible for human beings of individually limited capacity to design and produce ever more complex machines and programs. As a bonus, the resulting modular designs could change in unpredicted, yet coordinated ways. They could be improved via a decentralized, value-seeking process, which we have called *design evolution.*

In the United States, the modularity of computer designs came to be reflected in the surrounding systems of engineering design and industrial organization. First, in the 1960s and 1970s, beginning with System/360, computer system designs became "truly modular" for the first time. These systems, we have shown, were laden with valuable opportunities to design new hidden modules. As is natural in an advanced market economy, the perception of opportunities quickly led to the formation of new firms and new product markets. Thus, in the 1970s, clusters of firms and markets grew up around every successful modular computer system. Soon, moreover, the various separate clusters began to merge and overlap.

By 1980, what had emerged was a complex, expanding set of artifacts that reflected a dynamically evolving set of designs. The value of those designs in turn was lodged in two places. One place was IBM, a classic "modern corporation." The other was a large and growing modular cluster, made up almost entirely of new firms, which sold new types of products. Investors at that time judged that about half the capital market value of the industry belonged to the shareholders of IBM; the other half belonged to the shareholders of firms in the modular cluster.[1]

Once the computer industry attained this structure, value-seeking could begin to operate in new ways at many levels of the economic system. In a highly concentrated industry, the key players must be conscious of one another's actions, and must move together. However, a far more dynamic pattern of experimentation, entry, exit, and restructuring is possible within a modular cluster. First, as we have said, the designs of modules may evolve swiftly along independent paths. Second, new systems may be constructed opportunistically from preexisting modular components. Finally, firms that make modules and markets that trade them can also develop swiftly and independently of one another.

In volume 2, we shall argue that the modular cluster type of industry structure gave rise to an especially rapid and turbulent form of design and industry evolution. Design evolution generated new artifacts, designs, and task structures; industry evolution generated new enterprises, contracts, and industry structures. Each of these decentralized, dynamic processes affected the other. In biology, this type of interaction is called *coevolution.*[2] When complex designs are organized as modular structures, and enterprises are organized into modular clusters, then the designs and the industry that supplies them will coevolve. The coevolution of designs and enterprises in turn may span all levels of the economic system, as is shown in the figure.

---

[1] This calculation of value share includes only the value of publicly traded firms then in existence. The value of yet-to-be founded firms is not counted. This creates a potential asymmetry between the valuation of IBM (which includes future opportunities) and that of the modular cluster (which excludes many future opportunities).

[2] Lumsden and Wilson (1981); Edelman (1992); Plotkin (1993); Kauffman (1993); Nelson (1995); Deacon (1997); Wilson (1998).

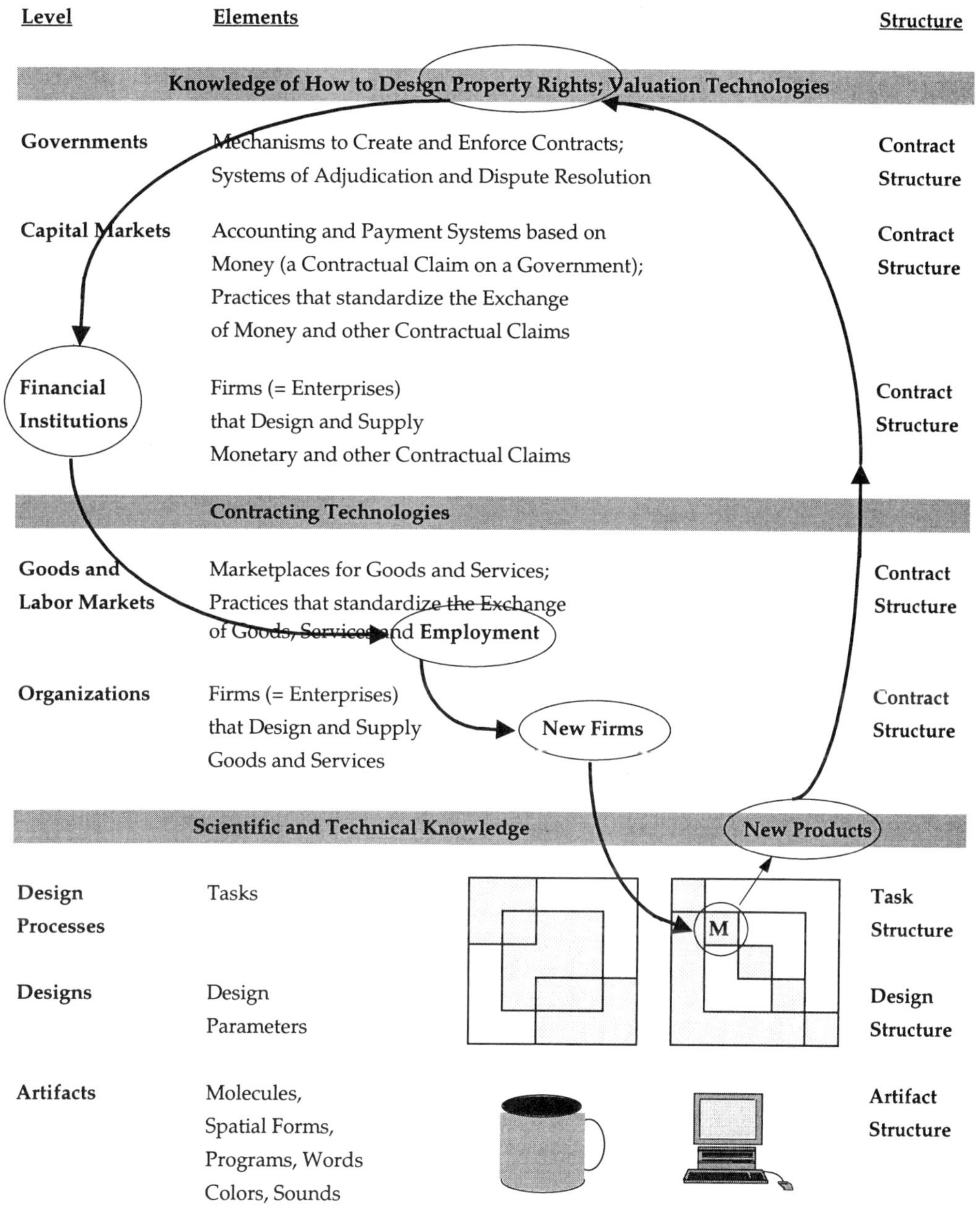

Figure 1 Coevolutionary dynamics across the levels of the economic system.

The figure depicts an interrelated cascade of potential adaptations. First, through splitting or augmentation, a new module arises in one or more computer systems. Once module tests are sufficiently well defined, we have said, that module can become a product in the larger economic system. If circumstances are favorable, the arrival of a new product can lead to the formation of new firms. First, however, *property rights* to the module and its designs must be clarified. (Property rights to the pieces of modular systems are inherently ambiguous; thus legal processes must be invoked to answer questions about who owns what.)

New firms usually need financing. Hence if they are large enough and special enough as a group, their needs may justify the creation of new financial contracts, institutions, and markets. Taken together, a specialized set of contracts, institutions, and markets forms a *conduit* between investors and opportunities. Thus the formation of new firms may initiate the collateral formation of new financial conduits in the capital markets.

New firms also need people to carry out their work (to implement their task structures). Firms compete for people in a variety of labor markets, and they can gain advantage by creating employment and service contracts that match the desires of the people they want to employ and retain. Hence the formation of new firms will also create an environment that rewards experimentation and innovation in employment and service contracts.

Finally, firms compete on the basis of how well they work. Thus they have reason to create and try out new guidance mechanisms for collective enterprises. These guidance mechanisms, we have said, include include management tools, such as incentive and information systems, and methods of building up technical and scientific knowledge. The purpose of guidance mechanisms is to enable people working in enterprises that are insulated from day-to-day market forces to see and seek market value more effectively.

This brings the process of "adaptive coevolution" full cycle. As we have sketched it, the process begins with a *new* artifact—a *module*—which, by virtue of its design, can be created separately from, but can still function within, a larger system. The process ends with *new firms* that make and sell modules; they in turn will be supported by new markets, new contracts, new financing conduits, and, possibly, newly defined property rights. These new firms will then compete on the basis of their product designs, contract designs, and overall enterprise designs.

In volume 2, we will argue that, in an advanced market economy, powerful dynamic forces can arise through the interaction of modular artifact designs with the "modular cluster" type of industry structure. In the computer industry, however, these forces had just begun to operate in 1980. Ahead lay new artifacts: personal computers, workstations, networks, and the Internet. Also in the future were new patterns of competition, including competition among firms that owned visible information (architects), and competition among architects and hidden module designers. Finally, at the financial level, the future path of the industry would lead to the formation of a cluster of over a thousand firms, valued (in 1996) at slightly less than $1 trillion.

Beginning in the 1980s, and continuing through today, financial conduits, property rights, employment contacts, and management practices all came to be adapted to work with the new modular artifact designs and corresponding task structures. So well adapted are the organizations to the artifacts today that it is tempting to declare that the cluster form of organization is "ideal" for the job of allocating resources and dividing knowledge, effort, and rewards among the many contributors to a large and complex modular design.

Indeed, the evolved industry structure *is* well adapted to the underlying artifact designs. However, the artifact designs were themselves influenced by the evolving industry structure. And so in the U.S. computer industry, the designs, task structures, contract structures, and industry structures have all evolved together, and each is in some measure a product of the others.

In this volume, we have developed the theory of design evolution in some detail, but we have only made a start at understanding competition and industry evolution in modular clusters. Thus, our aim in volume 2 is to explain the elemental structures of the modular cluster and how and why those structures have changed over the last two decades. As in volume 1, we will continue to view the phenomenon through the theoretical lens of complex adaptive systems, and we will continue to assume that human designers are capable of "seeing and seeking" value as it is defined in the greater economic system.

Overall, we believe that *modularity* and *evolution* are the two main tenets that together explain the designs of computers and the shape of the computer industry today. These tenets describe an architecture and a dynamic process that our society uses to organize value-seeking in what is a very large and complex space of artifact and enterprise designs. The architecture of modularity partitions designers' efforts, and efficiently coordinates their actions. The process of evolution directs their searches, and appropriately rewards their efforts. Today, this highly evolved, decentralized social framework allows tens of thousands of computer designers to work on complex, interrelated systems in a decentralized, yet highly coordinated way. These designers create a never-ending, ever-changing stream of new artifacts, which are testimony to the power of modularity in the world.

Bibliography

Abernathy, W. J. (1978). *The Productivity Dilemma.* Baltimore: Johns Hopkins University Press.

Abernathy, W. J., and K. B. Clark (1985). "Innovation: Mapping the Winds of Creative Destruction." *Research Policy* 14: 3–22.

Abernathy, W. J., and J. M. Utterback (1978). "Patterns of Industrial Innovation." *Technology Review* 80: 40–47.

Adams, W., and J. W. Brock (1982). "Integrated Monopoly and Market Power: System Selling, Compatibility Standards, and Market Control." *Quarterly Review of Economics and Business* 22(4): 29–42.

Aghion, P., and J. Tirole (1994). "On the Management of Innovation." *Quarterly Journal of Economics* 109(4): 1185–1209.

Alchian, A. (1950). "Uncertainty, Evolution and Economic Theory." *Journal of Political Economy* 58: 211–221.

Alchian, A. (1969). "Corporate Management and Property Rights in Economic Policy and the Regulation of Corporate Securities," in ed. H. G. Manne, *Economic Policy and Regulation of Corporate Securities,* Washington, D.C.: American Enterprise Institute for Public Policy Research, 337–360.

Alchian, A. A., and H. Demsetz (1972). "Production, Information Costs, and Economic Organization." *American Economic Review* 62(5): 777–895.

Alchian, A., and H. Demsetz (1973). "The Property Rights Paradigm." *Journal of Economic History* 33: 16–27.

Alexander, C. (1964). *Notes on the Synthesis of Form.* Cambridge, Mass.: Harvard University Press.

Alic, J. A., M. H. Harris, and R. R. Miller. (1983). "Microelectronics: Technological and Structural Change." Paper presented at the Conference on Microelectronics in Transition: Transformation and Social Change, University of California, Santa Cruz.

Alic, J. A., L. Branscomb, H. Brooks, A. Carter, and G. Epstein (1992). *Beyond Spinoff: Military and Commercial Technologies in a Changing World.* Boston: Harvard Business School Press.

Amdahl, G. M., G. A. Blaauw, and F. P. Brooks, Jr. (1964). "The Architecture of the IBM System/360." *IBM Journal of Research and Development* 8(April): 87–101.

Amram, M., and N. Kulatilaka (1998). *Real Options: Managing Strategic Investment in an Uncertain World.* Boston, Mass.: Harvard Business School Press.

Anand, B. N., and T. Khanna (1997). "Intellectual Property Rights and Contract Structure." Boston: *Harvard Business School Working Paper* 97–016.

Anchordoguy, M. (1989). *Computers Inc.* Cambridge, Mass.: Harvard University Press.

Anderson, P., and M. L. Tushman (1990). "Technological Discontinuities and Dominant Designs: A Cyclical Model of Technological Change." *Administrative Science Quarterly* 35: 604–633.

Andrews, M. (1994). *C++ Windows NT Programming.* New York: M&T Books.

Argyris, C. (1990). *Overcoming Organizational Defenses.* Boston: Allyn and Bacon.

Argyris, E. (1991). "Teaching Smart People How to Learn." *Harvard Business Review.* (May/June): 99–109.

Aron, J. D., F. P. Brooks, B. O. Evans, J. W. Fairclough, A. Finerman, B. A. Galler, W. P. Heising, W. T. Johnson, and N. Stern (1983). "Discussion of the SPREAD Report, June 23, 1982." *Annals of the History of Computer* 5(1): 27–44.

Arrow, K. J. (1964). "The Role of Securities in the Optimal Allocation of Risk Bearing." *Review of Economic Studies* 31: 91–96.

Arthur, W. B. (1988). "The Economy and Complexity" in *Lectures in the Sciences of Complexity: SFI Studies in the Sciences of Complexity,* ed. D. Stein. Reading, Mass.: Addison-Wesley.

Arthur, W. B. (1989). "Competing Technologies: Increasing Returns, and Lock-In by Historical Events." *Economic Journal* 99(394): 116–131.

Ashby, W. R. (1952, 1960). *Design for a Brain.* New York: John Wiley and Sons.

Aspray, W., and A. Burks, eds. (1987). *The Papers of John von Neumann.* Cambridge, Mass.: MIT Press.

Athey, S., and A. Schmutzler (1995). "Product and Process Flexibility in an Innovative Environment." *RAND Journal of Economics* 26(4): 557–574.

Athey, S., and S. Stern (1996). "An Empirical Framework for Testing Theories About Complementaries in Organizational Design." Cambridge, Mass.: National Bureau of Economic Research.

Ayres, I., and P. Klemperer (1995). *Are Uncertainty and Delay in Intellectual Property Litigation Desirable?* New Haven, Conn.: Yale Law School.

Ayres, I., and B. J. Nalebuff (1997). *Common Knowledge as a Barrier to Negotiation.* New Haven, Conn.: Yale Law School.

Bachelier, L. (1900). *Theory of Speculation.* Faculty of Sciences. Paris, France, Academy of Paris.

Baetjer, H. J. (1998). *Software as Capital.* Los Alimitos, Calif.: IEEE Computer Society Press.

Baker, G. P., R. Gibbons, K. J. Murphy (1997). *Implicit Contracts and the Theory of the Firm.* Cambridge, Mass.: National Bureau of Economic Research.

Baker, G. P., M. C. Jensen, and K. J. Murphy (1988). "Compensation and Incentive: Theory and Practice." *Journal of Finance* 43:593–616.

Baker, G. P., and G. D. Smith (1998). *The New Financial Capitalists: Kohlberg, Kravis, Roberts and the Creation of Corporate Value.* Cambridge, U.K.: Cambridge University Press.

Baldwin, C. Y. (1983). "Productivity and Labor Unions: An Application of the Theory of Self-Enforcing Contracts." *Journal of Business* 56(2): 155–185.

Baldwin, C. Y., and K. B. Clark (1992). "Capabilities and Capital Investment: New Perspectives on Capital Budgeting." *Journal of Applied Corporate Finance* 5(2): 67–82.

Baldwin, C. Y., and K. B. Clark (1993). "Modularity and Real Options." *Harvard Business School Working Paper* 93–026.

Baldwin, C. Y., and K. B. Clark (1994). "Capital Budgeting Systems and Capabilities Investments in U.S. Companies after World War II." *Business History Review* 68: 73–109.

Baldwin, C. Y., and K. B. Clark (1997a). "Managing in an Age of Modularity." *Harvard Business Review* (September/October): 84–93.

Baldwin, C. Y., and K. B. Clark (1997b). "Sun Wars—Competition within a Modular Cluster," in *Competing in the Age of Digital Convergence,* ed. D. B. Yoffie. Boston: Harvard Business School Press.

Band, J. (1995). "Proprietary Rights." *Computer Lawyer* 12(6): 1–8.

Barnard, C. I. (1938). *The Functions of an Executive.* Cambridge, Mass.: Harvard University Press.

Bashe, C. J., L. R. Johnson, J. H. Palmer, and E. W. Pugh (1986). *IBM's Early Computers.* Cambridge, Mass.: MIT Press.

Bauer, R. A., E. Collar, and V. Tang with J. Wind and P. Houston (1992). *The Silverlake Project: Transformation at IBM.* New York: Oxford University Press.

Bean, A. S., D. Schiffel, and M. Mogee (1975). "The Venture Capital Market and Technological Innovation." *Research Policy* 4: 380–408.

Beggs, A., and P. Klemperer (1992). "Multi-Period Competition with Switching Costs." *Econometrica* 60(3): 651–666.

Bell, C. G., and J. C. Mudge (1978). "The Evolution of the PDP-11," in *Computer Engineering: A DEC View of Hardware Systems Design,* ed. C. G. Bell, J. C. Mudge, and J. E. McNamara. Maynard, Mass.: Digital Press.

Bell, C. G., and J. E. McNamara (1978). "The PDP-8 and Other 12-Bit Computers," in *Computer Engineering: A DEC View of Computer Engineering,* ed. C. G. Bell, J. C. Mudge, and J. E. McNamara. Maynard, Mass.: Digital Press.

Bell, C. G., R. Cady, H. McFarland, B. A. Delagi, J. F. O'Loughlin, R. Noonan, and W. A. Wulf (1970). "A New Architecture for Minicomputers—The DEC PDP-11,'' reprinted in C. G. Bell, J. C. Mudge, and J. E. McNamara, eds. (1978), *Computer Engineering: A DEC View of Hardware Systems Design.* Maynard, Mass.: Digital Press.

Bell, C. G., G. Butler, R. Gray, J. E. McNamara, D. Vonada, and R. Wilson (1978). "The PDP-1 and Other 18-Bit Computers," in *Computer Engineering: A DEC View of Hardware Systems Designs,* ed. C. G. Bell, J. C. Mudge, and J. E. McNamara. Maynard, Mass.: Digital Press.

Bell, C. G., and A. Newell, eds. (1971). *Computer Structures: Readings and Examples.* New York: McGraw-Hill.

Bell, C. G., J. C. Mudge, and J. E. McNamara, eds. (1978). *Computer Engineering: A DEC View of Hardware Systems Design.* Maynard, Mass.: Digital Press.

Benavides, D., J. Duley, and B. Johnson (1998). "Financial Evaluation of Alternative Semiconductor Manufacturing Capacity Design and Deployment Policies: Summary of Results." Mimeo.

Bender, D. (1995). "Lotus v Borland Appeal—On-Screen Program Menus Not Copyright-Protected." *Computer Law & Practice* 11(3): 71–83.

Bergin, T. J., and R. G. Gibson, eds. (1996). *History of Programming Languages.* Reading, Mass.: Addison-Wesley.

Berk, J., R. Green, and V. Naik (1997). "Valuation and Return Dynamics of R&D Ventures." Mimeo.

Berle, A. A., and G. C. Means (1932). *The Modern Corporation and Private Property.* New York: Macmillan.

Bernanke, B. S. (1983). "Irreversibility, Uncertainty, and Cyclical Investment." *Quarterly Journal of Economics* 300: 85–106.

Bernstein, P. L. (1992). *Capital Ideas: The Improbable Origins of Modern Wall Street.* New York: Free Press.

Bernstein, P. L. (1996). *Against the Gods: The Remarkable Story of Risk.* New York: John Wiley and Sons.

Blaauw, G. A., and F. P. Brooks, Jr. (1964). "The Structure of System/360: Part I—Outline of the Logical Structure," reprinted in C. G. Bell and A. Newell, eds. (1971), *Computer Structures: Readings and Examples.* New York: McGraw-Hill, 588–601.

Black, F., and M. Scholes (1973). "The Pricing of Options and Corporate Liabilities." *Journal of Political Economy* 81: 637–654.

Blout, E., ed. (1996). *The Power of Boldness: Ten Master Builders of American Industry Tell Their Success Stories.* Washington, D.C.: Joseph Henry Press.

Bodie, Z., and R. C. Merton (1995). "The Informational Role of Asset Prices," in *The Global Financial System,* D. B. Crane et al., eds. Boston: Harvard Business School Press.

Bodie, Z., and R. C. Merton (1999). *Finance.* New Jersey: Prentice-Hall.

Bodo, M. (1994). *Hard Drive Bible,* 7th ed. Santa Clara, Calif.: Corporate Systems Center.

Bohn, R. E., and R. Jaikumar (1986). "The Development of Intelligent Systems for Industrial Use: An Empirical Investigation," in *Research on Technological Innovation, Management and Policy,* ed. R. S. Rosenbloom. Greenwich, Conn.: JAI Press.

Bolton, P., and D. Scharfstein (1990). "A Theory of Predation Based on Agency Problems in Financial Contracting." *American Economics Review* 53: 94–106.

Bower, J. L. (1970). *Managing the Resource Allocation Process.* Homewood, Ill.: Richard D. Irwin.

Braudel, F. (1979a). *The Perspective of the World.* New York: Harper & Row.

Braudel, F. (1979b). *The Structures of Everyday Life: The Limits of the Possible.* New York: Harper & Row.

Braudel, F. (1979c). *The Wheels of Commerce.* New York: Harper & Row.

Breeden, D. T. (1979). "An Intertemporal Asset Pricing Model with Stochastic Consumption and Investment Opportunities." *Journal of Financial Economics* 7: 265–296.

Bresnahan, T. F. (1999). "New Modes of Competition and the Future Structure of the Computer Industry," in *Competition, Convergence, and the Microsoft Monopoly:* A Kluwer Press Progress and Freedom Foundation Volume.

Bresnahan, T. F., and S. Greenstein (1995). "The Competitive Crash in Large-Scale Commercial Computing," in *Growth and Development, The Economics of the 21st Century,* ed. R. Landau, N. Rosenberg, and T. Taylor. Stanford, Calif.: Stanford University Press.

Bresnahan, T. F., and S. M. Greenstein (1997). "Technical Progress and Co-Invention in Computing and the Use of Computers." Washington, D.C.: *Brookings Papers on Economics Activity,* Micro Economics.

Bresnahan, T. F., and S. M. Greenstein (1999). "Technological Competition and the Structure of the Computer Industry." *Journal of Industrial Economics* 47(1): 1–40.

Bresnahan, T. F., and P. C. Reiss (1994). "Measuring the Importance of Sunk Costs." *Annales D'Economie et de Statistique* 31: 183–217.

Brock, G. W. (1975). *The U.S. Computer Industry: A Study of Market Power.* Cambridge, Mass.: Ballinger.

Brooks, F. P., Jr. (1995). *The Mythical Man-Month,* 2nd ed. Reading, Mass.: Addison-Wesley.

Brooks, F. P., Jr. (1987). "No Silver Bullet." *IEEE Computer 20*(April): 10–19.

Brown, G. S. (1969). *Laws of Form.* London: George Allen and Unwin.

Brown, K. H., and S. M. Greenstein (1995). *How Much Better is Bigger, Faster & Cheaper? Buyer Benefits From Innovation in Mainframe Computers in the 1980s.* Mimeo. Stanford, Calif.: Stanford University Center for Economic Policy Research.

Bucciarelli, L. L. (1994). *Designing Engineers.* Cambridge, Mass.: MIT Press.

Burgelman, R. A. (1994). "Fading Memories: A Process Theory of Strategic Business Exit in Dynamic Environments." *Administrative Science Quarterly* 39(March): 24–57.

Burks, A. W. (1987). Foreword to *The Papers of John von Neumann,* ed. W. Aspray and A. Burks. Cambridge, Mass.: MIT Press.

Burks, A. W., H. H. Goldstine, and J. von Neumann (1946). "Preliminary Discussion of the Logical Design of an Electronic Computing Instrument," reprinted in C. G. Bell and A. Newell, eds. (1971), *Computer Structures: Readings and Examples.* New York: McGraw-Hill, pp. 92–119.

Bygrave, W. D., and J. A. Timmons (1992). *Venture Capital at the Crossroads.* Boston: Harvard Business School Press.

Chandler, A. D., Jr. (1962). *Strategy and Structure.* Cambridge, Mass.: MIT Press.

Chandler, A. D., Jr. (1977). *The Visible Hand: The Managerial Revolution in American Business.* Cambridge, Mass.: Harvard University Press.

Chandler, A. D., Jr. (1990). *Scale and Scope: The Dynamics of Industrial Capitalism.* Cambridge, Mass.: Harvard University Press.

Chandler, A. D., Jr. (1997). "The Computer Industry: The First Half Century," in *Competing in the Age of Digital Convergence,* ed. D. B. Yoffie. Boston: Harvard Business School Press.

Chesbrough, H. W. (1999). "The Differing Organizational Impact of Technological Change: A Comparative Theory of National Institutional Factors," *Industrial and Corporate Change,* 8(3): forthcoming.

Chesbrough, H. W. (1999). "Assembling the Elephant: A Review of Empirical Studies on the Impact of Technical Change on Competing Firms." Boston: *Harvard Business School Working Paper* 99–104.

Chou, C., and O. Shy (1990). "Network Effects Without Network Externalities." *International Journal of Industrial Organization* 8: 259–270.

Christensen, C. M. (1993). "The Rigid Disk Drive Industry: A History of Commercial and Economic Turbulence." *Business History Review* 67: 531–588.

Christensen, C. M. (1997). *The Innovator's Dilemma: When New Technologies Cause Great Firms to Fail.* Boston: Harvard Business School Press.

Christensen, C. M., and R. S. Rosenbloom (1995). "Explaining the Attacker's Advantage: Technological Paradigms, Organizational Dynamics, and the Value Network." *Research Policy* 24: 233–257.

Church, J., and N. Gandal (1992). "Network Effects, Software Provision, and Standardization." *Journal of Industrial Economics* 40(1): 85–103.

Church, J., and N. Gandal (1993). "Complementary Network Externalities and Technological Adoption." *International Journal of Industrial Organization* 11: 239–260.

Clark, K. B. (1985). "The Interaction of Design Hierarchies and Market Concepts in Technological Evolution." *Research Policy* 14(5): 235–251.

Clark, K. B. (1987). "Investment in New Technology and Competitive Advantage," in *The Competitive Challenge,* ed. D. J. Teece. Cambridge, Mass.: Ballinger, 59–81.

Clark, K. B. (1988a). "Knowledge, Problem Solving and Innovation in the Evolutionary Firm." Unpublished manuscript, Harvard Business School.

Clark, K. B. (1988b). "Managing Technology in International Competition: The Case of Product Development in Response to Foreign Entry," in *International Competitiveness,* ed. A. M. Spence and H. A. Hazard. Cambridge, Mass.: Ballinger, 27–84.

Clark, K. B. (1989). "Project Scope and Project Performance: The Effect of Parts Strategy and Supplier Involvement on Product Development." *Management Science* 35(10): 1247–1263.

Clark, K. B. (1995). "Notes On Modularity in Design and Innovation in Advanced Ceramics and Engineering Plastics," Paper presented at the Meetings of the American Economics Association, New York, N. Y.

Clark, K. B., and T. Fujimoto (1989). "Lead Time in Automobile Product Development: Explaining the Japanese Advantage." *Journal of Engineering and Technology Management* 6: 25–58.

Clark, K. B., and R. M. Henderson (1990). "Architectural Innovation: The Reconfiguration of Existing Systems and the Failure of Established Firms." *Administrative Science Quarterly* 35: 9–30.

Clark, K. B., and E. Rothman (1986). "Management and Innovation: The Evolution of Ceramic Packaging for Integrated Circuits," in *High Technology Ceramics: Past, Present and Future,* ed. W. D. Kingery, The American Ceramic Society, 335–349.

Clarke, D. L. (1978). *Analytical Anchaeology* 2nd ed. rev. by B. Chapman. New York: Columbia University Press.

Coase, R. H. (1937). "The Nature of the Firm." *Economica* 4(November): 386–405.

Constantine, L. L., G. J. Myers, and W. P. Stevens (1974). "Structured Design." *IBM Systems Journal* 13(2): 115–139.

Cortada, J. W. (1987). *Historical Dictionary of Data Processing—Organizations.* New York: Greenwood Press.

Cox, J. C., and S. A. Ross (1976). "The Valuation of Options for Alternative Stochastic Processes." *Journal of Financial Economics* 3(January-March): 145–146.

Cox, J. C., J. E. Ingersoll, and S. A. Ross (1985a). "An Intertemporal General Equilibrium Model of Asset Prices." *Econometrica* 53(March): 363–384.

Cox, J. C., J. E. Ingersoll, and S. A. Ross (1985b). "A Theory of the Term Structure of Interest Rates." *Econometrica* 53(March): 385–408.

Crane, D. B., K. A. Froot, S. P. Mason, R. C. Merton, A. F. Perold, E. R. Sirri, and P. Tufano (1995). *The Global Financial System: A Functional Perspective.* Boston: Harvard Business School Press.

Crosby, A. W. (1997). *The Measure of Reality: Quantification and Western Society 1250–1600.* New York: Cambridge University Press.

Cusumano, M. A. (1991). *Japan's Software Factories—A Challenge to US Management.* New York: Oxford University Press.

Cusumano, M. A., and R. W. Selby (1995). *Microsoft Secrets.* New York: Free Press.

Dam, K. W. (1995). "Some Economic Considerations in the Intellectual Property Protection of Software." *Journal of Legal Studies* 24: 321–378.

Darwin, C. (1859). *On the Origin of Species by Means of Natural Selection.* London. Reprint, 1990, The Folio Society.

Davis, R., P. Samuelson, M. Kapor, and J. Reichman (1996). "A New View of Intellectual Property and Software." *Communications of the ACM* 39(3): 21–30.

Dawkins, R. (1989). *The Selfish Gene,* 2nd ed. New York: Oxford University Press.

Dawkins, R. (1996). *The Blind Watchmaker: Why the Evidence of Evolution Reveals a Universe without Design.* New York: W. W. Norton.

Deacon, T. W. (1997). *The Symbolic Species: The Co-Evolution of Language and the Brain.* New York: W. W. Norton.

Debreu, G. (1959). *Theory of Value.* New York: John Wiley and Sons.

DeLamarter, R. T. (1986). *Big Blue: IBM's Use and Abuse of Power.* New York: Dodd, Mead.

Demsetz, H. (1967). "Toward a Theory of Property Rights." *American Economic Review* 57: 347–359.

Demsetz, H. (1997). "The Firm in Economic Theory: A Quiet Revolution." *American Economic Review* 87(2): 426–429.

Deutsch, D. (1997). *The Fabric of Reality.* New York: Penguin Books.

Dixit, A. (1989). "Entry and Exit Decisions under Uncertainty." *Journal of Political Economy* 97(3): 620–638.

Dixit, A. K., and R. S. Pindyck (1994). *Investment under Uncertainty.* Princeton, N.J.: Princeton University Press.

Dixit, A. K., and R. S. Pindyck (1995). "The Options Approach to Capital Investment." *Harvard Business Review* 105–115.

Dominguez, J. R. (1974). *Venture Capital.* Lexington, Mass.: Lexington Books.

Dorfman, N. S. (1986). *Innovation and Market Structure: Lessons from the Computer and Semiconductor Industries.* Cambridge, Mass.: Ballinger.

Dosi, G. (1986a). "Sources, Procedures and Microeconomic Effects of Innovation," *Journal of Economic Literature* 26: 1120–1171.

Dosi, G. (1986b). "Technological Paradigms and Technological Trajectories: A Suggested Interpretation of the Determinants and Directions of Technical Change." *Research Policy* 11: 147–162.

Dubinskas, F. A. (1986). *VLSI Technology, Inc.: Automating ASIC Design.* Boston: Harvard Business School Press.

Dyson, G. B. (1997). *Darwin among the Machines.* Reading, Mass.: Addison-Wesley.

Economides, N. (1996a). "Network Externalities, Complementarities, and Invitations to Enter," *European Journal of Political Economy* 12: 211–232.

Economides, N. (1996b). "The Economics of Networks." *International Journal of Industrial Organization* 14(6): 673–699.

Economides, N. (1999). "Quality Choice and Vertical Integration." *International Journal of Industrial Organization,* forthcoming.

Economides, N., and S. C. Salop (1992a). "Competition and Integration among Complements and Network Market Structure." *Journal of Industrial Economics* 40(1): 105–123.

Edelman, G. M. (1992). *Bright Air, Brilliant Fire.* New York: Basic Books.

Einhorn, M. A. (1992). "Mix and Match Compatibility with Vertical Product Dimensions." *RAND Journal of Economics* 23(4): 535–547.

Eppinger, S. D. (1991). "Model-Based Approaches to Managing Concurrent Engineering." *Journal of Engineering Design* 2(4).

Eppinger, S. D. (1997). "A Planning Method for Integration of Large-Scale Engineering Systems." Presented at International Conference on Engineering Design.

Eppinger, S. D., D. E. Whitney, R. P. Smith, and D. Gebala (1994). "A Model-Based Method for Organizing Tasks in Product Development." *Research in Engineering Design* 6(1): 1–13.

Eppinger, S. D., M. V. Nukala, and D. E. Whitney (1997). "Generalised Models of Design Iteration Using Signal Flow Graphs." *Research in Engineering Design* 9: 112–123.

Evans, R. O. (1983). "Introduction to the SPREAD Report." *Annals of the History of Computing* 5(1): 4–5.

Fama, E. F. (1965). "The Behavior of Stock Market Prices." *Journal of Business* 38: 34–105.

Fama, E. F. (1970). "Efficient Capital Markets: A Review of Theory and Empirical Work." *Journal of Finance* 25: 383–417.

Fama, E. F. (1978). "The Effects of a Firm's Investment and Financing Decisions on the Welfare of Its Securityholders." *American Economic Review* 68: 272–284.

Fama, E. F. (1980). "Agency Problems and the Theory of the Firm." *Journal of Political Economy* 88: 288–307.

Fama, E. F., and M. C. Jensen (1983a). "Separation of Ownership and Control." *Journal of Law and Organization* 26(2): 301–326.

Fama, E. F., and M. C. Jensen (1983b). "Agency Problems and Residual Claims." *Journal of Law and Economics* 26(2): 327–329.

Fama, E. F., and M. H. Miller (1972). *The Theory of Finance.* New York: Holt, Rinehart and Winston.

Farrell, J., and G. Saloner (1986a). "Economic Issues in Standardization," in *Telecommunications and Equity: Policy Research Issues,* ed. J. Miller, Amsterdam: Elsevier Science. 165–177.

Farrell, J., and G. Saloner (1986b). "Installed Base and Compatibility: Innovation, Product Preannouncements, and Predation." *American Economic Review* 76(5): 940–955.

Farrell, J., and C. Shapiro (1988). "Dynamic Competition with Switching Costs." *RAND Journal of Economics* 19(1): 123–137.

Farrell, J., and C. Shapiro (1990). "Asset Ownership and Market Structure in Oligopoly." *RAND Journal of Economics* 21(2): 275–292.

Farrell, J., H. K. Monroe, and G. Saloner (1998). "The Vertical Organization of Industry: Systems Competition versus Component Competition." *Journal of Economics & Management Strategy* 7(2): 143–182.

Fenn, G., N. Liang, and S. Prowse (1995). *The Economics of the US Private Equity Market.* Washington, D.C.: Board of Governors of the Federal Reserve System.

Ferguson, C. H. (1985). *American Microelectronics in Decline: Evidence, Analysis and Alternatives.* Cambridge, Mass.: MIT Press.

Ferguson, C. H. (1990). "Computers and the Coming of the US Keiretsu." *Harvard Business Review* 90(4): 56–80.

Ferguson, C. H., and C. R. Morris (1993). *Computer Wars: How the West Can Win in a Post-IBM World.* New York: Times Books.

Fershtman, C., and K. L. Judd (1987). "Equilibrium Incentives in Oligopoly." *American Economic Review* 77(5): 927–940.

Fine, C. H. (1998). *Clockspeed: Winning Industry Control in the Age of Temporary Advantage* Reading, Mass.: Perseus Books.

Fisher, F., J. J. McGowan, and J. E. Greenwood (1983). *Folded, Spindled and Mutilated: Economic Analysis and U.S. vs. IBM.* Cambridge, Mass.: MIT Press.

Fisher, M., K. Ramdas, and K. Ulrich (1997). "Component Sharing in the Mangement of Product Variety." *Wharton School Working Paper.*

Fishman, K. D. (1981). *The Computer Establishment.* New York: Harper & Row.

Flamm, K. (1988). *Creating the Computer: Government, Industry, and High Technology.* Washington, D.C.: Brookings Institution.

Foster, R. J. (1986). *Innovation: The Attackers Advantage.* New York: Summit Books.

Fowler, T. (1990). *Value Analysis in Design.* New York: Van Nostrand Reinhold.

Frank, R. H., and P. J. Cook (1992). *Winner-Take-All Markets.* Stanford, Calif.: Center for Advanced Study in Behavioral Sciences.

Freeman, C. (1982). *The Economics of Innovation.* Cambridge, Mass.: MIT Press.

Freeman, J., and J. Hannan (1993). *Organizational Ecology.* Boston: Harvard University Press.

Fruhan, W. E. (1979). *Financial Strategy: Studies in the Creation, Transfer and Destruction of Shareholder Value.* Homewood, Ill.: Richard D. Irwin.

Fudenburg, D., R. Gilbert, J. Stiglitz, and J. Tirole (1983). "Preemption, Leapfrogging and Competition in Patent Races." *European Economic Review* 22: 3–31.

Fuller, S. L. (1974). "Minimal-Total-Processing Time Drum and Disk Scheduling Disciplines." *Communications of the ACM* 17(7): 376–381.

Futia, C. (1980). "Schumpeterian Competition." *Quarterly Journal of Economics* 94(June): 675–695.

Gabriel, R. P. (1996). *Patterns of Software: Tales from the Software Community.* New York: Oxford University Press.

Gallini, N. T., and M. Trebilcock (1995). *Intellectual Property Rights and Competition Policy.* Toronto: University of Toronto.

Gandal, N., S. Greenstein, and D. Salant (1995). *Adoptions and Orphans in the Early Microcomputer Market.* Ramat Aviv, Israel: The Foerder Institute for Economic Research, Tel-Aviv University.

Garud, R., and A. Kumaraswamy (1993). "Changing Competitive Dynamics in Network Industries: An Exploration of Sun Microsystems' Open Systems Strategy." *Strategic Management Journal* 14: 351–369.

Gates, B., N. Myhrvold, and P. H. Rinearson (1996). *The Road Ahead.* New York: Penguin.

Gavetti, G. and D. Levinthal (1998). *Looking Forward and Looking Backward: Cognitive and Experiential Search.* Philadelphia: The Wharton School, University of Pennsylvania.

Ghemawat, P. (1991). *Commitment: The Dynamic of Strategy.* New York: Free Press.

Gilder, G. (1989). *Microcosm.* New York: Simon & Schuster.

Gomes-Casseres, B. (1994a). "Group Versus Group: How Alliance Networks Compete." *Harvard Business Review* 62–84.

Gomes-Casseres, B. (1997). *Collective Competition: International Alliances in High Technology.* Boston: Harvard University Press.

Gompers, P. A. (1994). "The Rise of Venture Capital." *Business and Economic History* 23: 1–24.

Gompers, P. A. (1995). "Optimal Investment Monitoring, and the Stage of Venture Capital." *Journal of Finance* 50: 1461–1489.

Gompers, P. A., and J. Lerner (1998a). "Venture Capital Distributions: Short-Run and Long-Run Reactions." *Journal of Finance* 53: 2161–2184.

Gompers, P. A., and J. Lerner (1999). "An Analysis of Compensation in the U.S. Venture Capital Industry." *Journal of Financial Economics* 51(1): 3–44.

Gould, S. J., and R. C. Lewontin (1984). "The Spandrels of San Marco and the Panglossian Paradigm: A Critique of the Adaptationist Programme," in *Conceptual Issues in Evolutionary Biology,* ed. E. Sober. Cambridge, Mass.: MIT Press: 252–270.

Greenstein, S. (1993). "Markets, Standards, and the Information Infrastructure." *IEEE Micro* 36–51.

Greenstein, S. (1994). *From Superminis to Supercomputers: Estimating Surplus in the Computer Market.* Boston: National Bureau of Economic Research.

Greenstein, S. M. (1994). *Did Computer Technology Diffuse Quickly?: Best and Average Practice in Mainframe Computers, 1968–83.* Boston: National Bureau of Economic Research.

Grenadier, S. R. (1995). "Valuing Lease Contracts: A Real-Options Approach." *Journal of Financial Economics* 38: 297–331.

Grenadier, S. R., and A. M. Weiss (1994). *Optimal Migration Strategies for Firms Facing Technological Innovations: An Option Pricing Approach.* Stanford, Calif.: Stanford University.

Groenewold, G. (1996). "Software patents: A new order?" *UNIX Review* 14(10): 89.

Grossman, S. J., and O. D. Hart (1986). "The Costs and Benefits of Ownership: A Theory of Vertical Integration." *Journal of Political Economy* 94(4): 691–819.

Grossman, G. M., and C. Shapiro (1987). "Dynamic R&D Competition." *The Economic Journal* 97: 372–387.

Grove, A. (1983). *High Output Management.* New York: Random House.

Haanstra, J. W., B. O. Evans, J. D. Aron, F. P. Brooks, Jr., J. W. Fairclough, W. P. Heising, H. Hellerman, W. H. Johnson, M. J. Kelly, D. V. Newton, B. G. Oldfield, S. A. Rosen, and J. Svigals (1983). "Final Report of the SPREAD Task Group, December 28, 1961." *Annals of the History of Computing* 5(1): 6–26.

Hagedorn, J. (1993). "Strategic Technology Partnering during the 1980's: Trends, Networks and Corporate Patterns in No-Core Technologies." *Research Policy* 24: 207–231.

Hall, B. J., and J. B. Liebman (1998). "Are CEO's Really Paid Like Bureaucrats?" *Quarterly Journal of Economics* 113(3): 653–690.

Hamel, G., and C. K. Prahalad (1994). *Competing for the Future.* Boston: Harvard Business School Press.

Hannan, M. T., and J. Freeman (1989). *Organizational Ecology.* Cambridge, Mass.: Harvard University Press.

Hansmann, H. (1996). *The Ownership of Enterprise.* Cambridge, Mass.: Harvard University Press.

Hare, C., E. Dulaney, G. Eckel, S. Lee, L. Ray, and S. William (1996). *Inside Unix,* 2nd ed. Indianapolis: New Riders Publishing.

Hart, O., and J. Moore (1990). "Property Rights and the Nature of the Firm." *Journal of Political Economy* 98(6): 1119–1158.

Hayek, F. A. (1945). "The Uses of Knowledge in Society." *American Economic Review* 35(4).

Hayek, F. A. (1988). *The Fatal Conceit: The Errors of Socialism.* Chicago: University of Chicago Press.

Hayes, R. H., and W. J. Abernathy (1980). "Managing Our Way to Economic Decline." *Harvard Business Review*(July/August): 67–87.

Hayes, R. H., S. Wheelwright, and K. B. Clark (1988). *Dynamic Manufacturing: Creating the Learning Organization.* New York: Free Press.

He, H. and R. S. Pindyck (1992). "Investments in Flexible Production Capacity." *Journal of Economic Dynamics and Control* 16: 575–599.

Helpman, E., and M. Trajtenberg (1994). *A Time to Sow and a Time to Reap; Growth Based on General Purpose Technologies. Working Paper Series.* Cambridge, Mass.: National Bureau of Economic Research.

Henderson, R. M., and K. B. Clark (1990). "Architectural Innovation: The Reconfiguring of Existing Product Technologies and the Failure of Established Firms." *Administrative Science Quarterly* 35(March): 9–30.

Henderson, R. (1993). "Underinvestment and Incompetence as Responses to Radical Innovation: Evidence from the Photolithographic Alignment Equipment Industry." *RAND Journal of Economics* 24(2): 248–270.

Henderson, Y. K. (1989). "The Emergence of the Venture Capital Industry." New England Economic Review(July/August): 64–89.

Hennessy, J. L., and D. A. Patterson (1990). *Computer Architecture: A Quantitative Approach.* San Mateo, Calif.: Morgan Kaufmann.

Hicks, J. R. (1946). *Value and Capital.* London: Oxford University Press.

Hodder, J. E., and A. J. Triantis (1992). "Valuing Flexibility: An Impulse Control Framework." Annals of Operations Research 45: 109–130.

Hofstadter, D. R. (1979). *Gödel, Escher and Bach: An Eternal Golden Braid.* New York, Random House.

Hofstadter, D. R., *The Fluid Analogies Research Group Fluid Concepts and Creative Analogies: Computer Models of the Fundametnal Mechanisms of Thought.* New York: Basic Books.

Holland, J. H. (1992). *Adaptation in Natural and Artificial Systems,* 2nd ed. Ann Arbor, Mich.: University of Michigan Press.

Holland, J. H. (1995). *Hidden Order: How Adaption Builds Complexity.* Reading, Mass.: Addison-Wesley.

Holland, J. H. (1998). *Emergence: From Chaos to Order.* Reading, Mass.: Perseus Books.

Holmstrom, B., and P. Milgrom (1991). "Multitask Principal-Agent Analyses: Incentive Contracts, Asset Ownership, and Job Design." *Journal of Law, Economics and Organization* 7SP: 24–52.

Holmstrom, B., and P. Milgrom (1994). "The Firm as an Incentive System." *American Economic Review* 84(4): 972–991.

Holmstrom, B., and J. Tirole (1989). "The Theory of the Firm," in *Handbook of Industrial Organization,* ed. R. Schmalensee and R. D. Willij. Amsterdam: Elsevier Science.

Holmstrom, B., and J. Tirole (1991). "Transfer Pricing and Organizational Form." *Journal of Law, Economics, & Organization* 7(2): 201–228.

Hounshell, D. A. (1984). *From the American System to Mass Production 1800–1932: The Development of Manufacturing Technology in the United States.* Baltimore: Johns Hopkins University Press.

Hounshell, D. A. (1987). *Du Pont: The World War II Years and Post-War Expansion of Research.* Boston: Harvard Business School Press.

Hubbard, R. G. (1994). "Investment Under Uncertainty: Keeping One's Options Open." *Journal of Economic Literature* 32(December): 1816–1831.

Hubbard, R. G., and D. Palia (1998). "A Re-Examination of the Conglomerate Merger Wave in the 1960s: An Internal Capital Markets View." Cambridge, Mass.: National Bureau of Economic Research.

Hubka, V., and W. E. Eder (1988). *Theory of Technical Systems.* New York: Springer-Verlag.

Hyde, A. (1998). "Real Human Capital: The Economics and Law of Shared Knowledge." New York: Columbia Law School Sloan Project.

Iansiti, M. (1994). "Technology Integration: Managing Technological Evolution in a Complex Environment." *Research Policy* 24: 521–542.

Iansiti, M. (1995). "Science-Based Product Development: An Empirical Study of the Mainframe Computer Industry." *Production and Operations Management* 4(4): 335–359.

Iansiti, M. (1997a). "Managing Chaos: System Focused Product Development in the Computer and Multimedia Environment," in *Competing in an Age of Digital Convergence,* ed. D. B. Yoffie. Boston: Harvard Business School Press, pp. 413–444.

Iansiti, M. (1997b). *Technology Integration: Making Critical Choices in a Dynamic World.* Boston: Harvard Business School Press.

Iansiti, M., and T. Khanna (1995). "Technological Evolution, System Architecture and the Obsolescence of Firm Capabilities." *Industrial and Corporate Change* 4(2): 333–361.

Iansiti, M., and K. B. Clark (1993). "Integration and Dynamic Capability: Evidence From Product Development in Automobiles and Mainframe Computers." *Industrial and Corporate Change* 3(3): 557–605.

Iansiti, M., and A. MacCormack (1997). "Developing Products on Internet Time," *Harvard Business Review* (September/October): 108–117.

Iansiti, M., and J. West (1997). "Technology Integration: Turning Great Research into Great Products." *Harvard Business Review*(May/June): 69–89.

Jaikumar, R., and R. E. Bohn (1986). "The Development of Intelligent Systems for Industrial Use: A Conceptual Framework," in *Research on Technological Innovation, Management and Policy,* vol. 3, ed. R. S. Rosenbloom. Greenwich, CT, JAI Press, 169–211.

Janowski, J. E. (1990). *National Patterns of R&D Resources.* National Science Foundation, Washington, D.C.: U.S. Government Printing Office.

Jelinek, M., and G. B. Schoonhoven (1990). *The Innovation Marathon: Lessons from High Technology Firms.* Oxford, U.K.: Basil Blackwell.

Jensen, M. C. (1986). "Agency Costs of Free Cash Flow: Corporate Finance and Takeovers." *American Economic Review* 76: 323–329.

Jensen, M. C. (1988). "Takeovers: Their Causes and Consequences." *Journal of Economic Perspectives* 2(1): 35–44.

Jensen, M. C. (1989). "Eclipse of the Public Corporation." *Harvard Business Review* 67(5): 61–84.

Jensen, M. C. (1993). "The Modern Industrial Revolution, Exit and Failure of Internal Control Systems." *Journal of Finance* 48(3): 831–880.

Jensen, M. C. (1998). *Foundations of Organizational Strategy.* Cambridge, Mass.: Harvard University Press.

Jensen, M. C., and W. H. Meckling (1976). "The Theory of the Firm: Managerial Behavior, Agency Costs, and Capital Structure." *Journal of Financial Economics* 3: 305–360.

Jensen, M. C. and W. H. Meckling (1992). "Specific and General Knowledge, and Organizational Structure," in *Contract Economics,* ed. L. Werin and H. Wijkander. Oxford, U.K.: Basil Blackwell, 251–274.

John, R. R. (1997). "Elaborations, Revisions and Dissents: Alfred D. Chandler, Jr.'s *The Visible Hand* after Twenty Years." *Business History Review* 71: 151–200.

Johnson, H. T., and R. S. Kaplan (1987). *Relevance Lost.* Boston: Harvard Business School Press.

Jones, J. C. (1970). *Design Methods: Seeds of Human Futures.* London: Wiley-Interscience.

Karjala, D. S., and P. S. Menell (1995). "Applying Fundamental Copyright Principles to Lotus Development Corp. v. Borland International, Inc." *High Technology Law Journal* 10: 177–192.

Katz, M. L., and C. Shapiro (1986). "Technology Adoption in the Presence of Network Externalities." *Journal of Political Economy* 94(4): 823–841.

Katz, M. L., and C. Shapiro (1992). "Product Introduction with Network Externalities." *Journal of Industrial Economics* 40(1): 55–83.

Kauffman, S. (1989). *Adaptation on a Rugged Fitness Landscape.* Reading, Mass.: Addison-Wesley.

Kauffman, S. A. (1989). "Principles of Adaptation in Complex Systems," in *Lectures in the Sciences of Complexity,* ed. D. L. Stein. Reading, Mass.: Addison-Wesley, 619–812.

Kauffman, S. A. (1993). *The Origins of Order: Self-Organization and Selection in Evolution.* New York: Oxford University Press.

Kauffman, S. (1995). *At Home in the Universe.* New York: Oxford University Press.

Kelley, A. J., F. B. Campanella, and J. McKiernan (1972). *Venture Capital: A Guidebook for New Enterprises.* Washington, D.C.: U.S. Government Printing Office.

Kelly, K. (1994). *Out of Control.* Reading, Mass.: Addison-Wesley.

Kernighan, B. W., and R. Pike (1999). *The Practice of Programming.* Reading, Mass.: Addison-Wesley.

Kernighan, B. W., and P. J. Plauger (1976). *Software Tools.* Reading, Mass.: Addison-Wesley.

Kernighan, B. W., and P. J. Plauger (1978). *The Elements of Programming Style.* New York: McGraw-Hill.

Khanna, T. (1995). "Racing Behavior: Technological Evolution in the High-End Computer Industry." *Research Policy* 24(6): 933–958.

Khanna, T., and M. Iansiti (1997). "Firm Asymmetries and Sequential R&D: Theory and Evidence from the Mainframe Computer Industry." *Management Science* 43(4): 405–421.

Khanna, T., and K. G. Palepu (1996). "Corporate Strategy and Institutional Context: An Empirical Analysis of Diversified Indian Business Groups." Boston: *Harvard Business School Working Paper* 96–051.

Khanna, T., and Palepu, K. G. (1997). "Why Focused Strategies May Be Wrong for Emerging Markets." *Harvard Business Review* (July/August).

Kidder, T. (1981). *The Soul of a New Machine.* New York: Avon Books.

Kirzner, I. M. (1997). "Entrepreneurial Discovery and the Competitive Market Process: An Austrian Approach." *Journal of Economic Literature* 35: 60–85.

Klein, B., and K. M. Murphy (1997). "Vertical Integration as a Self-Enforcing Contractual Arrangement." *American Economics Review* 87(2): 415–420.

Klein, B., A. Crawford, and A. A. Alchian (1978). "Vertical Integration, Appropriable Rents, and the Competitive Contracting Process." *Journal of Law and Economics* 21(October): 297–326.

Klepper, S. (1992). "Entry, Exit, Growth, and Innovation over the Product Life Cycle." *American Economic Review* 86(3): 562–583

Klevorick, A. K., R. C. Levin, R. R. Nelson, and S. G. Winter (1993). "On the Sources and Significance of Interindustry Differences in Technological Opportunities." *Research Policy* 24: 185–205.

Knight, K. E. (1966). "Changes in Computer Performance: A Historical View." *Datamation* 40–42, 45–49, 54.

Knuth, D. E. (1992). *Literate Programming.* Center for the Study of Language and Information, Stanford University, Stanford, CA.

Koehn, N. F. (1995). "Josiah Wedgwood and the First Industrial Revolution." HBS Case 1-796-079, Boston: Harvard Business School.

Kogut, B., and E. H. Bowman (1995). "Modularity and Permeability as Principles of Design," in *Redesigning the Firm,* ed. E. H. Bowman and B. Kogut, New York: Oxford University Press, 243–260.

Korson, T. D., and V. K. Vaishnavi (1986). *An Empirical Study of the Effect of Modularity on Program Modifiability.* Milledgeville, Ga.: Georgia College State University, Department of Computer Information Systems.

Krafcik, J. F. (1988). "Triumph of the Lean Production System." *Sloan Management Review* 30: 41–52.

Krishnan, V., S. D. Eppinger, and D. E. Whitney (1997). "A Model-based Framework to Overlap Product Development Activities." *Management Science* 43(4): 437–451.

Laffont, J. (1990). "Analysis of Hidden Gaming in a Three-Level Hierarchy." *Journal of Law, Economics, and Organization* 6(2): 301–324.

Landes, D. S. (1987). *Revolution in Time.* New York: W. W. Norton.

Landes, D. S. (1998). *The Wealth and Poverty of Nations.* New York: W. W. Norton.

Langlois, R. N. (1992a). "Capabilities and Vertical Integration in Process Technology: The Case of Semiconductor Fabrication Equipment," Working Paper, Consortium on Competitiveness and Cooperation, University of California, Berkeley (November).

Langlois, R. N. (1992). "External Economies and Economic Progress: The Case of the Microcomputer Industry." *Business History Review* 66(1): 1–51.

Langlois, R. N. (1997). "Cognition and Capabilities: Opportunities Seized and Missed in the History of the Computer Industry," in *Technological Innovation: Oversights and Foresights,* eds. R. Garud, P. Nayyar, Z. Shapira. New York: Cambridge University Press.

Langlois, R. N., and P. L. Robertson (1992). "Networks and Innovation in a Modular System: Lessons from the Microcomputer and Stereo Component Industries." *Research Policy* 21: 297–313.

Leonard-Barton, D. (1995). *Wellsprings of Knowledge: Building and Sustaining Sources of Innovation.* Boston: Harvard Business School Press.

Lerner, J. (1993). "Venture Capitalists and the Decision to Go Public." *Journal of Financial Economics* 35: 293–316.

Lerner, J. (1994a). "The Importance of Patent Scope: An Empirical Analysis." *RAND Journal of Economics* 25(2): 319–333.

Lerner, J. (1994b). "The Syndication of Venture Capital Investments." *Financial Management* 23(3): 16–27.

Lerner, J. (1995a). "Patenting in the Shadow of Competitors." *Journal of Law and Economics* 38(October): 563–595.

Lerner, J. (1995b). "Pricing and Financial Resources: An Analysis of the Disk Drive Industry, 1980–88." *Review of Economics and Statistics* 77: 585–598.

Lerner, J. (1995c). "Venture Capitalists and the Oversight of Private Firms." *Journal of Finance* 50(1): 301–318.

Lerner, J. (1999). "The Government as Venture Capitalist: The Long-Run Impact of the SBIR Program." *Journal of Business* 72 (July): 285–318.

Lerner, J. (1997). "An Empirical Exploration of a Technology Race." *RAND Journal of Economics* 25: 228–247.

Levin, R. C. (1982). "The Semiconductor Industry," in *Government and Technical Progress: A Cross Industry Analysis.* ed. R. R. Nelson. New York: Pergamon Press.

Levinthal, D. (1992). *Surviving Schumpeterian Environments: An Evolutionary Perspective.* Philadelphia: The Wharton School, University of Pennsylvania.

Levinthal, D. (1994a). "Adaptation on Rugged Landscapes." *Management Science* 43: 934–950.

Levinthal, D. (1994b). *Strategic Management and the Exploration of Diversity.* Philadelphia: Wharton School, University of Pennsylvania.

Levinthal, D., and M. Warglien (1997). *Landscape Design: Designing for Local Action in Complex Worlds.* Philadelphia: Wharton School, University of Pennsylvania.

Levy, J. (1978). "Buses: The Skeleton of Computer Structures," in *Computer Engineering: A DEC View of Hardware Systems,* ed. C. G. Bell, J. C. Mudge, and J. E. McNamara. Maynard, Mass.: Digital Press.

Liebowitz, S. J., and S. E. Margolis (1995). "Path Dependence, Lock-In, and History." *Journal of Law, Economics & Organization* 11(1): 205–226.

Liles, P. R. (1974). *New Business Ventures and the Entrepreneur.* Homewood, Ill.: Richard D. Irwin.

Lindgren, B. W. (1968). *Statistical Theory.* New York: Macmillan.

Lintner, J. (1965a). "The Valuation of Risk Assets and the Selection of Risky Investments in Stock Portfolios and Capital Budgets." *Review of Economics and Statistics* 47(February): 13–37.

Lintner, J. (1965b). "Security Prices, Risk and Maximal Gains from Diversification." *Journal of Finance* 20(December): 587–615.

Lovejoy, W. S. (1992). "Rationalizing the Design Process." Presented at Conference on Design Management, 17–8 September, Anderson Graduate School of Managment, University of California at Los Angeles.

Lumsden, C. J., and E. O. Wilson (1981). *Genes, Mind and Culture: The Coevolutionary Process.* Cambridge, Mass.: Harvard University Press.

Machlup, F. (1962). *The Production and Distribution of Knowledge in the United States.* Princeton, N.J.: Princeton University Press.

Macready, W. G., A. Siapas, and S. A. Kauffman (1996). "Criticality and Parallelism in Combinatorial Optimization." *Science* 271: 56–58.

Manes, S., and P. Andrews (1993). *Gates: How Microsoft's Mogul Reinvented an Industry and Made Himself the Richest Man in America.* New York: Doubleday.

Manne, H. G. (1965). "Mergers and the Market for Corporate Control." *Journal of Political Economy* 73(April): 110–120.

Markowitz, H. (1952). "Portfolio Selection." *Journal of Finance* 7(March): 77–91.

Markowitz, H. (1959). *Portfolio Selection: Efficient Diversification of Investment.* New York: John Wiley and Sons.

Marples, D. L. (1961). "The Decisions of Engineering Design." *IEEE Transactions on Engineering Management* 2: 55–81.

Mason, S. P., and R. Merton (1985). "The Role of Contingent Claims Analysis in Corporate Finance," in *Recent Advances in Corporate Finance,* ed. E. Altman and M. Subrahmanyam. Homewood, Ill.: Richard D. Irwin.

Massa, M. (1998). *Information and Financial Innovations: Is Less Information Better than More?* Unpublished manuscript, New Haven, Conn.: Yale University.

Matutes, C., and P. Regibeau (1988). "'Mix and Match': Product Compatibility Without Network Externalities." *RAND Journal of Economics* 19(2): 221–234.

Mauer, D.C., and A. J. Triantis (1994). "Interactions of Corporate Financing and Investment Decisions: A Dynamic Framework." *Journal of Finance* 49(4): 1253–1277.

Mayle, N. (1996). *Strategic Issues of Software Components: Market Failures in the Silver Bullet Manufacturing Industry.* Unpublished manuscript, Cambridge, Mass.: Massachusetts Institute of Technology.

Maynard-Smith, J. (1989). *Evolutionary Genetics.* New York: Oxford University Press.

McClelland, J. L., and D. E. Rumelhart (1995). *Parellel Distributed Processing.* Cambridge, Mass.: MIT Press.

McCord, K. R., and S. D. Eppinger (1993). "Managing the Integration Problem in Concurrent Engineering." Sloan School of Management Working Paper #3594–93-MSA, Cambridge, Mass.: MIT (August).

McCullough, M. (1996). *Abstracting Craft, The Practiced Digital Hand.* Cambridge, Mass.: MIT Press.

McGahan, A. M. (1999). "The Performance of U.S. Corporations: 1981–1994." *Journal of Industrial Economics* (in press).

McGahan, A. M. (1999). "Competition, Strategy and Business Performance 1981–1997," *California Management Review* 41(3): 74–101.

McGahan, A. M., and M. E. Porter (1997). "The Emergence and Sustainability of Abnormal Profits." Boston: *Harvard Business School Working Paper* 97–103.

McGahan, A., D. Yoffie, and L. Vadasz (1994). "Creating Value and Setting Standards: The Lessons of Consumer Electronics for Personal Digital Assistants," in *Competing in the Age of Digital Convergence.* ed. D. B. Yoffie. Boston: Harvard Business School Press.

McGrath, R. G. (1997) "A Real Options Logic for Initiating Technology Positioning Investments." *Academy of Management Review* 22(4): 974–996.

Mead, C., and L. Conway (1980). *Introduction to VLSI Systems.* Reading, Mass.: Addison-Wesley.

Merton, R. C. (1973a). "Theory of Rational Option Pricing." *Bell Journal of Economics and Management Science* 4: 141–183.

Merton, R. C. (1973b). "An Intertemporal Capital Asset Pricing Model." *Econometrica* 41: 867–887.

Merton, R. C. (1990). *Continuous-Time Finance.* Cambridge, Mass.: Basil Blackwell.

Merton, R. C. (1997a). "A Model of Contract Guarantees for Credit-Sensitive, Opaque Financial Institutions." *European Finance Review* 1(1): 1–13.

Merton, R. C. (1998a). "Applications of Option Pricing: Twenty-Five Years Later." *American Economics Review* 88(3): 323–349.

Merton, R. C., and Z. Bodie (1995). "A Conceptual Framework for Analyzing the Financial Environment," in *The Global Financial System: A Functional Perspective.* Boston: Harvard Business School Press.

Merton, R. C., and A. F. Perold (1993). "The Theory of Risk Capital in Financial Firms." *Journal of Applied Corporate Finance* 5: 16–32.

Meurer, M. J. (1995). *The Nonobviousness Standard and the Optimal Probability of Patent Validity.* Buffalo: State University of New York–Buffalo Law School.

Meyer, R. F. (1970). "On the Relationship among the Utility of Assets, the Utility of Consumption, and Investment Strategy in an Uncertain, but Time-Invariant World," in *OR69: Proceedings of the Fifth International Conference on Operational Research,* ed. J. Lawrence. London: Tavistock.

Milgrom, P., and J. Roberts (1990). "The Economics of Manufacturing: Technology, Strategy and Organization." *American Economic Review* 80(3): 511–528.

Milgrom, P., and J. Roberts (1992). *Economics, Organization and Management.* Englewood Cliffs, N.J.: Prentice Hall.

Milgrom, P., and J. Roberts (1994). "Comparing Equilibria." *American Economic Review* 84(3): 441–459.

Milgrom, P., and J. Roberts (1995). "Complementarities and Fit: Strategy, Structure, and Organizational Change in Manufacturing." *Journal of Accounting and Economics* 19: 179–208.

Miller, R. R., and J. A. Alic (1986). "Financing Expansion in an International Industry: The Case of Electronics." *International Journal of Technology Management* 1:101–117.

Miller, M. H., and F. Modigliani (1961). "Dividend Policy, Growth and the Valuation of Shares." *Journal of Business* 34: 411–433.

Mills, D. Q., and G. B. Friesen (1996). *Broken Promises: An Unconventional View of What Happened at IBM.* Boston: Harvard Business School Press.

Mises, L. (1949). *Human Actions.* New Haven, Conn.: Yale University Press.

Mishina, K. (1993). "A Study of Toyota's Approach to Product Variety." Boston: *Harvard Business School Working Paper* 93–076.

Mithen, S. (1996). *The Prehistory of the Mind.* London: Thames and Hudson.

Modigliani, F., and M. H. Miller (1958). "The Cost of Capital, Corporation Finance, and the Theory of Investment." *American Economic Review* 48 (June): 261–297.

Moore, G. E. (1996). "Intel—Memories and the Microprocessor." *Daedalus* 125(2): 55.

Morelli, M. D., S. D. Eppinger, and R. Gulati (1995). "Predicting Technical Communication in Product Development Organizations." *IEEE Transactions on Engineering Management* 42(3): 215–222.

Morris, C. R., and C. H. Ferguson (1993). "How Architecture Wins Technology Wars." *Harvard Business Review* (March/April): 86–96.

Morris, P. A., E. O. Teisberg, and L. Kolbe (1991). "When Choosing R&D Projects, Go with Long Shots." *Research/Technology Management* 35–40.

Motta, M. (1992). "Cooperative R&D and Vertical Product Differentiation." *International Journal of Industrial Organization* 10: 643–661.

Mudge, J. C. (1978). "Design Decisions for the PDP-11/60 Mid-Range Minicomputer," in *Computer Engineering: A DEC View of Hardware Systems,* ed. C. G. Bell, J. C. Mudge, and J. E. McNamara. Maynard, Mass.: Digital Press, 315–316.

Myer, T. H., and I. E. Sutherland (1968). "On the Design of Display Processors." *Communications of the ACM* 11(6): 410–414.

Myers, G. J. (1975). *Reliable Software through Composite Design.* New York: Van Nostrand Reinhold.

Myers, G. J. (1979). *The Art of Software Testing.* New York: John Wiley and Sons.

Myers, S. C., and S. Majd (1983). *Calculating Abandonment Value Using Option Pricing Theory.* Cambridge, Mass.: MIT Sloan School of Management.

Nadler, D. A., and M. L. Tushman (1997). *Competing by Design: The Power of Organizational Architecture.* New York: Oxford University Press.

Nanda, A. (1994). "Strategy, Organization, and Performance in the U.S. Semiconductor Industry." Boston: *Harvard Business School Working Paper* 94–047.

Nelson, R. R. (1993). "Uncertainty, Learning, and the Economics of Parallel Research and Development Efforts." *Review of Economics and Statistics* 43: 351–364.

Nelson, R. R. (1995). "Recent Evolutionary Theory about Economic Change." *Journal of Economic Literature* 33: 48–90.

Nelson, R. R., and S. G. Winter (1978). "Forces Generating and Limiting Concentration under Schumpeterian Competition." *Bell Journal of Economics* 9(Winter): 524–548.

Nelson, R. R., and S. G. Winter (1982). *An Evolutionary Theory of Economic Change.* Cambridge, Mass.: Harvard University Press.

Nevins, J. L., and D. V. Whitney (1989). *Concurrent Design of Products and Processes: A Strategy for the Next Generation in Manufacturing.* New York: McGraw-Hill.

Novak, S., and S. D. Eppinger (1998). "Sourcing by Design: Product Architecture and the Supply Chain." Sloan School of Management Working Paper 4045, Cambridge, Mass.: MIT (November).

Olsen, K. H. (1957). "Transistor Circuitry in the Lincoln TX-2," reprinted in C. G. Bell, J. C. Mudge, and J. E. McNamara, eds. (1978), *Computer Engineering: A DEC View of Hardware Systems Design.* Maynard, Mass.: Digital Press.

Opitz, H. (1970). *A Classification System to Describe Workpieces.* London: Pergamon Press.

Osborne, S. M. (1993). *Product Development Cycle Time Characterization Through Modeling of Process Iteration.* Unpublished Masters Thesis, Cambridge: MIT Sloan School of Management.

Pahl, G., and W. Beitz (1984). *Engineering Design.* London: The Design Council.

Parnas, D. L. (1972a). "Information Distribution Aspects of Design Methodology." *Information Processing* 71: 339–344.

Parnas, D. L. (1972b). "On the Criteria to Be Used in Decomposing Systems into Modules." *Communications of the ACM* 15(12): 1053–1058.

Parnas, D. L., P. Clements, and D. Weiss (1985). "The Modular Structure of Complex Systems." *IEEE Transactions on Software Engineering* 11(3): 259–66.

Patterson, D. A., and J. L. Hennessy (1994). *Computer Organization and Design.* San Mateo, CA: Morgan Kaufmann Publishers.

Penrose, R. (1989). *The Emperor's New Mind.* New York: Oxford University Press.

Penrose, R. (1994). *Shadows of the Mind.* New York: Oxford University Press.

Perold, A. F. (1995). "The Payment System and Derivative Instruments," in *The Global Financial System,* D. B. Crane et al., eds. Boston: Harvard Business School Press.

Perold, A. F. (1997). *Capital Allocation in Financial Firms.* Boston: Harvard Business School manuscript.

Petroski, H. (1996). *Invention by Design: How Engineers Get from Thought to Thing.* Cambridge, Mass.: Harvard University Press.

Pfister, G. F. (1998). *In Search of Clusters.* Upper Saddle River, N.J.: Prentice Hall.

Pimmler, T. U., and S. D. Eppinger (1994). "Integration Analysis of Product Decompositions." *Design Theory and Methodology* 68: 343–351.

Pindyck, R. S. (1991). "Irreversibility, Uncertainty, and Investment." *Journal of Economic Literature* 29: 1110–1152.

Pindyck, R. S. (1992). *Investments of Uncertain Cost.* Cambridge, Mass.: National Science Foundation Grant No. SES-8618502 and the MIT Center for Energy Policy Research.

Pinker, S. (1994). *The Language Instinct.* New York: William Morrow.

Pirenne, H. (1917). *A History of Europe.* Garden City, N.Y.: Doubleday.

Pirenne, H. (1925). *Medieval Cities: Their Origins and the Revival of Trade.* Princeton, N.J.: Princeton University Press.

Plotkin, H. (1993). *Darwin Machines.* Cambridge, Mass.: Harvard University Press.

Polanyi, M. (1958). *Personal Knowledge.* Chicago: University of Chicago Press.

Popek, G. J., and R. P. Goldberg (1974). "Formal Requirements for Virtualizable Third Generation Architectures." *Communications of the ACM* 17(7): 412–421.

Popper, K. R. (1963). *Conjectures and Refutations.* New York: Harper Collins.

Popper, K. R. (1972). *Objective Knowledge: An Evolutionary Approach.* New York: Oxford University Press.

Porter, M. E. (1980). *Competitive Strategy: Techniques for Analyzing Industries and Competition.* New York: Free Press.

Porter, M. E. (1985). *Competitive Advantage: Creating and Sustaining Superior Performance.* New York: Free Press.

Porter, M. E. (1990). *The Competitive Advantage of Nations.* New York: Free Press.

Premus, R. (1985). *Venture Capital and Innovation: A Study.* Washington, D.C.: Joint Economic Committee, Congress of the United States.

Pugh, E. W., L. R. Johnson, and J. H. Palmer (1991). *IBM's 360 and Early 370 Systems.* Cambridge, Mass.: MIT Press.

Pugh, E. W. (1995). *Building IBM: Shaping an Industry and Its Technology.* Cambridge, Mass.: MIT Press.

Quinn, J. B. (1985). "Managing Innovation: Controlled Chaos." *Harvard Business Review* (May/June): 73–84.

Radner, R. (1992). "Hierarchy: The Economics of Managing." *Journal of Economic Literature* 30(September): 1382–1415.

Rajan, R. G., and L. Zingales (1998a). "Financial Dependence and Growth." *American Economics Review* 88: 559–586.

Rajan, R. G., and L. Zingales (1998b). "Power in a Theory of the Firm." *Quarterly Journal of Economics,* 113: 387–432.

Reid, R. H. (1997). *Architects of the Web.* New York: John Wiley and Sons.

Rifkin, G., and G. Hanon (1988). *The Ultimate Entrepreneur: The Story of Ken Olsen and Digital Equipment Corporation.* New York: Contemporary Books.

Rinderle, J. R., and N. P. Suh (1982). "Measures of Functional Coupling in Design." *ASME Journal of Engineering for Industry* (November): 383–388.

Ritchie, D. M. (1984). "The Evolution of the UNIX Time-Sharing System." *AT&T Bell Labs Technical Journal* 63(6): 1577–1593.

Ritchie, D. M., and K. Thompson (1974). "The UNIX Time Sharing System." *Communications of the ACM* 17(7): 365–375.

Rivkin, J. (1998a). "Optimally Local Search on Rugged Landscapes." Boston: *Harvard Business School Working Paper* 98–068.

Rivkin, J. (1998b). "Imitation of Complex Strategies." Boston: Harvard Business School Working Paper 98–068.

Robertson, D., and K. Ulrich (1998). "Platform Product Development." *Sloan Mangement Review* 39(4).

Robertson, P. L., and R. N. Langlois (1995). "Innovation, Networks, and Vertical Integration." *Research Policy* 24: 543–562.

Roller, L.-H., and M. M. Tombak (1990). "Strategic Choice of Flexible Production Technologies and Welfare Implications." *Journal of Industrial Economics* 38(4): 417–431.

Romer, P. M. (1996). "Why, Indeed, In America? Theory, History, and the Origins of Modern Economic Growth." *New Growth Theory and Economic History* 86(2): 202–206.

Rosenberg, N. (1982). *Inside the Black Box: Technology and Economics.* New York: Cambridge University Press.

Rosenkopf, L., and M. L. Tushman (1999). "The Co-Evolution of Community Networks and Technology: Lessons from the Flight Simulation Industry." *Industrial and Corporate Change* (in press).

Ross, S. A. (1976). "Arbitrage Theory of Capital Asset Pricing." *Journal of Economic Theory* 13(December): 341–360.

Rothwell, R., and P. Gardiner (1988). "Re-Innovation and Robust Designs: Producer and User Benefits." *Journal of Marketing Management* 3(3): 372–387.

Sah, R. K., and J. E. Stiglitz (1986). "The Architecture of Economic Systems: Hierarchies and Polyarchies." *American Economic Review* 76(4): 716–826.

Sahlman, W. A. "Aspects of Financial Contracting in Venture Capital." *Journal of Applied Corporate Finance* 23–36.

Sahlman, W. A. (1990). "The Structure and Governance of Venture-Capital Organizations." *Journal of Financial Economics* 27(2): 473–524.

Sahlman, W. A. (1999). "The Horse Race between Capital and Opportunity." *National Venture Capital Journal* (in press).

Sahlman, W. A., and H. H. Stevenson (1985). "Capital Market Myopia." *Journal of Business Venturing* 1: 7–30.

Salinger, M. A. (1984). "Tobin's q, Unionization, and the Concentration-Profits Relationship." *RAND Journal of Economics* 15(2): 159–170.

Salter, M. S. (1997). "Reversing History: The Economics and Politics of Vertical Disintegration at General Motors." Boston: Harvard Business School Working Paper 97-001.

Salter, M. S. (1998). "Value Creation and Destruction in the Management of Organizations." Boston: Harvard Business School Working Paper 99-035.

Saltzer, J. H. (1974). "Protection and the Control of Information Sharing in Multics." *Communications of the ACM* 17(7): 388–402.

Salus, P. H. (1994). *A Quarter Century of UNIX.* Reading, Mass.: Addison-Wesley.

Samuelson, P. A. (1973). "Proof that Properly Discounted Prices Vibrate Randomly." *Bell Journal of Economics and Management Science* 4: 369–374.

Sanchez, R. A. (1991). "Strategic Flexibility, Real Options and Product-based Strategy," Unpublished Ph.D. dissertation, MIT, Cambridge, Mass.

Sanchez, R., and J. T. Mahoney (1996a). "Modularity and Dynamic Capabilities," in *Schools of Thought in Strategic Management,* ed. T. Elfring and H. Volberda. Thousand Oaks, Calif.: Sage.

Sanchez, R., and J. T. Mahoney (1996b). "Modularity, Flexibility, and Knowledge Management in Product and Organization Design." *Strategic Management Journal* 17(Winter): 63–76.

Sanchez, R., and D. Sudharshan (1992). "Real-Time Market Research: Learning by Doing in the Development of New Products." in *Proceedings of the International Conference on Product Development Managment,* Brussels, Belgium, Institute of Advanced Studies in Managment.

Sanderson, S. W. (1991). "Cost Models for Evaluating Virtual Design Strategies in Multicycle Product Families." *Journal of Engineering and Technology Management* 8: 339–358.

Sanderson, S., and M. Uzumeri (1995). "Managing Product Families: The Case of the Sony Walkman," *Research Policy* 24: 761–782.

Saxenian, A. (1994). *Regional Advantage: Culture and Competition in Silicon Valley.* Cambridge, Mass.: Harvard University Press.

Schach, S. R. (1993). *Software Engineering.* Burr Ridge, Ill.: Richard D. Irwin and Aksen Associates.

Schaefer, S. (1997). "Product Design Partitions with Complementary Components," Unpublished manuscript, Kellogg School of Management, Northwestern University, Evanston, Ill.

Schary, M. A. (1991). "The Probability of Exit." *RAND Journal of Economics* 22(3): 339–353.

Scherer (1992). "Schumpeter and Plausible Capitalism." *Journal of Economic Literature* 30: 1416–1433.

Scherer, F. M. (1994). *Learning-By-Doing and International Trade in Semiconductors.* Cambridge, Mass.: Harvard University John F. Kennedy School of Government.

Schoemaker, P. J. H. (1991). "The Quest for Optimality: A Positive Heurisitic of Science?" *Behavioral and Brain Sciences* 14(2): 205–245.

Scholes, M. S. (1998). "Derivatives in a Dynamic Environment." *American Economic Review* 88(3): 350–370.

Schonberger, R. J. (1982). *Japanese Manufacturing Techniques: Nine Hidden Lessons in Simplicity.* New York: Free Press.

Schumpeter, J. A. (1976). *Capitalism, Socialism and Democracy.* New York: Harper & Row.

Schwartz, J. T. (1980). "Ultracomputers." *ACM Transactions on Programming Languages and Systems* 2(4): 484–521.

Selten, R. (1975). "A Reexamination of the Perfectness Concept for Equilibrium Points in Extensive Games." *International Journal of Game Theory* 4(1): 25–55.

Selten, R. (1978). "The Chain-Store Paradox." *Theory and Decision* 9: 127–159.

Shapiro, C., and H. R. Varian (1998). *Information Rules: A Strategic Guide to the Network Economy.* Boston: Harvard Business School Press.

Sharpe, W. F. (1964). "Capital Asset Prices: A Theory of Market Equilibrium under Conditions of Risk." *Journal of Finance* 19(September): 425–442.

Shelanski, H. A., and P. G. Klein (1995). "Empirical Research in Transaction Cost Economics: A Review and Assessment." *Journal of Law, Economics & Organization* 11(2): 335–361.

Shirley, G. V. (1990). "Models for Managing the Redesign and Manufacture of Product Sets." *Journal of Manufacturing and Operations Management* 3: 85–104.

Shleifer, A., and M. Boycko (1993). "The Politics of Russian Privatization," *Post-Communist Reform: Pain and Progress.* Cambridge, Mass.: MIT Press.

Simon, H. A. (1957). *Models of Man.* New York: Wiley.

Simon, H. A. (1969). *The Sciences of the Artificial.* Cambridge, Mass.: MIT Press.

Simon, H. A. (1976). *Administrative Behavior* 3rd ed. New York: Macmillan.

Skinner, W. (1969). "Manufacturing—Missing Link in Corporate Strategy." *Harvard Business Review*(May/June): 136–145.

Slytowsky, A. J. (1996). *Value Migration: How to Think Several Moves ahead of the Competition.* Boston: Harvard Business School Press.

Smith, J. K. (1985). "The Ten-Year Invention: Neoprene and Du Pont Research, 1930–1939." *Technology and Culture* 34–55.

Smith, J. K. (1986). *The Rise of Industrial Research in America.* Boston: Mass.: Harvard Business School.

Smith, J. K., and D. A. Hounshell (1985). "Wallace H. Carothers and Fundamental Research at Du Pont." *Science.* 229: 436–442.

Smith, N. P. (1997). "Computing in Japan: From Cocoon to Competition. Computer:" *IEEE Computer* 30(1): 26–33.

Smith, R. P., and S. D. Eppinger (1997a). "Identifying Controlling Features of Engineering Design Iteration." *Management Science* 43(3): 276–292.

Smith, R. P., and S. D. Eppinger (1997b). "A Predictive Model of Sequential Iteration in Engineering Design." *Management Science* 43(8): 1104–1120.

Spear, S. (1999). "Managing Complex Technical/Social Production Systems." Unpublished manuscript: Harvard Business School, Boston, Mass.

Stalk, G., Jr., and T. M. Hout (1990). *Competing Against Time: How Time-based Competition is Replacing Global Markets.* New York: Free Press.

Stein, J. C. (1994). *Waves of Creative Destruction: Customer Bases and the Dynamics of Innovation.* Cambridge, Mass.: National Bureau of Economic Research.

Stevens, W. P., G. J. Myers, and L. L. Constantine (1974). "Structured Design." *IBM Systems Journal* 13(2): 115–139.

Stevens, W. Y. (1964). "The Structure of System/360: Part II—System Implemenation," reprinted in C. G. Bell and A. Newell, eds. (1971). *Computer Structures: Reading and Examples.* New York: McGraw-Hill, 602–606.

Steward, D. V. (1981a). "The Design Structure System: A Method for Managing the Design of Complex Systems." *IEEE Transactions in Engineering Management* 28(3): 71–84.

Steward, D. V. (1981). *Systems Analysis and Management: Structure, Strategy and Design.* New York: Petrocelli Books.

Stigler, G. S. (1951). "The Division of Labor Is Limited by the Extent of the Market." *Journal of Political Economy* 59(3): 185–193.

Stole, L. A., and J. Zwiebel (1996). "Organizational Design and Technology Choice under Interfirm Bargaining." *American Economic Review* 86(1): 195–222.

Stulz, R. M. (1982). "Options on the Minimum or the Maximum of Two Risky Assets." *Journal of Financial Ecomomics* 10: 161–185.

Suh, N. P. (1990). *The Principles of Design.* New York: Oxford University Press.

Sull, D. N., R. S. Tedlow, and R. S. Rosenbloom (1997). "Managerial Commitments and Technological Change in the U.S. Tire Industry." *Industrial and Corporate Change* 6(2): 461–501.

Sullivan, K. J., P. Chalasani, and S. Jha (1997). *Software Design Decisions as Real Options.* Charlottesville, VA, University of Virginia School of Engineering and Applied Science.

Sutton, J. (1991). *Sunk Costs and Market Structure: Price, Competition, Advertising and the Evolution of Concentration.* Cambridge, Mass.: MIT Press.

Sutton, J. (1998). *Technology and Market Structure: Theory and History.* Cambridge, Mass.: MIT Press.

Suzue, T., and A. Kohdate (1990). *Variety Reduction Program: A Production Strategy for Product Diversification.* Cambridge, Mass.: Productivity Press.

Tedlow, R. S. (1994). *New and Improved: A History of Mass Marketing in America.* Boston: Harvard Business School Press.

Tedlow, R. S. (1998). "Thomas J. Watson, Sr. and American Salesmanship." Unpublished manuscript, Harvard Business School, Boston, Mass.

Teece, D., G. Pisano and A. Shuen (1992). "Dynamic Capabilities and the Concept of Strategy." University of California at Berkeley Working Paper.

Teece, D. J. (1993). "The Dynamics of Industrial Capitalism: Perspectives on Alfred Chandler's Scale and Scope." *Journal of Economic Literature* 31: 199–225.

Teisberg, E. O. (1989). *An Option Valuation Analysis of Investment Choices by a Regulated Firm.* Boston: Harvard University Graduate School of Business Administration.

Telser, L. G. (1980). "A Theory of Self-Enforcing Agreements." *Journal of Business* 53: 27–44.

Thomke, S. (1998). "Simulation, Learning and R&D Performance: Evidence from Automotive Development." *Research Policy* (May): 55–74.

Thomke, S. (1998). "Managing Experimentation in the Design of New Products." *Management Science* 44: 743–762.

Thomke, S., E. von Hippel, and R. Franke (1998). "Modes of Experimentation: An Innovation Process—and Competitive—Variable." *Research Policy* 27(July): 315–332.

Thompson, D. W. (1942). *On Growth and Form.* Cambridge, U.K.: Cambridge University Press.

Tirole, J. (1988). *The Theory of Industrial Organization.* Cambridge, Mass.: MIT Press.

Tobin, J. (1958). "Liquidity Preference as Behavior Towards Risk." *Review of Economic Studies* 25: 68–85.

Triantis, A. J., and J. E. Hodder (1990). "Valuing Flexibility as a Complex Option." *Journal of Finance* 45(2): 549–565.

Trigeorgis, L. (1996). *Real Options: Managerial Flexibility and Strategy in Resource Allocation.* Cambridge, Mass.: MIT Press.

Tufano, P. (1989). "Financial Innovation and First Mover Advantages," *Journal of Financial Economics* 25: 213–240.

Tufano, P. (1995). "Securities Innovations: A Historical and Functional Perspective." *Journal of Applied Corporate Finance* 7(4): 90–104.

Tufano, P. (1996). "Business Failure, Judicial Intervention, and Financial Innovation: Restructuring U.S. Railroads in the Nineteenth Century." *Business History Review* (Spring): 1–40.

Turing, A. M. (1950). "Computing Machinery and Intelligence." in *Mind Design II: Philosophy, Psychology and Artificial Intelligence.* ed. J. Haugeland. Cambridge, Mass.: MIT Press.

Tushman, M. L., and P. Anderson (1986). "Technological Discontinuities and Organizational Environments." *Administrative Sciences Quarterly* 31: 439–465.

Tushman, M. L., and C. A. O'Reilly III (1996). "The Ambidextrous Organization: Managing Evolutionary and Revolutionary Change." *California Management Review* 38(4): 8–30.

Tushman, M. L., and C. A. O'Reilly III (1997). *Winning through Innovation.* Boston: Harvard Business School Press.

Tushman, M. L., and J. P. Murmann (1998). "Dominant Designs, Technology Cycles, and Organizational Outcomes." *Research in Organizational Behavior* 20: 231–266.

Tushman, M. L., and L. Rosenkopf (1992). "Organizational Determinants of Technological Change: Toward a Sociology of Organizational Behavior." in *Research in Organizational Behavior,* ed. B. Staw and L. Cummings, Greenwich, Conn.: JAI Press. pp. 311–347.

Ullman, D. G. (1992). *The Mechanical Design Process.* New York: McGraw-Hill.

Ulrich, K. T. (1995a). "Fundamentals of Product Modularity." *Management of Design: Engineering and Management Perspectives,* Boston: Kluwer Academic Publishers, 219–231.

Ulrich, K. T. (1995). "The Role of Product Architecture in the Manufacturing Firm." *Research Policy* 24: 419–440.

Ulrich, K. T., and D. J. Ellison (1999). "Holistic Customer Requirements and the Design-Select Decision." *Management Science* (Forthcoming).

Ulrich, K. T., and S. D. Eppinger (1995). *Product Design and Development.* New York: McGraw-Hill.

Ulrich, K. T., and W. P. Seering (1989). "Synthesis of Schematic Descriptions in Mechanical Design." *Research in Engineering Design* 1: 3–18.

Ulrich, K. T., and W. P. Seering (1990). "Function Sharing in Mechanical Design." *Design Studies* 11(4): 233–234.

Ulrich, K. T., and K. Tung (1991). "Fundamentals of Product Modularity," in *Proceedings of the 1991 ASME Winter Symposium on Issues in Design/Manufacturing Integration,* Atlanta.

Ulrich, K. T., T. Randall, M. Fisher, and D.Reibstein (1998). "Managing Product Variety: A Study of the Bicycle Industry." in *Managing Product Variety,* ed. T. Ho. and C. Tang, Boston: Kluwer.

Utterback, J. (1994). *Mastering the Dynamics of Innovation: How Companies Can Seize Opportunities in the Face of Technological Change,* Boston: Harvard Business School Press.

Utterback, J. M., and F. F. Suarez (1993). "Innovation, Competition, and Industry Structure." *Research Policy* 22: 1–21.

Uzumeri, M., and S. Sanderson (1995). "A Framework for Model and Product Family Competition." *Research Policy* 24: 583–607.

Veblen, T. (1994). *Theory of the Leisure Class.* Reprint, New York: Penguin Classics.

Venezia, I. (1983). "A Bayesian Approach to the Optimal Growth Period Problem: A Note." *Journal of Finance* 38(1): 237–246.

Venezia, I., and M. Brenner (1979). "The Optimal Duration of Growth Investments and Search." *Journal of Business* 52(3): 393–407.

Vincenti, W. (1990). *What Engineers Know and How They Know It.* Baltimore: Johns Hopkins University Press.

Von Hippel, E. (1988). *The Sources of Innovation.* New York: Oxford University Press.

Von Hippel, E. (1990). "Task Partitioning: An Innovation Processing Variable." *Research Policy* (19): 407–418.

Von Neumann, J. (1987). *Papers of John von Neumann.* Cambridge, Mass.: MIT Press.

Von Neumann, J., and O. Morgenstern (1953). *Theory of Games and Economic Behavior.* Princeton, N.J.: Princeton University Press.

Waddington, C. H. (1977). *Tools for Thought.* New York: Basic Books.

Ward, A., J. K. Liker, J. J. Cristiano, and D. Sobek II (1995). "The Second Toyota Paradox: How Delaying Decisions Can Make Better Cars Faster." *Sloan Management Review* (Spring): 43–61.

Watson, T. J., Jr. (1990). *Father, Son & Company.* New York: Bantam Books.

Weitzman, M. L. (1996). "Hybridizing Growth Theory." *American Economics Review* 86 (May): 207–212.

Wernerfelt, B. (1997). "On the Nature and Scope of the Firm: An Adjustment-Cost Theory." *Journal of Business* 70(4): 489–514.

West, J., and M. Iansiti (1998). Experience, Experimentation and the Accumulation of Knowledge: An Empirical Study of R&D in the Semiconductor Industry. Boston: Harvard Business School (paper under review).

Wheelwright, S. C., and K. B. Clark (1992). *Revolutionizing Product Development.* New York: Free Press.

Wheelwright, S. C. (1978). "Reflecting Corporate Strategy in Manufacturing Decisions." *Business Horizons* 57–66.

Whitney, D. (1989). *Concurrent Design of Products and Processes.* New York: McGraw-Hill.

Whitney, D. E. (1990). "Designing the Design Proces," *Research in Engineering Design* 2: 3–13.

Whitney, D. E. (1993). "Nippondenso Co. Ltd: A Case Study of Strategic Product Design." *Research in Engineering Design* 5: 1–20.

Whitney, D. E. (1996). "Why Mechanical Design Cannot be Like VLSI Design." *Research in Engineering Design* 8:125–138.

Wilkes, M. V. (1985). *Memoirs of a Computer Pioneer.* Cambridge, Mass.: MIT Press.

Wilkes, M. V., and J. B. Stringer (1953). "Microprogramming and the Design of Control Circuits in an Electronic Digital Computer," reprinted in G. C. Bell and A. N. Newell, eds. (1971). *Computer Structures: Readings and Examples.* New York: McGraw-Hill, 335–340.

Wilkes, M. V. (1951). "The Best Way to Design an Automatic Calculating Machine," in *Proceedings of the Manchester University Computer Inauguaral Conference,* London: Ferranti, Ltd.

Williams, G. C. (1966). *Adaptation and Natural Selection.* Princeton, N.J.: Princeton University Press.

Williamson, O. E. (1975). *Markets and Hierarchies: Analysis and Antitrust Implications.* New York: Free Press.

Williamson, O. E. (1981). "The Modern Corporation: Origins, Evolution, Attributes." *Journal of Economic Literature* 19: 1537–1568.

Williamson, O. E. (1985). *The Economic Institutions of Capitalism: Firms, Markets, Relational Contracting.* New York: Free Press.

Wilson, E. O. (1998). *Consilience.* New York: Alfred A. Knopf.

Wise, T. A. (1966a). "IBM's $5,000,000 Gamble," *Fortune* (September).

Wise, T. A. (1966b). "The Rocky Road to the Marketplace," *Fortune* (October).

Wood, S. C. (1994). *Adaptable Manufacturing Systems for Semiconductor Integrated Circuit Production.* Stanford, Calif.: Stanford Electronics Laboratories, Department of Electrical Engineering, Stanford University.

Wruck, K. H. (1994). "Financial Policy, Internal Control and Performance: Sealed Air Corporation's Leveraged Special Dividend," *Journal of Financial Economics* 36: 157–192.

Wruck, K. H. (1998). "Releasing Value Held Hostage: Financial Policy, Organizational Strategy and the Externalization of Discipline and Control," Unpublished manuscript, Harvard Business School, Boston, Mass.

Yoffie, D. B. (1989). *The Global Semiconductor Industry, 1987.* Boston: Harvard Business School.

Yoffie, D. B., ed. (1997). *Competing in the Age of Digital Convergence,* Boston: Harvard Business School Press.

Young, G. (1985). *Venture Capital in High Tech Companies.* Westport, Conn.: Quorum Books.

Zachary, G. P. (1994). *Showstopper!—The Breakneck Race to Create Windows NT and the Next Generation at Microsoft.* New York: Free Press.

Zaltman, G. (1997). "Rethinking Market Research: Putting People Back In." *Journal of Marketing Research* 34 (November).

Zaltman, G., and R. Coulter (1995). "Seeing the Voice of the Customer: Metaphor-Based Advertising Research." *Journal of Advertising Research* 35(4): 35–51.

Index

Note: Figures are indicated by an italic *f* after the page number, and tables by an italic *t*.